Acoustics for Engineers

Ning Xiang · Jens Blauert

Acoustics for Engineers

Troy Lectures

Third Edition

Ning Xiang
Program in Architectural Acoustics
Rensselaer Polytechnic Institute (RPI)
Troy, NY, USA

Jens Blauert
Institute of Communication Acoustics
Ruhr-Universität Bochum (RUB)
Bochum, Germany

ISBN 978-3-662-63344-1   ISBN 978-3-662-63342-7   (eBook)
https://doi.org/10.1007/978-3-662-63342-7

Jointly published with ASA Press

1st & 2nd editions: © Springer-Verlag Berlin Heidelberg 2008, 2009
3rd edition: © Springer-Verlag GmbH Germany, part of Springer Nature 2021

This Springer imprint is published by the registered company Springer-Verlag GmbH, DE part of Springer Nature.
The registered company address is: Heidelberger Platz 3, 14197 Berlin, Germany

The Acoustical Society of America

On 27 December 1928 a group of scientists and engineers met at Bell Telephone Laboratories in New York City to discuss organizing a society dedicated to the field of acoustics. Plans developed rapidly, and the Acoustical Society of America (ASA) held its first meeting on 10–11 May 1929 with a charter membership of about 450. Today, ASA has a worldwide membership of about 7000.

The scope of this new society incorporated a broad range of technical areas that continues to be reflected in ASA's present-day endeavors. Today, ASA serves the interests of its members and the acoustics community in all branches of acoustics, both theoretical and applied. To achieve this goal, ASA has established Technical Committees charged with keeping abreast of the developments and needs of membership in specialized fields, as well as identifying new ones as they develop.

The Technical Committees include acoustical oceanography, animal bioacoustics, architectural acoustics, biomedical acoustics, engineering acoustics, musical acoustics, noise, physical acoustics, psychological and physiological acoustics, signal processing in acoustics, speech communication, structural acoustics and vibration, and underwater acoustics. This diversity is one of the Society's unique and strongest assets since it so strongly fosters and encourages cross-disciplinary learning, collaboration, and interactions.

ASA publications and meetings incorporate the diversity of these Technical Committees. In particular, publications play a major role in the Society. *The Journal of the Acoustical Society of America* (JASA) includes contributed papers and patent reviews. *JASA Express Letters* (JASA-EL) and *Proceedings of Meetings on Acoustics* (POMA) are online, open-access publications, offering rapid publication. *Acoustics Today*, published quarterly, is a popular open-access magazine. Other key features of ASA's publishing program include books, reprints of classic acoustics texts, and videos. ASA's biannual meetings offer opportunities for attendees to share information, with strong support throughout the career continuum, from students to retirees. Meetings incorporate many opportunities for professional and social interactions, and attendees find the personal contacts a rewarding experience. These experiences result in building a robust network of fellow scientists and engineers, many of whom become lifelong friends and colleagues.

From the Society's inception, members recognized the importance of developing acoustical standards with a focus on terminology, measurement procedures, and criteria for determining the effects of noise and vibration. The ASA Standards Program serves as the Secretariat for four American National Standards Institute Committees and provides administrative support for several international standards committees.

Throughout its history to present day, ASA's strength resides in attracting the interest and commitment of scholars devoted to promoting the knowledge and practical applications of acoustics. The unselfish activity of these individuals in the development of the Society is largely responsible for ASA's growth and present stature.

Preface

This book provides the material for an introductory course in engineering acoustics for students with basic knowledge of mathematics. The contents are based on extensive teaching experience at the university level.

Under the guidance of an academic teacher, the book is sufficient as the sole textbook for the subject. Each chapter deals with a well-defined topic and represents the material for a two-hour lecture. The chapters alternate between more theoretical and more application-oriented concepts.

For self-study, we advise our readers to consult complementary introductory material. Chapter 16 lists several textbooks for this purpose.

Thanks go to various colleagues and graduate students who most willingly helped with corrections, proofreading, and stylistic improvement, and last but not the least, to the reviewers of the first edition, in particular, to Profs. *Gerhard Sessler* and *Dominique J. Chéenne*. Nevertheless, the authors assume full responsibility for all contents. For the current edition, we reversed the authors' order. Ning Xiang is now the corresponding author, and Jens Blauert acts as the co-author.

In this (third) edition, we corrected recognized errors and typos, and edited several figures, notations, and equations to increase the clarity of the presentation. Also, we made some appropriate amendments.

At every chapter's end, we offer exercise problems. Chapter 15 proposes approaches to solving them. The problems provide our readers with the opportunity to explore the underlying mathematical background in more detail. However, the study of the problems and their proposed solutions is no prerequisite for comprehending the material presented in the book's main body.

Troy, NY, USA   Ning Xiang
Bochum, Germany   Jens Blauert
April 2021

Contents

About the Authors

Ning Xiang, Ph.D. is professor of Acoustics and Signal Processing at the Rensselaer Polytechnic Institute (RPI) in Troy, New York. He is the director of RPI's Program in Architectural Acoustics.

Jens Blauert, Dr.-Ing., Dr.-Tech. h.c. is emeritus professor of Acoustics and Electrical Engineering at the Ruhr-University Bochum (RUB) in Bochum, Germany. He is the founder and former director of RUB's Institute of Communication Acoustics.

Chapter 1
Introduction

Human beings are usually considered to predominantly perceive their environment through the visual sense—in other words, humans are conceived as *visual beings*. However, this is certainly not true for inter-individual communication.

It is audition and not vision that is the most relevant social sense of human beings. The auditory system is their most prominent communication organ—particularly in speech communication. Take as proof that it is much easier to educate blind people than deaf ones. Also, when watching TV, an interruption of the sound is much more distracting than an interruption of the picture. Particular attributes of audition compared to vision are the following.

- In audition, communication is compulsory. The ears cannot close by reflex like the eyes can.
- The field of hearing extends to regions all around the listener—in contrast to the visual field. Further, it is possible to listen behind optical barriers and in darkness.

These special features, among other things, lead many engineers and physicists, especially those in the field of communication technology, to a particular interest in acoustics. A further reason for the affinity of engineers and physicists to acoustics is based on the fact that many physical and mathematical foundations of acoustics are usually well known to them, such as mechanics, electrodynamics, vibrations, waves, and fields.

1.1 Definition of Three Basic Terms

When working your way into acoustics, you will usually start with the phenomenon of hearing. The term *acoustics* is derived from the Greek verb $\alpha\kappa o\acute{u}\epsilon\iota\nu$ [akúin], which means *to hear*. We thus start with the following definition.

© Springer-Verlag GmbH Germany, part of Springer Nature 2021
N. Xiang and J. Blauert, *Acoustics for Engineers*,
https://doi.org/10.1007/978-3-662-63342-7_1

Auditory event ... An auditory event is something that exists as heard. It
becomes actual in the act of hearing. Frequently used synonyms are *auditory object, auditory percept,* and *auditory sensation.*

Consequently, the question arises of when auditory events appear? As a rule, we hear
something when our auditory system interacts via the ears with a medium that moves
mechanically in the form of vibrations and/or waves. Such a medium may be a fluid
like air or water, or a solid like steel or wood. The phenomenon of hearing usually
requires the presence of mechanic vibration and/or waves. The following definition
follows this line of reasoning.

Sound ... Sound is mechanic vibration and/or mechanic waves in elastic media.

According to this definition, sound is a purely physical phenomenon. Be warned,
however, that the term *sound* is also sometimes used for auditory events, particularly
in sound engineering and sound design. Such an ambiguous usage of the term is
avoided in this book.

It it worthy noting that vibrations and waves are often mathematically expressible as differential equations—see Chap. 2. Vibrations require common differential
equations with the dependent variable being a function of time, while waves require
partial ones because the dependent variable is a function of both time and space.
Further, note that, although rare, auditory events may happen without sound being
present, such as with tinnitus. In turn, there may be no auditory events in the presence
of sound, for example, for deaf people, or when the frequency range of the sound
is not in the reach of hearing. Sounds are categorized in terms of their frequency
ranges—listed in Table 1.1.

The interrelation of auditory events and sound is captured by the following definition of acoustics.

Acoustics ... Acoustics is the science of sound and of its accompanying auditory
events.

This book deals with *engineering acoustics*. Synonyms for engineering acoustics
are *applied acoustics* and *technical acoustics*.

Table 1.1 Sound categories by frequency range

Sound category	Frequency range
Audible sound	$\approx 16\,\mathrm{Hz}$–$16\,\mathrm{kHz}$
Ultrasound	$>16\,\mathrm{kHz}$
Infrasound	$<16\,\mathrm{Hz}$
Hypersound	$>1\,\mathrm{GHz}$

1.2 Specialized Areas within Acoustics

Figure 1.1a presents a schematic of a transmission system as is often used in communication technology. A source renders information that is fed into a sender in coded form and transmitted over a channel. At the receiving end, a receiver picks up the transmitted signals, decodes them, and delivers the information to its final destination, that is, the information sink.

Figure 1.1b depicts a modified schematic to describe the receiving end of a transmission chain with acoustics involved. This delineation helps to distinguish major areas within engineering acoustics. The transmission channel delivers signals that are essentially chunks of electric energy. The receiver picks up these signals and feeds them into an energy transducer which transforms the electric energy into mechanic (acoustic) energy. The acoustic signals are then sent out into a sound field where they propagate to the listener. The listener receives them, decodes them, and processes the information. Note that undesired noise may enter the system at different points in addition to the desired signals.

The main areas of acoustics are as follows. The field that deals with the transduction of acoustic energy into electric energy, and vice versa, is called *electroacoustics*. The field that deals with the radiation, the propagation, and the reception of acoustic energy is called *physical acoustics*. The fields that deal with sound reception and auditory information processing by human listeners are called *psychoacoustics* and *physiological acoustics*. The first of these two fields focuses on the relationship between the sound and the auditory events associated with it, and the second one deals with sound-induced physiological processes in the auditory system and brain.

Acoustics as a discipline is usually further differentiated due to practical considerations. The following cover labels of the sessions at a major acoustics conference illustrate the broadness of the field.

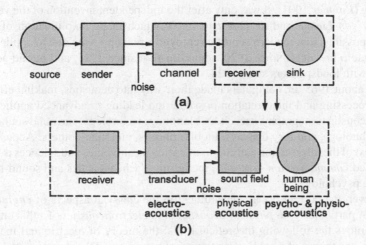

Fig. 1.1 Schematic of a transmission system. (**a**) General form. (**b**) Electroacoustic transmission system—receiving end

Acoustic and auditory signal processing, acoustic-measurement engineering, active acoustic systems, aircraft noise, architectural acoustics, artificial intelligence in acoustics, audio technology, audiological acoustics, bioacoustics, cognitive systems in acoustics, effects of noise, electroacoustics, evaluation of noise, flow acoustics, hydro-acoustics, medical acoustics, musical acoustics, noise propagation, noise protection, numerical acoustics, philosophy in acoustics, physical acoustics, physiological acoustics, psychoacoustics, room acoustics, speech-and-language processing, structure-borne sound, ultrasound, vehicle acoustics, vibration technology, virtual auditory reality.

Accordingly, there is a wide variety of professions dealing with acoustics, including diverse engineers, such as audio, biomedical, civil, electrical, environmental, and mechanical engineers. Further, for example, administrators, architects, audiologists, designers, ear-nose-and-throat-doctors, lawyers, managers, musicians, computer scientists, patent attorneys, physicists, physiologists, psychologists, sociologists and linguists.

1.3   About the History of Acoustics

Acoustics is an ancient science. *Pythagoras* already knew, around 500 BC, of the quantitative relationship between the length of a string and the pitch of its accompanying auditory event. In 1643, *Torricelli* demonstrated the vacuum experimentally and showed that there is no sound propagation in it. In 1802, *Chladni*, in his German-language book "Die Akustik" introduced acoustics as a separate branch into physics. At the end of the 19th century, classical physical acoustics had matured. The book "The Theory of Sound" by *Rayleigh* 1896, is considered a substantial reference even today.

At about the same time, basic inventions in acoustical communication technology were made, including the telephone (*Reis* 1867), television (*Nipkow* 1884) and tape recording (*Ruhmer* 1901). It was only after the independent invention of the vacuum triode by *von Lieben* and *de Forest* in 1910, which made amplification of weak currents possible, that modern acoustics enjoyed a real up-swing due to applications such as radio broadcast since 1920, sound-on-film since 1928, and public-address systems with loudspeakers since 1924.

From about 1965 on, computers made their way into acoustics, making effective signal processing and interpretation possible and leading to advanced applications such as acoustic tomography, speech-and-language technology, surround sound, binaural technology, auditory displays, mobile phones, and many others. Acoustics in the context of the information and communication technologies and sciences is nowadays called *communication acoustics*, including psychoacoustics and sound-related cognitive psychology.

However, the current book concentrates on the classical aspects of *engineering acoustics*, particularly on *physical acoustics* and *electroacoustics*. To this end, this book employs the following theoretical tools: the theory of electric and magnetic processes, the theory of signals, vibrations and systems, and the theory of waves and fields.

1.4 Relevant Quantities in Acoustics

The following quantities are of particular relevance in acoustics.[1]

- *Displacement, elongation*

  $\vec{\xi}$, in [m] ... displacementor of an oscillating particle or element from its resting position

- *Particle velocity*[2]

  $\vec{v}$, in [m/s] ... alternating velocity of an oscillating particle or oscillating element. In this context, it is simply termed *velocity*.

- *Acceleration*

  $\vec{a}$, in [m/s^2] ... alternating acceleration of an oscillating particle or element

- *Sound pressure*

  p, in [N/m^2 = Pa] ... alternating pressure as caused by particle oscillation[3]

- *Sound intensity*

  $\vec{I}$, in [W/m^2] ... sound power per effective area, $A_\perp$, that is the area component perpendicular to the direction of energy propagation

- *Speed of sound*

  $\vec{c}$, in [m/s] ... propagation speed of a sound wave[4]

The superscribed arrows denote vectors, but in this book we use them only when the vector quality is of relevance. Otherwise the magnitude, $c = |\vec{c}|$, is used.

Since sound is essentially vibrations and waves, the quantities ξ, v, a, and p are often periodically alternating quantities. *Fourier* analysis allows to decompose them into sinusoidal components. These components are notable in complex form—see Sect. 16.1. In this book, we underline quantities that are known to be complex.

In acoustics, there are the following three definitions of impedances.

- *Field impedance*

$$\underline{Z}_f = \frac{\underline{p}}{\underline{v}}, \quad \text{in} \quad \left[\frac{\text{Ns}}{\text{m}^3}\right] \tag{1.1}$$

---

[1] The units used here are consistent with a subclass of the SI–system (système international d'unités, namely, the m/kg/s/A–system.

[2] In acoustics often called sound velocity or just velocity.

[3] $1\,\text{Pa} = 1\,\text{N/m}^2 = 1\,\text{kg/(ms}^2) = 1\,(\text{Ws})/\text{m}^3$.

[4] Warning: $\vec{c}$ must not be mistaken as a particle velocity!

- *Mechanic impedance*

$$\underline{Z}_{\mathrm{mech}} = \frac{\underline{F}}{\underline{v}}, \quad \mathrm{in} \ \left[\frac{\mathrm{Ns}}{\mathrm{m}}\right] \tag{1.2}$$

- *Acoustic impedance*

$$\underline{Z}_{\mathrm{a}} = \frac{\underline{p}}{\underline{q}}, \quad \mathrm{in} \ \left[\frac{\mathrm{Ns}}{\mathrm{m}^5}\right] \tag{1.3}$$

with $\underline{F}$ being the force, $\underline{F} = \underline{p}\, A$, and $\underline{q}$ being the so-called volume velocity, $\underline{q} = \underline{v}\, A$. The different kinds of impedances are convertible into each other, provided that the effective radiation area of the sound source, A, is known.

Note that impedances represent the complex ratio of two quantities, the product of which forms a power-related quantity.

1.5   Some Numerical Examples

To derive some illustrative numerical examples, this section considers a plane wave in air.[5] A plane wave is a wave where all quantities are invariant across areas perpendicular to the direction of wave propagation. The field impedance in a plane wave is a quantity that is specific to the medium and is called the *characteristic field impedance*, $\underline{Z}_{\mathrm{w}}$—see Sect. 7.4. Disregarding dissipation, this is a real quantity. In air, we find $Z_{\mathrm{w,air}} \approx 412\,\mathrm{Ns/m}^3$ under standard conditions.

- *Sound pressure*

    – Sound pressure at the threshold of discomfort (maximum sound pressure),

$$p_{\mathrm{max,rms}} \approx 10^2\,\mathrm{N/m}^2 = 100\,\mathrm{Pa} \tag{1.4}$$

    – Sound pressure at a normal conversation level at a 1-m distance from the talker (normal sound pressure),

$$p_{\mathrm{normal,rms}} \approx 0.1\,\mathrm{N/m}^2 = 100\,\mathrm{mPa} \tag{1.5}$$

    – Sound pressure at 1 kHz at the threshold of hearing (minimum sound pressure),

$$p_{\mathrm{min,rms}} \approx 2 \cdot 10^{-5}\,\mathrm{N/m}^2 = 20\,\mu\mathrm{Pa} \tag{1.6}$$

For reference: The static atmospheric pressure under normal conditions is about $10^5\,\mathrm{N/m}^2 = 1000\,\mathrm{hPA} \mathrel{\widehat{=}} 1\,\mathrm{bar}$.

---

[5] The following synopsis presents rms-values rather than peak values to account for all kinds of finite-power sounds, such as noise, speech, and music, besides sinusoidal sounds. See Sect. 16.1 for the definition of "rms". Peak values of sinusoidal signals (magnitudes)—as usually used in complex notations throughout this book—exceed their rms-values by a factor of $\sqrt{2}$, that is $\hat{x} = \sqrt{2}\, x_{\mathrm{rms}}$.

- *Particle velocity*

  The following particle velocities appear with the above sound pressures, considering the relationship $\underline{p} = Z_{w,air} \cdot \underline{v}$.

  – Maximum particle velocity, $v_{max,rms} \approx 0.25$ m/s

  – Normal particle velocity, $v_{normal,rms} \approx 25 \cdot 10^{-5}$ m/s

  – Minimum particle velocity, $v_{min,rms} \approx 5 \cdot 10^{-8}$ m/s

For reference: The speed of sound in air is $c \approx 340$ m/s.

- *Particle displacement*

  The relationship between particle velocity and particle displacement is frequency dependent since $\xi(t) = \int v(t)\, dt$, or, in complex notation, $\underline{\xi} = \underline{v}/j\omega$. A comparison thus requires the selection of a specific frequency, say, 1 kHz here. With this presupposition, it follows

  – Maximum particle displacement, $\xi_{max,rms} \approx 4 \cdot 10^{-5}$ m

  – Normal particle displacement, $\xi_{normal,rms} \approx 4 \cdot 10^{-8}$ m

  – Minimum particle displacement, $\xi_{min,rms} \approx 8 \cdot 10^{-12}$ m

For reference: The diameter of hydrogen atoms is 10^{-10} m. Indeed, for the small displacements near the threshold of hearing it becomes questionable whether consideration of the medium as a continuum is still valid.

It is also worth noting here that the particle displacements due to the *Brownian* molecular motion are only one order of magnitude smaller than those induced by sound at the threshold of hearing. Thus, the auditory system works definitely at the brink of what makes sense physically. If the system were only slightly more sensitive, one would "hear the grass growing".

1.6 Logarithmic Level Ratios and Logarithmic Frequency Ratios

Logarithmic Level Ratios

As shown above, the range of sound pressures to be handled in acoustics is at least $1:10,000,000$, which is $1:10^7$. This leads to unhandy numbers when describing sound pressures and sound-pressure ratios. For this reason, a logarithmic measure called the *level* is frequently used. Further reasons for its use are the following.

- Equal relative variations of the strength of a physical stimulus lead to equal absolute changes in the salience of the sensory events, which is called the *Weber-Fechner* law. This relationship leads to a logarithmic characteristic.
- When connecting two-port elements in a chain (cascade), the overall level reduction (attenuation) between input and output turns out to be the sum of the attenuations of each element.

The following level definitions are common in acoustics, with $\lg = \log_{10}$.

- *Sound-intensity level*

$$L_{\mathrm{I}} = 10 \lg \frac{|\vec{\underline{I}}|}{I_0} \, \mathrm{dB}, \quad \text{with } I_0 = 10^{-12} \, \mathrm{W/m^2} \text{ as reference} \qquad (1.7)$$

- *Sound-pressure level*

$$L_{\mathrm{p}} = 20 \lg \frac{p_{\mathrm{rms}}}{p_{0,\mathrm{rms}}} \, \mathrm{dB}, \quad \text{with } p_{0,\mathrm{rms}} = 2 \cdot 10^{-5} \, \mathrm{N/m^2} = 20 \, \mu\mathrm{Pa}$$
$$\text{as reference} \qquad (1.8)$$

- *Sound-power level*

$$L_{\mathrm{P}} = 10 \lg \frac{|\underline{P}|}{P_0} \, \mathrm{dB}, \quad \text{with } P_0 = 10^{-12} \, \mathrm{W} \text{ as reference} \qquad (1.9)$$

The reference levels are internationally standardized, and the first two roughly represent the threshold of hearing at $1 \, \mathrm{kHz}$. Other references may be used, but in such cases, the respective reference is indicated, for example, in the form $L = 15 \, \mathrm{dB}$ re $100 \, \mu\mathrm{Pa}$.

The symbol used to signify levels computed with the above definitions is [dB], which stands for deciBel, named after *Alexander Graham Bell*. Another unit-like symbol, based on the natural logarithm, $\log_{\mathrm{e}} = \ln$, the Neper [Np], is also in use to express level, particularly in electrical transmission theory. Levels in Neper convert into levels in deciBel with L [Np] $= 8.69 \, L$ [dB].[6]

For the cases of intensity and power levels, note that the levels describe ratios of the magnitudes of intensity and/or power. These magnitudes read as follows in complex notation, taking the intensity as an example—see Sect. 16.2.

$$|\vec{\underline{I}}| = \left| I \, \mathrm{e}^{\mathrm{j}\omega(\phi_{\mathrm{p}} - \phi_{\mathrm{v}})} \right| = \frac{1}{2} \left| \underline{p} \, \underline{v}^* \right|. \qquad (1.10)$$

For practical purposes, it is useful to learn some level differences by heart. A few representative examples are listed in Table 1.2. By knowing these values, it is easy to estimate level differences. For instance, the sound-pressure ratio of $1 : 2000$ $= (1 : 1000)(1 : 2)$ corresponds to $(-60 \, \mathrm{dB}) + (-6 \, \mathrm{dB}) = -66 \, \mathrm{dB}$.

---

[6] Note that deciBel [dB] and Neper [Np] are no units in the strict sense but letter symbols indicating a calculation process. When used in equations, their dimension is one.

Table 1.2 Some representative level differences

Ratio of sound pressure	Ratio of sound intensity or power
$\sqrt{2} : 1 \approx 3\,\text{dB}$	$\sqrt{2} : 1 \approx 1.5\,\text{dB}$
$2 : 1 \approx 6\,\text{dB}$	$2 : 1 \approx 3\,\text{dB}$
$3 : 1 \approx 10\,\text{dB}$	$3 : 1 \approx 5\,\text{dB}$
$5 : 1 \approx 14\,\text{dB}$	$5 : 1 \approx 7\,\text{dB}$
$10 : 1 = 20\,\text{dB}$	$10 : 1 = 10\,\text{dB}$

To compute the levels that add up when more than one sound source is active, we distinguish between, (a), sounds that are coherent, such as stemming from loudspeakers with the same input signals and, (b), those that are incoherent, such as signals originating from independent noise sources like vacuum cleaners. Coherent sounds interfere but incoherent ones do not. Consequently, we end up with the following two formulas for summation—here written for sinusoids.

• *Addition of coherent sounds*

$$L_{\Sigma} = 20 \lg \left(\frac{\frac{1}{\sqrt{2}} \left| \underline{p}_1 + \underline{p}_2 + \underline{p}_3 + \cdots + \underline{p}_n \right|}{p_{0,\,\text{rms}}} \right) \text{dB} \qquad (1.11)$$

• *Addition of incoherent sounds*

$$L_{\Sigma} = 10 \lg \left(\frac{|\vec{\underline{I}}_1| + |\vec{\underline{I}}_2| \cdots + |\vec{\underline{I}}_n|}{I_0} \right) \text{dB} \qquad (1.12)$$

Inter-signal phase differences need not be considered when the signals do not interfere, which is the case with incoherent signals.

Logarithmic Frequency Ratios

What holds for the magnitude of sound quantities, namely, that their range is enormous, also holds for the frequency range of the signal components. The audible frequency range roughly considered to extend from about 16 Hz to 16 kHz in young people, which is a range of $1 : 10^3$. Sensitivity to high frequencies decreases with age. With high-intensity sounds, some kind of hearing may even be experienced above 16 kHz.

Equal ratios between the fundamental frequencies of musical sounds lead to equal musical intervals of the perceived pitch. Therefore, a logarithmic ratio of frequencies has been introduced, called *logarithmic frequency interval*, Ψ, which is based on the *logarithmus dualis*, $\text{ld} = \log_2$. It also has the dimension one. With $f_2 > f_1$ the following four definitions are in use.

$$\Psi_{\text{oct}} = \text{ld}\,(f_2/f_1), \qquad \text{in [oct]}$$
$$\Psi_{\text{3rd oct}} = 3\,\text{ld}\,(f_2/f_1), \qquad \text{in [3rd oct]}$$

Table 1.3 Frequencies versus corresponding wavelengths in air

Octave-center frequency [Hz]	16	32	63	125	250	500	1k	2k	4k	8k	16k
Wave length in air [m]	20	10	5	2.5	1.25	0.63	0.32	0.16	0.08	0.04	0.02

$$\Psi_{semitone} = 12\,\mathrm{ld}\,(f_2/f_1), \qquad \text{in [semitone]}$$
$$\Psi_{cent} = 1200\,\mathrm{ld}\,(f_2/f_1), \qquad \text{in [cent]}$$

The first two logarithmic frequency intervals are often called *octave* and *third-octave*. The equations, $\Psi_{oct} = 1$, indicates one octave. $\Psi_{3rd\,oct} = 1$ indicates one third of an octave. The four logarithmic frequency intervals have the following relationship to each other, $1\,\mathrm{oct} = 3 \cdot 3\mathrm{rd}\,\mathrm{oct} = 12\,\mathrm{semitone} = 1200\,\mathrm{cent}$.

A bandpass filter, which is a filter that filters out certain regions from a frequency spectrum and blocks the rest, is called *octave filter* when the difference between the upper and lower limiting frequencies of the pass-band amounts to one octave. Accordingly, there are, for example, third-octave filters. Note that the specification of limiting frequencies is task-specific.

In communication engineering, decades (10 : 1) are sometimes preferred to octaves (2 : 1). Conversion is as follows: $1\,\mathrm{oct} \approx 0.3\,\mathrm{dec}$ or $1\,\mathrm{dec} \approx 3.3\,\mathrm{oct}$.

Wavelength, λ, and frequency, f, of an acoustic wave are linked by the relationship $c = \lambda \cdot f$. In air we have $c \approx 340\,\mathrm{m/s}$. In Table 1.3, a series of frequencies is presented with their corresponding wavelengths in air. The series is taken from a standardized octave series that is recommended for use in engineering acoustics.

In the audible frequency range, the wavelengths extend from a few centimeters to many meters. Because radiation, propagation, and reception of waves are characterized by the linear dimension of reflecting surfaces relative to the wavelength, a wide variety of different effects, including reflection, scattering, and diffraction, are experienced in acoustics.

1.7   Double-Logarithmic Plots

By plotting levels over logarithmic frequency intervals, one obtains a double-logarithmic graphic representation of the original quantities. This way of plotting has some advantages over linear representations and is quite popular in acoustics.[7] Figure 1.2a presents an example of a linear representation, and Fig. 1.2b shows its corresponding double-logarithmic plot.

In double-logarithmic plots, all functions that are proportional to ω^y appear as straight lines since

---

[7] In network theory, double-logarithmic graphic representations are known as *Bode* diagrams.

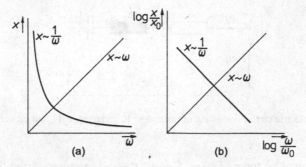

Fig. 1.2 Different representations of frequency functions. (**a**) Linear. (**b**) Double-logarithmic

$$x = a\,\omega^y \;\rightarrow\; \log x = \log a + y \log \omega . \tag{1.13}$$

For integer potencies, $y = \pm n$ with $n = 1, 2, 3, \ldots$, we arrive at slopes of $\pm n \cdot 6\,\mathrm{dB/oct}$ for sound pressure, displacement, and particle velocity, and of $\pm n \cdot 3\,\mathrm{dB/oct}$ for power and intensity. For decades the respective values are $\approx 20\,\mathrm{dB/dec}$ and $\approx 10\,\mathrm{dB/dec}$.

Functions with different potency of ω are quite frequent in acoustics. This results from the fact that differential equations of different degrees are used to describe vibrations and waves. The slope of the lines in the plot helps to estimate the order of the underlying oscillation processes.

1.8 Exercises

Recapitulation of Complex Notation

Problem 1.1 Use *Euler*'s formula to derive a real-valued sinusoidal expression for $z(t)$ from the complex-valued exponential time function. *Euler*'s formula reads as follows,

$$\underline{z} = \hat{z}\,\mathrm{e}^{\mathrm{j}(\omega t + \phi)} . \tag{1.14}$$

Problem 1.2 Given a sinusoidal electric voltage signal

$$u(t) = \hat{u}\cos(\omega t + \phi), \tag{1.15}$$

find the electric current, $i(t)$, in Fig. 1.3 by applying complex notation.

Problem 1.3 For experimental determination of the mechanic impedance, $\underline{Z}_{\mathrm{mech}}$, of a solid material (object) in practice, one may excite the object under test by

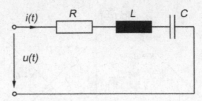

Fig. 1.3 Serial electric circuit consisting of resistance, R, inductance, L, and capacitance, C

applying a sinusoidal force, $F(t) = \hat{F} \cos(\omega t + \phi_F)$, and measuring the acceleration, $a(t) = \hat{a} \cos(\omega t + \phi_a)$, at the point of interest.

Determine the mechanic impedance, $\underline{Z}_{\text{mech}}$, at this point via the given excitation and the resulting acceleration responses.

Problem 1.4 Given a reference pressure of $p_{0,\text{rms}} = 20\,\mu\text{N/m}^2$,

(a) Find the sound-pressure level of $p_{\text{rms}} = 0.1\,\text{N/m}^2$ and $p_{\text{rms}} = 10^2\,\text{N/m}^2$, respectively.

(b) Evaluate the sound-pressure levels for the case that the sound pressures under (a) are three-times larger.

Problem 1.5 Which frequency ratios and which logarithmic frequency intervals are represented by three semitones and by twelve semitones?

Problem 1.6 $\Psi_{\text{oct}} = \text{ld}\,(f_2/f_1),$ in [oct]
$\Psi_{1/\text{3rd oct}} = 3\,\text{ld}\,(f_2/f_1),$ in $[\frac{1}{3}\,\text{oct}]$
$\Psi_{\text{semitone}} = 12\,\text{ld}\,(f_2/f_1),$ in [semitone]
$\Psi_{\text{cent}} = 1200\,\text{ld}\,(f_2/f_1),$ in [cent]

Given the third-octave-band interval, $\Psi_{\text{3rd oct}} = 1$, and the octave-band interval, $\Psi_{\text{oct}} = 1$.

– Determine the corresponding frequency ratios, $(f_2/f_1)_{\text{3rd oct}}$ and $(f_2/f_1)_{\text{oct}}$.
– Use these frequency ratios to find the two limiting frequencies, that is, the lower limiting and the upper limiting frequency, f_1, and f_u, of a 3rd-octave band and of an octave band with a given center frequency of f_c.

Problem 1.7 Establish a table for the addition of sound-pressure levels of multiple incoherent sound sources of equal level and equal distance to the receiving point. The number of sound sources is $1, 2, 3, 4, 5, 6, 8$, or 10, repectively.

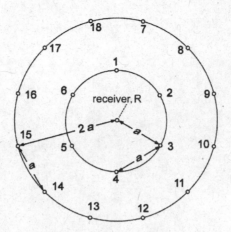

Fig. 1.4 Multiple sound sources of equal intensity in an open-space office

Problem 1.8 In an open-space office there are multiple noise sources (talking persons), spatially distributed as shown in Fig. 1.4. Sources 1–6 are located along a circle of radius a around the receiver, R, with a mutual distance of a between them. Sources 7–18 are located apart from each other by, again, a distance of a and along a circle of radius $2a$ around the receiver. The sound-pressure level is assumed to decrease by 5 dB when doubling the distance in this particular space.[8]

With the sound-pressure level being 65 dB(A) at a 1-m distance from any single source (talker) for all directions, find the sound-pressure level at the receiver as a function of the number of sound sources ($a = 4$ m).

---

[8] This, by the way, indicates that the sound-field in the specific space given here is not an entirely free field. For a perfectly free-field, we would expect a decrease of 6 dB/distance-doubling for spherical sound sources (4.25).

Chapter 2
Mechanic and Acoustic Oscillations

When physical or other quantities vary in a specific way as a function of time, we usually say they are *oscillating*. A common broad definition of oscillation is as follows.

Oscillation ... An oscillation is a process with attributes that are repeated regularly in time.

Oscillating processes are widespread in our world, and they are responsible for all wave propagation, such as sound, light, or radio waves. The time functions of oscillating quantities vary extensively. For example, oscillations are produced by intermittent sources like foghorns, sirens, the saw-tooth generator of an oscilloscope, or the blinking signal of a turning light.

A prominent category of oscillations is characterized by energy that is swinging between two complementary storages, namely, kinetic versus potential energy or electric versus magnetic energy. In many cases, these oscillating systems are approximately linear and constant in time, which categorizes them as so-called linear, time-invariant (LTI) systems.

The mathematical treatment of LTI systems is straightforward. A specific feature of these systems is that the *superposition principle* applies. Excitation of an LTI system by several individual excitation functions leads the system to respond according to the linear combination of the individual response to each excitation function. The superposition principle reads as follows in mathematical terms,

$$y(t) = \sum_k b_k\, y_k(t) = \mathcal{T}\left[\sum_k b_k\, x_k(t)\right], \quad \text{assuming } y_k(t) = \mathcal{T}[x_k(t)], \quad (2.1)$$

where $\mathcal{T}$ stands for a specified transformation by the LTI system concerned. $x_k(t)$ and $y_k(t)$ represent the kth excitation and response of the system, respectively. b_k is constant.

The general exponential function with the complex frequency, $\underline{s} = \breve{\alpha} + \mathrm{j}\,\omega$,

© Springer-Verlag GmbH Germany, part of Springer Nature 2021
N. Xiang and J. Blauert, *Acoustics for Engineers*,
https://doi.org/10.1007/978-3-662-63342-7_2

$$A \, e^{\underline{s}t} = A \, e^{\tilde{\alpha}t + j\omega t} = A \, e^{\tilde{\alpha}t}(\cos \omega t + j \sin \omega t) \,, \qquad (2.2)$$

is an *eigen-function* of LTI systems. This property means that excitation by a sinusoidal function results in a response that is a sinusoidal function of the same frequency, although generally with a different phase and amplitude. This unique feature of LTI systems is one of the reasons why sinusoidal functions play a prominent role in the analysis of LTI systems and linear oscillators.

Operations with LTI systems are often performed in what is called the *frequency domain*. To move from the time domain to the frequency domain, the time function of the excitation is decomposed by *Fourier* transform into sinusoidal components. Each component is then sent through the system, and the time function of the total response determined by summing up all the individual sinusoidal responses and performing the inverse *Fourier* transform.

The current book does not deal with *Fourier* transform in great detail, but the fact that all sounds are decomposable into sinusoidal components—and completely (re)composable from these—is a good argument for employing sinusoidal excitation in LTI systems.

2.1   Basic Elements of Linear, Oscillating, Mechanic Systems

Three elements are required to form a simple mechanic oscillator, namely, a *mass*, a *spring* and a fluidic *damper*(dashpot)—see Fig. 2.1.

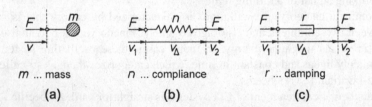

m ... mass n ... compliance r ... damping

(a)   **(b)**   **(c)**

Fig. 2.1 Basic elements of linear time-invariant mechanic oscillation systems. (**a**) Mass. (**b**) Spring. (**c**) Fluidic damper (dashpot)

For the introduction of these elements, we make three idealizing assumptions, namely,

(a) All relationships between the mechanic quantities displacement, ξ, velocity, v, force, F, and acceleration, a, are linear.

(b) The characteristic features of the elements are constant.

(c) This chapter considers one-dimensional motion only.

• *Mass*

An alternating force may be applied to a solid body with the mass m, as shown in Fig. 2.1a, so that *Newton*'s[1] law holds as follows,

$$F(t) = m\, a(t) = m\, \frac{\mathrm{d}v}{\mathrm{d}t} = m\, \frac{\mathrm{d}^2\xi}{\mathrm{d}t^2}\,. \tag{2.3}$$

For sinusoidal quantities this law reads in complex notation as follows,

$$\underline{F} = m\, \underline{a} = \mathrm{j}\,\omega\, m\, \underline{v} = -\omega^2 m\, \underline{\xi}\,. \tag{2.4}$$

Later in this chapter, we show that the mass stores kinetic energy. It is a *one-port* element in terms of network-theory because there is only one in/output port for transmitting power. The physical unit of the mass is [kg] or [N s^2/m]. The mechanic impedance of a mass is imaginary and expressed as

$$\underline{Z}_{\text{mech}} = \mathrm{j}\,\omega\, m\,. \tag{2.5}$$

• *Spring*

According to *Hooke*, the following applies to linear springs[2] with a compliance of n—as seen in Fig. 2.1b,

$$F(t) = \frac{1}{n}\,\xi_\Delta(t) = \frac{1}{n}\int v_\Delta(t)\,\mathrm{d}t = \frac{1}{n}\int\left[\int a_\Delta(t)\,\mathrm{d}t\right]\mathrm{d}t\,. \tag{2.6}$$

The physical unit of the compliance is [m/N]. For sinusoidal quantities in complex notation this is equivalent to

$$\underline{F} = \frac{1}{n}\,\underline{\xi}_\Delta = \frac{1}{\mathrm{j}\,\omega\, n}\,\underline{v}_\Delta = \frac{-1}{\omega^2 n}\,\underline{a}_\Delta\,. \tag{2.7}$$

The spring stores potential energy. It is a *two-port* element because it has both an input and an output port. The mechanic impedance of the spring is imaginary and equals

$$\underline{Z}_{\text{mech}} = \frac{1}{\mathrm{j}\,\omega\, n} \tag{2.8}$$

• *Damper* (Dashpot)

A dashpot is a damping element based on fluid friction due to a viscous medium—see Fig. 2.1c. At a dashpot with damping (mechanic resistance), r, the following holds,

---

[1] *Newton*'s law is valid in so-called inertial spatial coordinate systems. If no force is applied to a mass in these systems, the mass moves with constant velocity along a linear trajectory. As the origin of the coordinate system "ground" is usually defined to be a mass taken as infinite. Gravitation forces are not considered here.

[2] In acoustics, the compliance, n, is often preferred to its reciprocal, the stiffness, $k = 1/n$, as this leads to formula notations that engineers are more accustomed to—refer to Chap. 3.

$$F(t) = r\, v_\Delta(t) = r\, \frac{d\xi_\Delta}{dt} = r \int a_\Delta(t)dt . \tag{2.9}$$

In complex notation for sinusoidal quantities this is

$$\underline{F} = r\, \underline{v}_\Delta = j\omega\, r\, \underline{\xi}_\Delta = \frac{r}{j\omega}\, \underline{a}_\Delta . \tag{2.10}$$

The mechanic impedance of the damper is real and expressed as

$$\underline{Z}_{\text{mech}} = r . \tag{2.11}$$

The physical unit of the damper is [N s/m]. The dashpot does not store energy. It consumes energy through dissipation, which is a process of converting mechanic energy into thermodynamic energy, in other words, *heat*. The dashpot is a *two-port* element.

2.2   Parallel Mechanic Oscillators

This section now considers an arrangement where a mass, a spring, and a dashpot are connected in parallel by idealized, that is, rigid and massless rods—see Fig. 2.2.

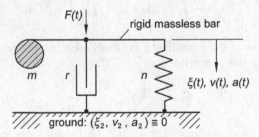

Fig. 2.2 Mechanic parallel oscillator, exited by an alternating force. The second port is grounded here for simplicity

The arrangement may be excited by an alternating force, $F(t)$, that is composed of three elements,

$$F(t) = F_{\text{m}}(t) + F_{\text{r}}(t) + F_{\text{n}}(t) \tag{2.12}$$

In this way, the following differential equation holds,

$$F(t) = m\, \frac{d^2\xi}{dt^2} + r\, \frac{d\xi}{dt} + \frac{1}{n}\xi \quad \text{or} \quad F(t) = m\, \frac{dv}{dt} + r\, v(t) + \frac{1}{n}\int v(t)\, dt . \tag{2.13}$$

As only one variable, ξ or v, is sufficient to describe the state of the system, it represents what is commonly called a *simple oscillator*.

For simplicity of the example, both the spring and the dashpot are connected to ground. In this way, the quantities ξ_2 and v_2 are set to zero at the output ports. Consequently, we omit the subscript Δ in the following.

2.3 Free Oscillations of Parallel Mechanic Oscillators

This section deals with the particular case that the oscillator is in a position away from its resting position, and the introduced force is set to zero, that is, $F(t) = 0$ for $t > 0$. The differential equation (2.13) then converts into a homogenous differential equation as follows,

$$m \frac{d^2\xi}{dt^2} + r \frac{d\xi}{dt} + \frac{1}{n}\xi = 0. \tag{2.14}$$

The solution of this equation is called *free oscillation* or *eigen-oscillation* of the system. A trial using $\underline{\xi} = e^{\underline{s}t}$ results in the characteristic equation[3]

$$m \underline{s}^2 + r \underline{s} + \frac{1}{n} = 0, \tag{2.15}$$

where $\underline{s}$ denotes the complex frequency. The general solution of this quadratic equation reads as

$$\underline{s}_{1,2} = -\frac{r}{2m} \pm \sqrt{\frac{r^2}{4m^2} - \frac{1}{mn}} \quad \text{or} \quad \underline{s}_{1,2} = -\delta \pm \sqrt{\delta^2 - \omega_0^2}, \tag{2.16}$$

where $\delta = r/2m$ is the damping coefficient and $\omega_0 = 1/\sqrt{mn}$ the characteristic angular frequency. This general form renders the three different types of solutions, namely,

Case (a) with $\delta = \omega_0$ ... critical damping, only one root, which is real.
Case (b) with $\delta > \omega_0$ ... strong damping, both roots real, s is negative.
Case (c) with $\delta < \omega_0$ ... weak damping, both roots are complex.

The differential equation for a simple oscillator is of second order, making it necessary to have two initial conditions to derive specific solutions. Three forms of general solutions exist, which are listed below. It remains to adjust them to the particular initial conditions to finally arrive at particular solutions.

- *Case* (a) $(\delta = \omega_0)$

$$\underline{\xi}(t) = (\underline{\xi}_1 + \underline{\xi}_2) e^{-\delta t}. \tag{2.17}$$

This case is at the brink of both periodic and aperiodic decay. Depending on the initial conditions, it may or may not render a single swing-over. It is called the *aperiodic limiting case*.

- *Case* (b) $(\delta > \omega_0)$

$$\underline{\xi}(t) = \underline{\xi}_1 e^{-(\delta - \sqrt{\delta^2 - \omega_0^2})t} + \underline{\xi}_2 e^{-(\delta + \sqrt{\delta^2 - \omega_0^2})t}. \tag{2.18}$$

---

[3] As noted in the introduction to this chapter, the general exponential function is an eigen-function of linear differential equations. This means that it stays to be an exponential function when differentiated or integrated.

This solution, called the *creeping case*, describes an aperiodic decay.

- *Case* (c) $(\delta < \omega_0)$

$$\underline{\xi}(t) = \underline{\xi}_1 e^{-\delta t} e^{-j\omega t} + \underline{\xi}_2 e^{-\delta t} e^{+j\omega t}, \quad \text{with} \quad \omega = \sqrt{\omega_0^2 - \delta^2}. \qquad (2.19)$$

This solution, called the *oscillating case*, describes a periodic, decaying oscillation. It represents indeed an oscillation as illustrated by looking at the particular special case of $\underline{\xi}_1 = \underline{\xi}_2 = \underline{\xi}_{1,2}/2$.

Substituting it into (2.19) yields

$$\xi(t) = \underline{\xi}_{1,2} \, e^{-\delta t} \cos(\omega t) \qquad (2.20)$$

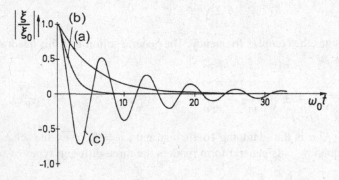

Fig. 2.3 Decays of a simple oscillator for different damping settings (schematic). **(a)** Aperiodic limiting case. **(b)** Aperiodic case. **(c)** Oscillating case

Figure 2.3 illustrates the three cases. The fastest-possible decay below a threshold—which, by the way, is the objective when tuning the suspension of road vehicles—is achieved with a slightly subcritical damping of $\delta \approx 0.6\,\omega_0$.

In addition to $\delta = r/2m$, the following two quantities are often used in acoustics to characterize the amount of damping in an oscillating system,

Q ... The sharpness-of-resonance factor, also known as quality factor.

T_d ... The decay time—similar to the reverberation time in Sect. 12.5.

The sharpness-of-resonance factor, Q, is defined as

$$Q = \frac{\omega_0}{2\,\delta} = \frac{\omega_0\,m}{r}, \qquad (2.21)$$

It is a measure of the width of the peak of the resonance curve—compare Fig. 2.6.

A more illustrative interpretation is possible in the time domain when one considers that after Q oscillations a mildly damped oscillation has decreased to 4 % of its starting value—which is at the brink of what is visually detectable on an oscilloscope screen.

Table 2.1 Typical Q values for various oscillators

Type of element	Quality factor
Electric oscillator of traditional construction (coil, capacitor, resistor)	$Q \approx 10^2-10^3$
Electromagnetic cavity oscillator	$Q \approx 10^3-10^6$
Mechanic oscillator, steel in vacuum	$Q \approx 5 \cdot 10^3$
Quartz oscillator in vacuum	$Q \approx 5 \cdot 10^5$

The decay time, T_d, measures how long it takes for an oscillation to decrease by 60 dB after the excitation has been stopped. At this level, the velocity and displacement have decayed to one-thousandth and the power to one-millionth of its original value. T_d and δ are related by $T_d \approx 6.9/\delta$—see Sect. 12.5.

Table 2.1 lists characteristic Q values for different kinds of technologically relevant oscillators. For comparison, in the aperiodic limiting case, Q has a value of 0.5.

2.4 Forced Oscillation of Parallel Mechanic Oscillators

For free oscillations, the exciting force is zero. Yet, in this section we now deal with the case where the oscillator is driven by an ongoing sinusoidal force, $F(t) = \hat{F} \cos(\omega t + \phi)$, with an angular frequency of $f = \omega/2\pi$, such that the oscillation of the system takes on a stationary state.[4] This mode of operation is termed force-driven or *forced oscillation*. The mathematical description leads to an inhomogeneous differential equation as follows,

$$\hat{F} \cos(\omega t + \phi) = m \frac{d^2\xi}{dt^2} + r \frac{d\xi}{dt} + \frac{1}{n} \xi. \tag{2.22}$$

For sinusoidal excitations, this equation reads as follows in complex form,

$$\underline{F} = -\omega^2 m \underline{\xi} + j \omega r \underline{\xi} + \frac{1}{n} \underline{\xi}. \tag{2.23}$$

Substitution of $\underline{\xi}$ by $\underline{v}$ yields

$$\underline{F} = j \omega m \underline{v} + r \underline{v} + \frac{1}{j \omega n} \underline{v}. \tag{2.24}$$

This equation directly admits the inclusion of the mechanic impedance, $\underline{Z}_{mech}$, as well as it's reciprocal, the mechanic admittance (also termed mobility), $\underline{Y}_{mech} = 1/\underline{Z}_{mech}$, so that

---

[4] Slowly varying frequencies are also in use, assuming that a stationary state has (approximately) been reached at each instant of observation.

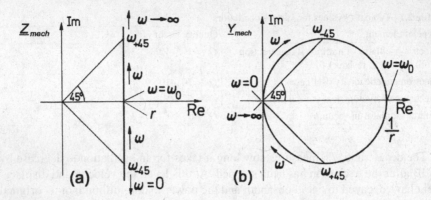

Fig. 2.4 Trajectories of mechanic impedance/admittance in the complex $\underline{Z}$ and $\underline{Y}$ planes as functions of angular frequency. (**a**) Mechanic impedance. (**b**) Mechanic admittance (mobility)

$$\underline{Z}_{\mathrm{mech}} = \frac{F}{\underline{v}} = j\omega m + r + \frac{1}{j\omega n} \quad \text{and} \tag{2.25}$$

$$\underline{Y}_{\mathrm{mech}} = \frac{\underline{v}}{\underline{F}} = \frac{1}{j\omega m + r + \frac{1}{j\omega n}}. \tag{2.26}$$

Figure 2.4 illustrates the trajectories of these two quantities in the complex plane as a function of frequency. The two quantities become real at the *characteristic frequency*, ω_0. At this frequency, the phase changes signs (jumps) from positive to negative values or vice versa.

Figure 2.5 schematically illustrates functions of $\xi(\omega)$ and $v(\omega)$ in the case of slow variations of the frequency of excitation. For simple oscillators these curves have a single peak. In this example, for a case of subcritical damping with $Q \approx 2$, the exciting force is kept constant over frequency.

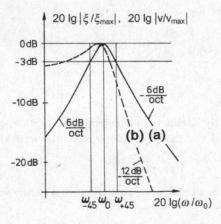

Fig. 2.5 Mechanic responses of a simple resonator as a function of the angular frequency for constant-amplitude forced excitation. (**a**) Velocity. (**b**) Displacement

For the velocity curve, the two points at -3 dB correspond to the frequency ω_{-45} and ω_{45}. The difference of these frequencies,

$$\omega_\Delta = \omega_{45} - \omega_{-45} \tag{2.27}$$

is termed (angular frequency) bandwidth.

The quality factor, Q, is determined by this bandwidth as

$$Q = \frac{\omega_0}{\omega_\Delta}. \tag{2.28}$$

For $Q \le 1/\sqrt{2}$, a displacement resonance does not occur. The course of calculations to arrive at these functions is as follows,

$$\frac{\underline{F}}{\underline{v}} = \frac{\underline{F}}{j\omega\underline{\xi}} = j\omega m + \frac{1}{j\omega n} + r, \tag{2.29}$$

$$\frac{\underline{\xi}}{\underline{F}} = \frac{1}{-\omega^2 m + \frac{1}{n} + j\omega r}, \tag{2.30}$$

$$\frac{|\underline{\xi}|}{|\underline{F}|} = \frac{1}{\sqrt{(\frac{1}{n} - \omega^2 m)^2 + (\omega r)^2}}, \tag{2.31}$$

$$\frac{|\underline{v}|}{|\underline{F}|} = \frac{1}{\sqrt{(\omega m - \frac{1}{\omega n})^2 + r^2}}. \tag{2.32}$$

Note that the phase of the velocity, v, is decreasing and passes zero at ω_0, while the phase of the displacement, ξ, is also decreasing but goes through $-\pi/2$ at this point—see Fig. 2.6. Furthermore, the position of the peak for the $|v(\omega)|$ curve is precisely at the characteristic frequency, while the peak of the $|\xi(\omega)|$ curve lies slightly lower—the higher the damping, the lower the frequency at this peak! Hence, this peak is called the *resonance*. Consequently, it should be distinguished appropriately between the terms resonance frequency and characteristic frequency.

Figure 2.6 shows the resonance curves for the velocity in a slightly different way to illustrate the role of the Q-factor concerning the form of these curves. The resonance peak becomes higher and more narrow with increasing Q. This is the reason that Q is termed sharpness-of-resonance factor, besides quality factor.

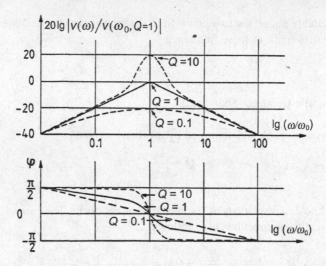

Fig. 2.6 Double-logarithmic plot of resonance curves of the velocity, illustrating the influence of the sharpness-of-resonance factor, Q

2.5   Energies and Dissipation Losses

To derive the energies and losses in the elements from which the oscillator is built, (2.13) is first multiplied with $v(t)$ to arrive at what is called *instantaneous power*, namely

$$P(t) = F(t)\, v(t) = m\, \frac{\mathrm{d}v}{\mathrm{d}t}\, v(t) + r\, v^2(t) + \frac{1}{n}\, v(t) \int \overbrace{v(t)\, \mathrm{d}t}^{\mathrm{d}\xi} \,. \qquad (2.33)$$

Integration over time then leads to a term with the dimension energy (work) as follows,

$$W_{0,\,t_1} = \int_0^{t_1} F(t)\, \overbrace{v\, \mathrm{d}t}^{\mathrm{d}\xi} = m \int_0^{t_1} v\, \frac{\mathrm{d}v}{\mathrm{d}t}\, \mathrm{d}t + r \int_0^{t_1} v^2 \mathrm{d}t + \frac{1}{n} \int_0^{t_1} \xi\, v\, \mathrm{d}t \,. \qquad (2.34)$$

For the case that the motion of the oscillator starts from its resting position, that is, for $\xi_{t=0} = 0$, this expression mutates to

$$\int_0^{\xi_1} F(t)\, \mathrm{d}\xi = \frac{1}{2}\, m\, v_1^2 + r \int_0^{t_1} v^2 \mathrm{d}t + \frac{1}{2}\, \frac{1}{n}\, \xi_1^2 \,. \qquad (2.35)$$

The left term denotes the energy that is fed into the system. The terms on the right side of the equality sign stand, from left to right, for the kinetic energy of the mass, the frictional losses (dissipation) in the dashpot, and the potential energy in the spring.

Our discussion starts with the case of no losses, that is, when $r \equiv 0$. In this case, the total energy in the system does not change. It simply swings between the mass and the spring. We express these relationships as follows,

$$W = \frac{1}{2} m\, v^2(t) + \frac{1}{2n}\, \xi^2(t)\,. \tag{2.36}$$

At the instant $\xi = 0$, all energy is kinetic, and when we have $v = 0$, all energy is potential. In mathematical terms, this is

$$W(\xi = 0) = \frac{1}{2} m\, \hat{v}^2 = W(v = 0) = \frac{1}{2n}\, \hat{\xi}^2\,. \tag{2.37}$$

When losses are present due to friction, that is, when $r \neq 0$, the stationary state is maintained by a driving force. Recall that this section discusses force-driven oscillation. To keep the oscillation amplitude constant, the system requires supplementary power. This power results from the middle term of (2.35) and amounts to

$$W_r = r \int_0^{t_1} v^2(t)\mathrm{d}t\,, \quad \text{and, thus} \quad P_r = r \frac{\mathrm{d}}{\mathrm{d}t} \int_0^{t_1} v^2(t)\mathrm{d}t\,. \tag{2.38}$$

Averaging over a full period, T, with the arbitrary phase, ϕ, we find

$$\overline{P} = r\, \frac{1}{T} \int_0^T \hat{v}^2 \cos^2(\omega t + \phi)\mathrm{d}t$$

$$= \frac{1}{2T}\, r\, \hat{v}^2\, T = \frac{1}{2}\, r\, \hat{v}^2 = \frac{1}{2}\, \hat{F}\, \hat{v} = F_{\mathrm{rms}}\, v_{\mathrm{rms}}\,. \tag{2.39}$$

At the dashpot, $\underline{v}$ and $\underline{F}$ are in phase, which means that the supplied power is purely resistive (active) power. This holds for the complete system when driven at its characteristic frequency. Off this frequency, additional (reactive) power is needed to keep the system stationarily oscillating.

2.6 Basic Elements of Linear, Oscillating, Acoustic Systems

In addition to the mechanic elements, there is a further class of elements for oscillators that are traditionally called *acoustic elements*. Note that the terms *mechanic* and *acoustic* are historic in this case. Since sound is mechanic, the oscillators built from both classes of elements are, to be sure, mechanic and acoustic at the same time.

The acoustic elements are formed by small cavities filled with fluid, that is, gas or liquid. To deal with these cavities as concentrated elements, their linear dimensions must be small compared to the wavelengths under consideration. To define the acoustic elements, this section uses the sound pressure, p, the sound-pressure difference, $p_\Delta = p_1 - p_2$, and the volume velocity,

$$q = \frac{\mathrm{d}V}{\mathrm{d}t} = A\, \frac{\mathrm{d}\xi}{\mathrm{d}t} = A\, v(t)\,. \tag{2.40}$$

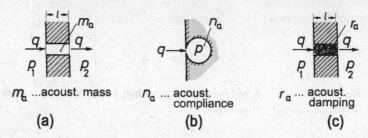

Fig. 2.7 Basic elements of linear acoustic oscillators. (**a**) Acoustic mass. (**b**) Acoustic spring.
(**c**) Acoustic damper

Figure 2.7 schematically illustrates the three acoustic elements—acoustic mass,
m_a, acoustic spring, n_a, and acoustic damper, r_a. Note that here the damper and the
mass are two-port elements while the spring has only one port.

The following equations define these elements.

• *Acoustic Mass*

$$p_\Delta(t) = m_a \frac{dq}{dt} \quad \text{or, in complex notation,} \quad \underline{p}_\Delta = j\omega\, m_a\, \underline{q} \qquad (2.41)$$

• *Acoustic Spring*

$$p(t) = \frac{1}{n_a} \int q\, dt \quad \text{or, in complex notation,} \quad \underline{p} = \frac{1}{j\omega\, n_a}\underline{q} \qquad (2.42)$$

• *Acoustic Damper*[5]

$$p_\Delta(t) = r_a\, q \quad \text{or, in complex notation,} \quad \underline{p}_\Delta = r_a\, \underline{q} \qquad (2.43)$$

2.7　The *Helmholtz* Resonator

The *Helmholtz* resonator is the best-known example of an oscillator with an acoustic
element. A *Helmholtz* resonator is commonly demonstrated by blowing over the open
end of a bottle to produce a musical tone. This is an auditory event with a distinct
pitch, which is adjustable by filling the bottle with some water.

---

[5] For the characteristic parameters of the acoustic elements, the following relations hold: $m_a = \varrho_- l/A$ with ϱ_- being density, $n_a = V/(\eta\, p_-) = V/c^2\, \varrho_-$ with V being volume, $\eta = c_p/c_v$, and
$r_a = \varXi\, l/A$ with $\varXi$ being flow resistivity—for details refer to Sect. 11.5.

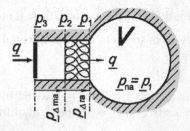

Fig. 2.8 *Helmholtz* resonator with friction

What happens when the bottle is blown on? The air in the bottleneck is a mass oscillating on the air inside the bottle, and the air inside the bottle acts as a spring.[6]

Figure 2.8 schematically illustrates the *Helmholtz* resonator with friction that causes damping. The three elements, mass, damping, and spring, are connected in cascade (chain), so that the total pressure results in

$$\underline{p}_\Sigma = \underline{p}_{\Delta m_a} + \underline{p}_{\Delta r_a} + \underline{p}_{\Delta n_a}. \tag{2.44}$$

Dividing $\underline{p}_\Sigma$ by the volume velocity, $\underline{q}$, delivers the acoustic impedance, $\underline{Z}_a$, namely,

$$\underline{Z}_a = \frac{\underline{p}_\Sigma}{\underline{q}} = j\,\omega\,m_a + r_a + \frac{1}{j\,\omega\,n_a}. \tag{2.45}$$

2.8 Exercises

Differential Equations, Free and Forced Oscillation, Resonance Curves, Complex Power in Mechanic Systems

Problem 2.1

(a) Given a (constant) mass, m, describe/sketch the mechanic impedance of the mass as a function of the angular frequency, ω. Determine the phase angle of its mechanic impedance.

(b) Given a (constant) compliance, n, describe/sketch the mechanic impedance of the spring as a function of the angular frequency, ω. Determine the phase angle of its mechanic impedance.

Problem 2.2 Given a mechanic parallel oscillator with mass, m, compliance, n, and damping, r.

(a) Establish a corresponding differential equation for a harmonic force excitation, $F_0 = \hat{F}_0 \cos \omega\,t$, expressed in a suitable quantity.

---

[6] Usually the spring characteristics of air are not directly experienced because the air evacuates. Yet, in this case, the effect is similar to operating a tire pump with the opening hole pressed closed.

(b) Find a solution to the differential equation if there is no external excitation.

(c) Find the possible types of solutions of the differential equation for excitation as given in (a).

Problem 2.3 A mechanic parallel oscillator be excited by a sinusoidal force of constant amplitude, $F_0 = const$, which is independent of frequency.

(a) Calculate and plot the trajectory of the velocity, $\underline{v}$, in the complex plane as a function of frequency, f.

(b) Find the relationship between the damping coefficient, $\delta = r/2m$, and the bandwidth, ω_Δ, as defined by the half-value (-3 dB) on the resonance curve.

(c) Quantitatively explain the following statement, which is applicable for the determination of the sharpness-of-resonance factor, Q, by inspection of an oscilloscopic plot.

"The quality factor" approximately represents the number of oscillations of a damped oscillation that precede a 96%-reduction of the starting amplitude.

Problem 2.4 A parallel mechanic oscillator contains a mass, m $= 0.63\,\text{Ns}^2/\text{m}$, a spring with a compliance of, n $= 0.027$ m/N, and a damper with r $= 6.02\,\text{Ns/m}$.

Determine for this parallel mechanic oscillator

- The resonance frequency, f_0
- The damping coefficient, δ
- The sharpness-of-resonance factor, Q and the decay time, T_d
- Will this system be oscillating or not?

Problem 2.5 Discuss why the complex power in a mechanic-oscillator system is

$$\underline{P} = \frac{1}{2}\,\underline{F}\,\underline{v}^*. \tag{2.46}$$

Problem 2.6 Given a lossy electric series resonator as shown in Fig. 1.3.

(a) Determine the following quantities using a graph in the complex plane: input impedance $\underline{Z}$, input admittance $\underline{Y}$, resonance frequency ω_0, characteristic resistance $\underline{Z}_0$.

(b) Derive the relation between bandwidth, ω_Δ, and sharpness-of-resonance factor, Q.

(c) Plot the magnitude of the input impedance, $\underline{Y}$, as a function of the angular frequency in double-logarithmic representation.

Problem 2.7 For a parallel mechanic oscillator, show that the phase of ξ, concerning the excitation, becomes $-\frac{\pi}{2}$ at the characteristic frequency of the forced oscillation, ω_0, when a sinusoidal excitation with constant amplitude is applied.

Chapter 3
Electromechanic and Electroacoustic Analogies

During our discussion of the simple mechanic and acoustic oscillators in Chap. 2, readers with some electrical engineering experience may have realized that many mathematical formulas are similar to those that appear when dealing with electric oscillators. There is a general isomorphism of the equations in mechanic, acoustic, and electric networks. This allows to describe mechanic and acoustic networks via analog electric ones. Formulation in electrical-coordinates is often to the advantage of those who are familiar with the theory of electric networks since analysis and synthesis methods from network theory are easily and figuratively applied.

There is more than one way to portray a mechanic or acoustic network by an analog electric one, depending on the coordinates used. There is never the best analogy but rather one which is optimal for the specific application considered. Also, note that analogies have limits of validity. If they mimicked the problem completely, they would cease to be analogies.

For electrical engineers, dealing with mechanic and acoustic networks in terms of their electric analogies means transforming uncommon problems into common ones, which is why they often prefer this method. Nevertheless, it is always possible to deal with the problems in their original form as well.

The following fundamental relations are to be considered when selecting coordinates for analog representations. The two terminals of an electric circuit serve to feed electric energy into the system or extracted it from it. In both cases, the two terminals form a *port*. By restricting ourselves to monofrequent (sinusoidal) signals, it is sufficient to consider complex power instead of energy.

Electric power is the complex product of the electric voltage, $\underline{u}$, and the electric current, $\underline{i}$—as derived in Sect. 16.2. Note that $\underline{u}$ and $\underline{i}$ denote electric coordinates in complex notation, with peak values as magnitudes. Thus, the complex electric power is

$$\underline{P}_{\mathrm{el}} = \frac{1}{2}\, \underline{u}\, \underline{i}^{*}, \qquad (3.1)$$

© Springer-Verlag GmbH Germany, part of Springer Nature 2021
N. Xiang and J. Blauert, *Acoustics for Engineers*,
https://doi.org/10.1007/978-3-662-63342-7_3

with the asterisk denoting the complex conjugate form. By applying the asterisk to $\underline{i}$ and not to $\underline{u}$, we have defined inductive reactive power as positive.

The terminals of mechanic elements and the openings of acoustic elements also form ports, but in these cases, in contrast to the electrical case, one terminal or opening forms a port by itself.

The *mechanic power* is defined as the complex product of force, $\underline{F}$, and particle velocity, $\underline{v}$, as follows,

$$\underline{P}_{\mathrm{mech}} = \frac{1}{2}\,\underline{F}\,\underline{v}^* \,. \tag{3.2}$$

Accordingly, we get the *acoustic power* as the complex product of sound pressure, $\underline{p}$, and volume velocity, $\underline{q}$,

$$\underline{P}_{\mathrm{a}} = \frac{1}{2}\,\underline{p}\,\underline{q}^*. \tag{3.3}$$

Since the asterisk has been applied to $\underline{v}$ and $\underline{q}$, the reactive power of mass is defined to be positive.

To arrive at isomorphisms, we use the electrical coordinates, $\underline{u}$ and $\underline{i}$, in analogy to the mechanic, $\underline{F}$ and $\underline{v}$, or the acoustic ones, $\underline{p}$ and $\underline{q}$. These analogies are restricted by the fact that the electrical coordinates are one-dimensional and thus only represent one dimension of the mechanic/acoustic coordinates. For the vectors $\overrightarrow{\underline{F}}$, $\overrightarrow{\underline{v}}$, and $\overrightarrow{\underline{q}}$, this means that only the spatial component that excites the terminal or opening in the normal direction is represented.

3.1   The Electromechanic Analogies

There are two kinds of analogies possible with mechanic networks. Analogy #1, usually called *impedance analogy*,[1] is expressed as

$$\underline{F} \,\hat{=}\, \underline{u} \quad \text{and} \quad \underline{v} \,\hat{=}\, \underline{i}, \tag{3.4}$$

and analogy #2, also known as *mobile analogy* or *dynamic analogy*, is expressed as

$$\underline{F} \,\hat{=}\, \underline{i} \quad \text{and} \quad \underline{v} \,\hat{=}\, \underline{u}\,. \tag{3.5}$$

Both kinds of electromechanic analogies are used in practice and are, therefore, discussed here. Figure 3.1 provides an overview.

---

[1] The names for the analogies are traditional but make sense in the light of our discussion in Sect. 3.6.

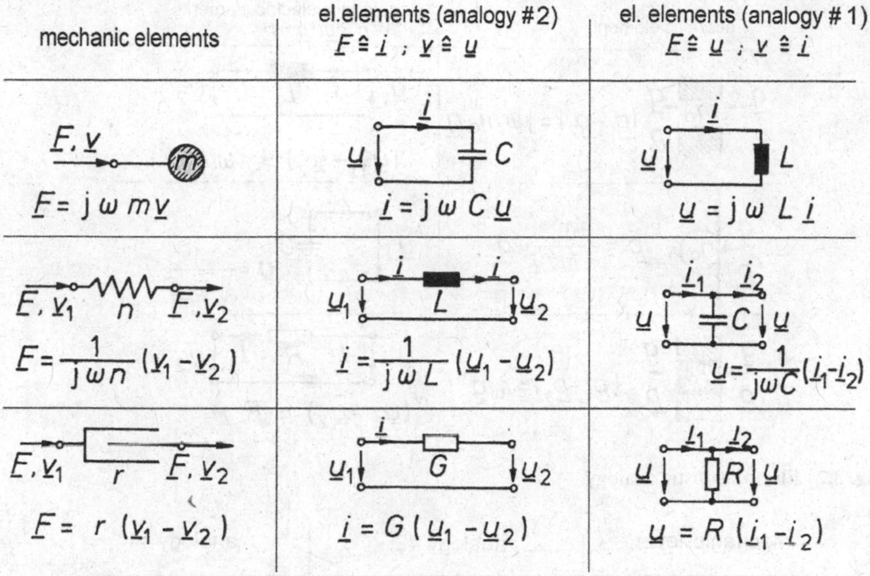

mechanic elements	el. elements (analogy #2) $\underline{F} \mathrel{\hat{=}} \underline{i}$; $\underline{v} \mathrel{\hat{=}} \underline{u}$	el. elements (analogy #1) $\underline{F} \mathrel{\hat{=}} \underline{u}$; $\underline{v} \mathrel{\hat{=}} \underline{i}$
$\underline{F}=j\omega m\underline{v}$	$\underline{i}=j\omega C\underline{u}$	$\underline{u}=j\omega L\underline{i}$
$\underline{F}=\dfrac{1}{j\omega n}(\underline{v}_1-\underline{v}_2)$	$\underline{i}=\dfrac{1}{j\omega L}(\underline{u}_1-\underline{u}_2)$	$\underline{u}=\dfrac{1}{j\omega C}(\underline{i}_1-\underline{i}_2)$
$\underline{F}=r(\underline{v}_1-\underline{v}_2)$	$\underline{i}=G(\underline{u}_1-\underline{u}_2)$	$\underline{u}=R(\underline{i}_1-\underline{i}_2)$

Fig. 3.1 Electromechanic analogies

3.2 The Electroacoustic Analogy

While both variants of electromechanic analogies are used in practice, this is not the case with the electroacoustic ones. Here only one of the two possible analogies is actually used, namely,

$$\underline{p} \mathrel{\hat{=}} \underline{u} \quad \text{and} \quad \underline{q} \mathrel{\hat{=}} \underline{i}. \tag{3.6}$$

Figure 3.2 presents the overview.

Note that all analogies dealt with in Sects. 3.1 and 3.2 relate to networks with lumped (concentrated) elements. This means that wave propagation is not considered. Accordingly, it is required that the acoustic elements are small compared to the wavelength of longitudinal waves across the dimensions of the elements.[2] We also assume that the individual elements are decoupled in every way except through their terminals.

3.3 Levers and Transformers

Besides m, n, r and L, C, R, respectively, there is an additional mechanic linear element that is frequently found in practical networks, namely, the mechanic lever. Its electric counterpart is the ideal, galvanically coupled (single-coil) transformer.

---

[2] An additional type of electroacoustic analogies covering waves will be introduced later in Sect. 8.5.

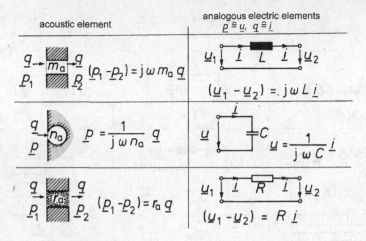

Fig. 3.2 Electroacoustic analogy

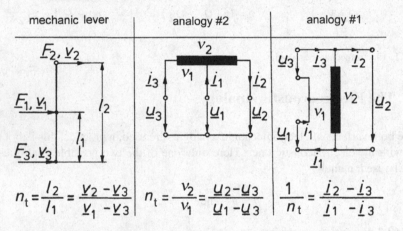

Fig. 3.3 Ideal one-coil transformers as electric analogies for the free-floating mechanic lever. l … lever length, ν … number of turns, n_{t} … transformation ratio

Both lever and single-coil transformers are triple-port elements. Figure 3.3 illustrates the isomorphic relationships for the free-floating lever in static equilibrium for both kinds of electromechanic analogies.

In the domain of electroacoustic analogies, a lever does not exist. The so-called *velocity transformer*—sketched in Fig. 3.4—is frequently mistaken for an acoustic lever, but it actually acts as a mechanic lever with one terminal fixed to ground.

Note that mass and compliance in the cavity are neglected and that the linear dimensions of the cone are small compared to the wavelength. With q and p being continuous, one gets

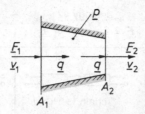

Fig. 3.4 Velocity transformer

$$q = \underline{v}_1 A_1 = \underline{v}_2 A_2 \quad \text{and} \quad \underline{p} = \frac{\underline{F}_1}{A_1} = \frac{\underline{F}_2}{A_2}. \tag{3.7}$$

Introducing $n_t = A_1/A_2$ for the area ratio leads to the following equations,

$$\underline{v}_2 = n_t \underline{v}_1 \quad \text{and} \quad \underline{F}_2 = \frac{1}{n_t} \underline{F}_1, \tag{3.8}$$

and, consequently,

$$\underline{Z}_{2,\,\text{mech}} = \frac{1}{n_t^2} \underline{Z}_{1,\,\text{mech}}. \tag{3.9}$$

Velocity transformers are applied as *impedance transformers* as exemplified by the compression chamber at the mouth of a horn loudspeaker—see Sect. 5.2 for details.

3.4 Rules for Deriving Analog Electric Circuits

When deriving the analog electric circuit of a mechanic or acoustic circuit, the mechanic or acoustic one-, two- or triple-port elements must be replaced by analog electric elements. When connecting those elements, the following rules apply.

For electromechanic analogies the following holds—refer to Fig. 3.5.

- *Cascading* of mechanic elements result in chains of electric elements. The masses or their analog single-port electric elements always form the end of a chain or of a branch.

- *Branching* of mechanic two-port elements leads to parallel branching in analogy #2, and to serial branching in analogy #1—in each case through rigid, massless rods. Again, the single-port elements form the end elements.

For electroacoustic analogies the following holds.

- Cascades of acoustic elements result in chains of electric elements. The single-port spring and its analog electric element form end elements.

- Parallel branching of acoustic elements leads to parallel branching of electric elements.

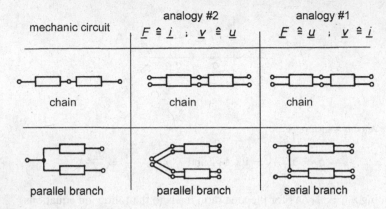

Fig. 3.5 Electromechanic analogies for mono-, dual- and triple-port elements

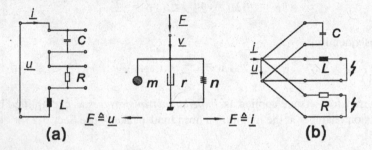

Fig. 3.6 Electric analogies of a simple mechanic parallel-branch oscillator. (**a**) Analogy #1. (**b**) Analogy #2. The lightning symbols denote shortcuts

3.5   Synopsis of Electric Analogies of Simple Oscillators

The schematic in Fig. 3.6 shows how the electric analogies are derived for mechanic *parallel-branch oscillators*, often simply called parallel oscillators, and further, how the electric analogies are derived for mechanic *serial-branch oscillators*, often simply called serial oscillators. Figure 3.7 provides a synopsis of the different possible analog relationships.

3.6   Circuit Fidelity, Impedance Fidelity and Duality

By looking at the electromechanic analogies given in Fig. 3.7, it becomes apparent that the circuits derived by analogy #2, namely, with $\underline{F} \,\hat{=}\, \underline{i}$ and $\underline{v} \,\hat{=}\, \underline{u}$, show the same topology as their mechanic counterpart. This behavior is called *circuit fidelity* or

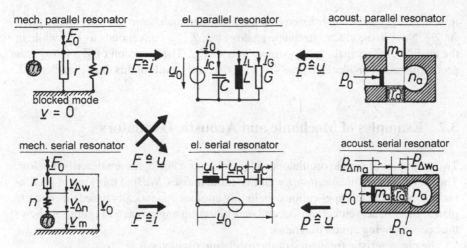

Fig. 3.7 Synopsis of the electric analogies of simple mechanic and acoustic oscillators

topological fidelity. Note that, in this case, impedances transform into admittances and vice versa.

However, those circuits derived with analogy #1, that is, with $\underline{F} \mathrel{\hat{=}} \underline{u}$ and $\underline{v} \mathrel{\hat{=}} \underline{i}$, result in a topology that is *dual* with respect to the mechanic original. In this case, the impedances lead to isomorphic expressions—what is called *impedance fidelity*. The circuit topologies transform into the dual ones.

In electrical networking terminology, the term *dual* refers to two circuits where one behaves in terms of voltages just as the other one behaves in terms of currents. We find that $\underline{Y} = \mathrm{const}^2 \cdot \underline{Z}$ for the elements of dual circuits. This means that the impedance of the one circuit is proportional via a real constant to the admittance of its dual pair. Important dual pairs include the elements capacitance, C, v.s. inductance, L, and resistance, R, v.s. conductance, G. Further, in dual networks, closed loops (meshes) in one correspond to nodes in the other one, and vice versa. Consequently, T-circuits correspond to π-circuits, and serial-branching circuits correspond to parallel-branching ones.

The electroacoustic analogy that we use possesses circuit fidelity as well as impedance fidelity, which is why the other possible analogy is never applied.

To understand the characteristic features discussed above, it is helpful to realize that the loop equation[3] holds for the quantities $\underline{u}$, $\underline{v}_\Delta$ and $\underline{p}_\Delta$, while the node equation holds for $\underline{i}$, $\underline{F}$, and $\underline{q}$.

In this book, we prefer the electromechanic analogy #2 for its topologic fidelity. Yet, this leads to a complication when mechanic and acoustic circuits are merged. To connect an acoustic circuit with a mechanic one, for example, through its input

---

[3] Recall that in electrical terms the loop equation is $\sum \underline{u}_n = 0$, with $n = 1, 2, 3, \ldots$, meaning that by completely circling a mesh we end at the same electric potential. The node equation is $\sum \underline{i}_n = 0$, with $n = 1, 2, 3, \ldots$, meaning that all electric charge that flows into a node must leave it at the same time.

impedance $\underline{Z}_a$, we start with deriving the equivalent mechanic impedance, $\underline{Z}_{mech} = A^2 \underline{Z}_a$. Now, in the electromechanic analogy #2, $\underline{Z}_{mech}$ corresponds to $\underline{Y}_{el}$, while in the electroacoustic analogy $\underline{Z}_a$ corresponds to $\underline{Z}_{el}$. The inversion of $\underline{Z}_{el}$ into $\underline{Y}_{el}$ is accomplished by an ideal *gyrator*—see Sect. 4.3 for details of this dual-port element.

3.7   Examples of Mechanic and Acoustic Oscillators

Two examples of simple oscillators and their electric analogies are described below. The first is a *mechanic oscillator* with two finite masses. We find this kind of oscillators in many practical applications, including engines dynamically based on concrete plates, ultrasound-source transducers, and vibrating engine parts. Figure 3.8 shows the accompanying circuit diagrams.

The characteristic frequency of the mechanic oscillator is

$$\omega_0 = \frac{1}{\sqrt{n\, m_\Sigma}} . \tag{3.10}$$

This relationship becomes evident by looking at the analog electric circuit and noting that the two capacitances are serially linked. Consequently, the effective mass is

$$m_\Sigma = \frac{m_1\, m_2}{m_1 + m_2} . \tag{3.11}$$

Figure 3.9 shows a simple *cavity resonator* with two finite compliances, otherwise known as an *acoustic oscillator*. Such closed-cavity resonators are, for example, applied for calibration of microphones because they are well insulated from external noise and have low power losses as may be caused by radiation. A vibrating piston excites the system and delivers an exactly known volume velocity of q_0.

The characteristic frequency of the acoustic resonator is

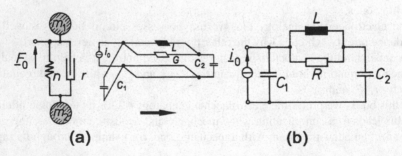

Fig. 3.8 Analog circuit diagram (analogy #2) of a two-mass mechanic oscillator. (a) Mechanic arrangement. (b) Analog electric circuit. The plot in the middle illustrates how the electric circuit is derived

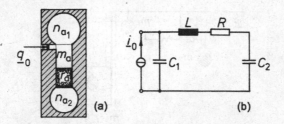

Fig. 3.9 Circuit diagram for a two-cavity acoustic oscillator. (**a**) Acoustic arrangement. (**b**) Analog electric circuit

$$\omega_0 = \frac{1}{\sqrt{m_a n_{a\Sigma}}}, \tag{3.12}$$

and its effective compliance is

$$n_{a\Sigma} = \frac{n_{a_1} n_{a_2}}{n_{a_1} + n_{a_2}}. \tag{3.13}$$

3.8 Exercises

Analog Electrical Circuits, Acceleration and Velocity Measurement

Problem 3.1 For the mechanic system shown in Fig. 3.10, construct an analog electric circuit using analogy #2, that is, $\underline{F} \mathrel{\hat=} \underline{i}$ and $\underline{v} \mathrel{\hat=} \underline{u}$.

Problem 3.2 For the mechanic system shown in Fig. 3.11, construct an analog electric network using analogy #1, that is, $\underline{F} \mathrel{\hat=} \underline{u}$ and $\underline{v} \mathrel{\hat=} \underline{i}$.

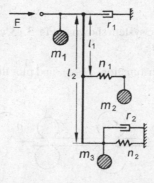

Fig. 3.10 Mechanic circuit to be converted into an equivalent electric one by analogy #2

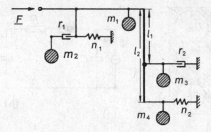

Fig. 3.11 Mechanic circuit to be converted into an equivalent electric one by analogy #1

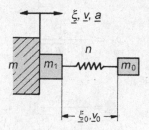

Fig. 3.12 Two-mass arrangement for vibration measurements

Problem 3.3 A mechanic two-mass oscillator—see Fig. 3.12—is to be used as a tool to determine the displacement, ξ, the velocity, $\underline{v}$, and the acceleration, $\underline{a}$, of an oscillating mass, m. To this end, the displacement, $\underline{\xi}_0$, and the velocity, $\underline{v}_0$, are measured between the two masses, m_0 and m_1.

(a) Derive the functions $\underline{\xi}_0 = f(\underline{v})$ and $\underline{v}_0 = f(\underline{v})$.

(b) Which oscillating quantities of the mass, m, are determinable by measurement of $\underline{\xi}_0$ and $\underline{v}_0$,

   – For $\omega \gg \omega_0$ (low-end tuning)?

   – For $\omega \ll \omega_0$ (high-end tuning)?

Problem 3.4 For the acoustic system shown in Fig. 3.13, construct an analog electric circuit using lumped elements.

(a) Determine the transfer function, $\underline{H}(\omega)$, and plot the course of

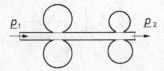

Fig. 3.13 Acoustic circuit to be converted into an equivalent electric one by appropriate analogy

$$|\underline{H}(\omega)| = \frac{|\underline{p}(\omega)_2|}{|\underline{p}(\omega)_1|} \tag{3.14}$$

(b) What are the properties of this network in terms of frequency response?
How to flatten the frequency curve?

Chapter 4
Electromechanic and Electroacoustic Transduction

In the preceding chapter, we dealt with simple linear, time-invariant mechanic and acoustic networks and their electric analogies. In general, these networks can be quite complicated and may assume any number of degrees of freedom. Yet, regardless of how sophisticated the networks are, the energy and power transported in these networks is either mechanic, acoustic, or electromagnetic. Acoustic power and energy are of mechanic nature. Thus the terminological distinction between mechanical, $\underline{F}$, $\underline{v}$, and acoustical coordinates, $\underline{p}$, $\underline{q}$, is purely operational. The current chapter presents the possibility of coupling electrical and mechanical domains, which results in a coupling of electric and mechanic energy and power. This topic is important for modern acoustics.

This coupling can be manifold. The coupling element, the *electromechanic coupler*, can contain its own power sources and may be either active or passive. The relationship between mechanic/acoustic and electric coordinates can be linear or nonlinear. The coupling may be bi-directional, that is, exist for both directions, electric-to-mechanic and vice versa, or only mono-directional. It may be retroactive or not.

In the current chapter, we restrict ourselves to examples of practical importance. On the mechanic/acoustic side, we use the coordinates $\underline{F}$ and $\underline{v}$, which transformes into $\underline{p}$ and $\underline{q}$, given that the effective area, $A_\perp$, is known. In most practical cases, a linear and time-invariant physical relationship between $\underline{F}$, $\underline{v}$ and $\underline{u}$, $\underline{i}$ is presumed. If this is not the case, we assume approximate linearity for small alternating quantities superimposed on large steady offsets.

© Springer-Verlag GmbH Germany, part of Springer Nature 2021
N. Xiang and J. Blauert, *Acoustics for Engineers*,
https://doi.org/10.1007/978-3-662-63342-7_4

4.1 Electromechanic Couplers as Two- or Three-Port Elements

A coupling element between mechanic and electric domains is generally represented as a three-port representation—shown in Fig. 4.1a. In a housing of mass, m, rigidly connected to the terminal 2, there is a *movable component* which can be operated through a rigid, massless rod. This rod penetrates the housing from the left, denoted terminal 1 in the figure.

The power that is transported into the movable component is

$$\underline{P}_\Delta = \frac{1}{2} \underline{F}_\Delta \, \underline{v}_\Delta^* \, . \tag{4.1}$$

The schema shown in Fig. 4.2 is the result of counting all masses rigidly blocked to the housing as part of the housing mass, m. Assuming lossless coupling, we have the following balance of power,

$$\underline{F}_\Delta \, \underline{v}_\Delta^* = \underline{F}_1 \, \underline{v}_1^* - \underline{F}_2 \, \underline{v}_2^* \, , \tag{4.2}$$

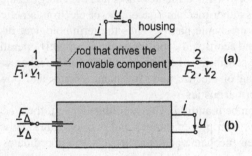

Fig. 4.1 Black-box representations of a coupling element. (**a**) Three-port representation. (**b**) Two-port representation

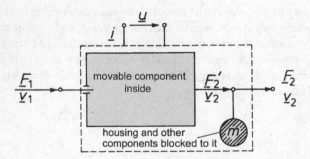

Fig. 4.2 Schematic representation of a coupling element, distinguishing mobile components and components blocked to the housing

which consequently leads to

$$\underline{F}_\Delta = \frac{\underline{F}_1 \, \underline{v}_1^* - \underline{F}_2 \, \underline{v}_2^*}{\underline{v}_1^* - \underline{v}_2^*} \quad \text{with} \quad \underline{v}_\Delta = \underline{v}_1 - \underline{v}_2. \tag{4.3}$$

When the housing is fixed to the ground, as is frequently the case, $\underline{v}_2$ becomes zero and yields $\underline{v}_\Delta = \underline{v}_1$ and $\underline{F}_\Delta = \underline{F}_1$. In this case, terminal 2 may be disregarded because no power passes through it.

Regardless of the specific case, the essential role of the electromechanic coupler is to couple the introduced mechanic power, $(1/2)\,\underline{F}_\Delta \, \underline{v}_\Delta^*$, at one port and the provided electric power, $(1/2)\,\underline{u}\,\underline{i}^*$, at the other—or vice versa. Figure 4.1b illustrates the situation in form of a two-port element. This figure is the basis of our discussion for the rest of this chapter.

4.2 The Carbon Microphone—A Controlled Coupler

An important class of electromechanic couplers consists of couplers where the signal-representing quantities in one network, mechanic or electric, accomplish the coupling by controlling elements of the other network.

An illustrative historical example is the carbon microphone, which was an important part of telephone technology for about a century and was, during that period, the most frequently-used microphone type worldwide.

The carbon microphone is a unidirectional coupler working in the mechanic-to-electric direction. It requires a DC-power supply and behaves as an active element at its output port. It performs a power amplification on the order of 30, which is about 15 dB. This property was the main reason for its widespread use when telephone terminals had not yet other built-in amplifiers.

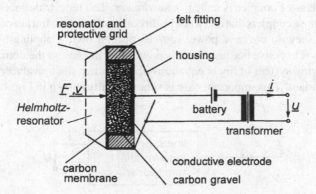

Fig. 4.3 Section view of a carbon microphone

Figure 4.3 illustrates a section through a carbon microphone. The electrically conducting membrane is excited by the sound-pressure impinging on it.

A conductive electrode—metal or carbone—is positioned behind the membrane, and the gap between this back electrode and the membrane is filled with fine-grain carbon gravel. Further, a *Helmholtz* resonator—compare Sect. 2.7—is arranged in front of the membrane to boost the sensitivity in the main frequency range of speech signals.

The electric resistance of the gravel, about $100\,\Omega$, varies according to the alternating pressure on it. If a DC-current is applied to this arrangement, the current superimposes on an alternating current as the resistor varies. The AC component of the current can be filtered out using a transformer—shown in Fig. 4.3. Typical carbon microphones have a sensitivity of $T_{up} \approx 500\,mV/Pa$, which corresponds to $500\,mV$ for a 94-dB sound.

Several disadvantages should be mentioned, though. First, carbon microphones produce many upper harmonics with a power of up to 25% of the fundamental harmonic. Second, they generate quite a bit of internal noise, and, finally, they are power consuming. For these reasons, these microphones have been replaced in communication technology with electret microphones. Such microphones can be miniaturizes and realized on a silicon chip—so-called *MEMS* microphones —compare Sect. 6.6.

Other examples of controlled couplers are the foil-strain gauge (a controlled resistor), the piezotransistor (a pressure-sensitive transistor), and the compressed-air loudspeaker in which an electromagnetic valve controls a stream of compressed air to generate sounds of up to 160 dB.

4.3 Fundamental Equations of Electroacoustic Transducers

From an application standpoint, the most important electromechanic couplers are those where electric power is directly transformed into mechanic power, or vice versa. This class of coupler is called *transducers*. The term transducer is usually reserved for those couplers that can work bi-directionally and are intrinsically passive, meaning that they do not have power sources of their own. Coupling by means of transducers is retroactive because there is power flowing across the domains.

The following system of linear equations applies when the transducers are linear and time-invariant. This important case is schematically shown in Fig. 4.4.

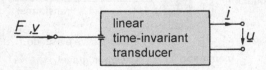

Fig. 4.4 Schematic plot of a linear, time-invariant transducer

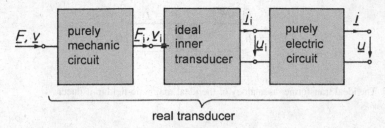

Fig. 4.5 A real transducer as a chain of three two-ports

$$\begin{pmatrix} \underline{v} \\ \underline{F} \end{pmatrix} = \begin{pmatrix} \underline{A}_{11} & \underline{A}_{12} \\ \underline{A}_{21} & \underline{A}_{22} \end{pmatrix} \begin{pmatrix} \underline{u} \\ \underline{i} \end{pmatrix},$$
(4.4)

where we omit the subscript Δ from now on for simplicity. This form of the fundamental transducer equations is called the primary form. The coefficients of the transfer matrix, A_{ik}, are called chain parameters.

We now concentrate on transducers that transform power using force effects in electromagnetic fields. In such transducers the *Lorentz* force is effective. This force is expressible in general form as

$$\overrightarrow{F} = Q_{el}\, \overrightarrow{E} + Q_{el}\, (\overrightarrow{v} \times \overrightarrow{B}),$$
(4.5)

or, in expanded form, with l being the path of i in the B-field,

$$\overrightarrow{F} = (C\, u)\overrightarrow{E} + i\, (\overrightarrow{l} \times \overrightarrow{B}).$$
(4.6)

Assuming the simple case in which the electric field strength, $\overrightarrow{E}$, and the magnetic-flux density, $\overrightarrow{B}$, are constant, we draw the following principles from these equations for the forces in transducers.[1]

– For purely electric fields the force, $\underline{F}$, is proportional to the voltage, $\underline{u}$.
– For purely magnetic fields the force, $\underline{F}$, is proportional to the current, $\underline{i}$.

Real transducers have additional components beyond the actual energy-transducing elements. Transducers working with magnetic fields always contain an inductance, and those working with electric fields always have a capacitance. We also have to expect a resistance, representing electric power losses. On the mechanical side, mass is unavoidable. A spring is required to provide a restoring force on the oscillating mass and to compensate for gravitation, and there is usually some mechanic damping as well.

It is possible to mathematically separate the real transducer into a chain of three two-port elements—depicted in Fig. 4.5. The *inner transducer* can be configured as

---

[1] In general, higher forces can be achieved with magnetic fields because the electric field-strength is limited by the danger of disruptive discharge. This is the reason why electric motors and generators usually use magnetic fields.

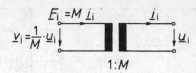

Fig. 4.6 The ideal transformer as analogy of the ideal magnetic-field transducer

ideal, that is, without any resistances, dampers, or reactances. The inner transducer cannot store energy in any way because it is lossless, and it is not directly accessible from outside. The complex power at its ports results as

$$\underline{F}_i\, \underline{v}_i^* = \underline{u}_i\, \underline{i}_i^*. \tag{4.7}$$

In the following, we derive the fundamental equations for inner transducers based on either magnetic or electric fields.

As noted above, $\underline{F}$ and $\underline{v}$ are proportional in magnetic-field transducers. The proportionality coefficient, M, is a real constant and specific to a particular transducer. With the movable component blocked, meaning $\underline{v}_i = 0$, and the electric port cut short, meaning $\underline{u}_i = 0$, we measure a force, $\underline{F}_i$, when applying a current, $\underline{i}_i$, as follows,

$$\underline{F}_i = M\, \underline{i}_i. \tag{4.8}$$

Since energy is neither lost nor stored in the inner transducer, the power at the two ports is identical by definition. Consequently, we know the relationship between $\underline{v}_i$ and $\underline{u}_i$ when $\underline{F}_i$ and $\underline{i}_i$ are zero, namely,

$$\underline{v}_i = \frac{1}{M}\, \underline{u}_i. \tag{4.9}$$

A combination of the two yields

$$\begin{pmatrix} \underline{v}_i \\ \underline{F}_i \end{pmatrix} = \begin{pmatrix} 1/M & 0 \\ 0 & M \end{pmatrix} \begin{pmatrix} \underline{u}_i \\ \underline{i}_i \end{pmatrix}. \tag{4.10}$$

By applying the electromechanic analogy #2, we identify the ideal transformer as analogy of the magnetic transducer shown in Fig. 4.6.

Electric-field transducers show proportionality of $\underline{F}$ and $\underline{u}$. The proportionality coefficient, N, is again a real constant, and specific to a particular transducer. With the movable component fixed, such that $v_i = 0$, the equation is

$$\underline{F}_i = N\, \underline{u}_i, \tag{4.11}$$

and, due to the identity of the power at the two ports, it follows that

$$\underline{v}_i = \frac{1}{N}\, \underline{i}_i, \tag{4.12}$$

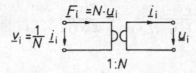

$$v_i = \frac{1}{N} \, i_i \qquad F_i = N \cdot u_i \qquad i_i$$

$$1:N$$

Fig. 4.7 The ideal gyrator as analogy of the ideal electric-field transducer

which results in the subsequent matrix equations,

$$\begin{pmatrix} \underline{v}_i \\ \underline{F}_i \end{pmatrix} = \begin{pmatrix} 0 & 1/N \\ N & 0 \end{pmatrix} \begin{pmatrix} \underline{u}_i \\ \underline{i}_i \end{pmatrix}. \tag{4.13}$$

The electromechanic analogy #2 renders the ideal gyrator as analogy for the electric-field transducer—see Fig. 4.7. The ideal gyrator is a two-port element which is dual to the ideal transformer.[2]

4.4 Reversibility

In mechanics as well as in electric networks, the principle of reciprocity may apply. For example, in mechanics, we observe reciprocity in experiments like the one shown in Fig. 4.8.

This experiment demonstrates that when we apply a force to a bending beam at position 2 and observe a deflection of the beam in position 1, the ratio of force and deflection is the same as would be observed if the force were applied at position 1 and the deflection were observed at position 2, assuming that all other forces are zero. This situation is illustrated by the following equation.

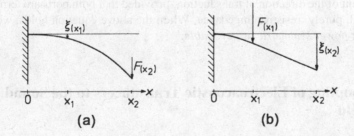

Fig. 4.8 Mechanic reciprocity experiment with a bending beam. (a) Force applied at position x_2, deflection at position x_1. (b) Vice versa

---

[2] When using the analogy #1, the transformer represents the electric-field transducer and the gyrator the magnetic-field transducer.

$$\left.\frac{\underline{F}(x_2)}{\underline{\xi}(x_1)}\right|_{\underline{F}(x_1)=0} = \left.\frac{\underline{F}(x_1)}{\underline{\xi}(x_2)}\right|_{\underline{F}(x_2)=0}. \tag{4.14}$$

The same principle holds in electric networks for the ratio of voltages at one port and currents at another, assuming that these networks are only constructed of inductances, capacitances, resistances, and ideal transformers.

Mathematically, the formulation of the principle of reciprocity is as follows,

$$\det |\mathcal{A}| = \underline{A}_{11}\underline{A}_{22} - \underline{A}_{12}\underline{A}_{21} = +1, \tag{4.15}$$

with chain parameters and by using the chain reference system for currents and voltages as depicted in Fig. 4.4.

To cover electric-field transducers, we include gyrators as additional elements. This requires us to modify Eq. (4.14). The requirement becomes clear when we use analogy #2 to check the reciprocity for magnetic-field and electric-field transducers. From (4.10) we find for magnetic-field transducers $\frac{1}{M} M - 0 = +1$, but from (4.13) for electric-field transducers $0 - \frac{1}{N} N = -1$. Compare Figs. 4.6 and 4.7 for more clarification.

The gyrator obviously introduces a polarity inversion. This inversion is solely introduced by the choice of the analogy and not by physical modification of the circuit! Fortunately, this a 180° phase shift is usually irrelevant from an application point of view.

Consequently, it is sufficient to require only that $\det|\mathcal{A}| \overset{!}{=} 1$, which leads to the following, more general equation,

$$\left.\left|\frac{\underline{u}}{\underline{F}}\right|\right|_{i=0} = \left.\left|\frac{\underline{v}}{\underline{i}}\right|\right|_{F=0}. \tag{4.16}$$

This equation essentially says that the ratio of the power at port 1 and port 2 is independent of the direction of transduction, provided that both ports are terminated with a real, purely resistive impedance. When the above equation holds, we call a transducer *power symmetric* or *reversible*.

4.5   Coupling of Electroacoustic Transducers to the Sound Field

For electromechanic transducers to work, they must be coupled to the sound field in such a way that they can either act as *receivers*[3] by withdrawing power from the

---

[3] Note that in hearing-aid technology the sound emitter is called receiver because it receives electrical signals.

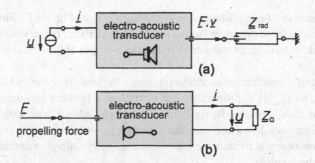

Fig. 4.9 Coupling electromechanic transducers to the sound field renders electroacoustic transducers

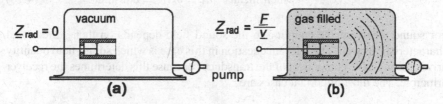

Fig. 4.10 Thought experiment to illustrate the radiation impedance. (a) Avacuated volume, no sound radiation. (b) Gas-filled volume, sound is radiated

field or as *emitters*[4] by delivering power to it. Figure 4.9 illustrates these roles. In this way, we have moved from electromechanic to electroacoustic transducers.

Let us consider the sound emitters first. The following thought experiment helps to better understand the sound-field coupling at the emitter's end—see Fig. 4.10. A sound emitter may be operated in a vacuum. In this case, coupling does not take place because there is no sound field. The mechanic output is idle, meaning that $F = 0$. An electric input impedance of $\underline{Z}_{el}|_{F=0} = \underline{u}/\underline{i}$ can be measured at the input of the transducer in this condition.

Now let air flow into the volume. A different input impedance is measured afterwards, which is explained by the fact that the emitter is no longer idle at the output port. This port is now terminated by a finite impedance called the *radiation impedance*, $\underline{Z}_{rad} = \underline{F}/\underline{v}$. The radiation impedance is a mechanic impedance. The emitter now delivers power to the sound field, namely,

$$\overline{P} = \frac{1}{2}\,\mathrm{Re}\{\underline{Z}_{rad}\}\,|\underline{v}^2| = \frac{1}{2}\,r_{rad}\,|\underline{v}|^2 = \frac{1}{2}\,\mathrm{Re}\{\underline{F}\,\underline{v}^*\}. \qquad (4.17)$$

The radiation impedance depends on the type of sound field. In the case of plane waves, it is a real quantity. We than have $\underline{Z}_{rad} = r_{rad}$, where r_{rad} is called *radiation resistance*. In a free field, plane waves are hard to realize. They are approximately

---

[4] Emitters are sometimes called *transmitters*, to denote that electrical signals are transmitted to the sound field.

given in the beam of a highly directional sound source—see Fig. 10.12. Further, plane waves are found at large distances from a sound source. Exact plane waves only exist in tubes with diameters small compared to the wave-length of the sounds—dealt with in Sect. 7.5.

In the case of omnidirectional radiation, characterized by spherical sound fields of zero order, $\mathrm{Re}\,\{\underline{Z}_{\mathrm{rad}}\}$ is proportional to ω^2 below a limiting frequency, ω'—see Sect. 9.1 for details. This holds, for example, for closed-box loudspeakers—refer to Sect. 5.2. One way to ensure that the radiated power does not decrease below the limiting frequency is to increase the volume velocity with decreasing frequency according to the formula

$$|v|^2 \propto \frac{1}{\omega^2}, \text{ which means } |a| = \omega\,|v| \overset{!}{=} \text{constant}. \tag{4.18}$$

For sound receivers, the coupling to the sound field depends on their directional characteristic as well. The relevant question in this case is which sound-field quantity drives the movable component of the transducer because this determines the receiver principle. The most important cases are,

– The driving force is proportional to the sound pressure, that is, $F \propto p$
– The driving force is proportional to the sound-pressure gradient, namely,
  $F \propto \partial p / \partial r$

We speak of *pressure receivers* in the former and *pressure-gradient receivers* in the latter case.

4.6 Pressure and Pressure-Gradient Receivers

Figure 4.11a schematically illustrates the construction of pressure receivers. There is a closed volume with a membrane of effective area, A, covering part of the surface. The complete arrangement is small compared to the wavelength of the sound, λ. The driving force, in this case, turns out to be $\underline{F} = A\,\underline{p}_1$.

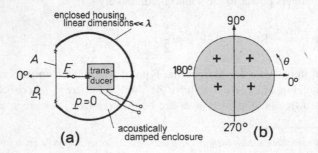

Fig. 4.11 Pressure receiver. (a) Construction. (b) Monopolar directional characteristic

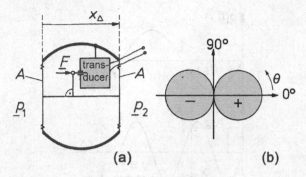

Fig. 4.12 Pressure-gradient receiver. (a) Construction, (b) Bipolar directional characteristic

This relation does not depend on frequency, that is, $\underline{F} \neq f(\omega)$. The sensitivity of the device has a spherical directional characteristic[5]—see Fig. 4.11b. Here we take the microphone axis normal to the membrane as reference direction, i.e. 0°. In other words, $\Gamma = |\underline{F}(\theta)|/|\underline{F}_{max}|$, with θ being the angle between the microphone axis and the sound-incidence direction. Note for all plots of directional characteristics that they have to be considered as three-dimensional, although only the vertical projection is shown here.

Figure 4.12a illustrates the construction of pressure-gradient receivers. As with the pressure receivers, the linear dimensions are small compared to the wavelength. There are two membranes, one on each side of the otherwise closed volume, and each having an effective area, A. The movable component is coupled to the membrane in such a way that its driving force is equal to the difference of the forces affecting each membrane. The same can, by the way, be achieved with only one membrane that is accessible for sound from both sides. The driving force, consequently, becomes

$$\underline{F} = A\,(\underline{p}_1 - \underline{p}_2) = A\,\underline{p}_\Delta. \tag{4.19}$$

For the pressure difference between two points in a sound field in the direction of wave propagation, we get

$$\underline{p}_\Delta = \frac{\partial \underline{p}}{\partial x}\,x_\Delta, \tag{4.20}$$

with $\partial \underline{p}/\partial x$ being a vector, called sound-pressure gradient, $\overrightarrow{\text{grad}\,p}$. For creating a pressure difference between the two membranes, only that portion of $\overrightarrow{\text{grad}\,p}$ is relevant which coincides with the microphone axis. This portion is

$$\underline{p}_\Delta = \overrightarrow{\text{grad}\,p}\,x_\Delta \cos\theta. \tag{4.21}$$

---

[5] The directional characteristic, Γ, is the ratio of the magnitude of the driving force taken for a sound incidence from a reference direction and the magnitude of the driving force in the direction of maximum sensitivity, $\Gamma = 1$.

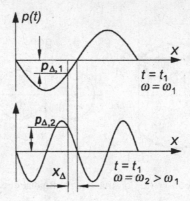

Fig. 4.13 Two sound-pressure waves of different frequencies

The directional characteristic—depicted in Fig. 4.12b—turns out to be

$$\Gamma = \cos\theta . \tag{4.22}$$

This is called the *figure-of-eight* characteristic. The plus signs and minus signs in the plot denote a 0°- or a 180°-phase difference (… polarity inversion), respectively, between the pressure-gradient signal and the electric output signal.[6]

We now consider two special cases, a plane sound field and a spherical one. For a diverging plane wave, we get in simplified form

$$\underline{p}(x) \propto e^{-j\beta x}, \quad \text{with} \quad \beta = \omega/c , \tag{4.23}$$

as derived in Sect. 7.3. Consequently, we obtain

$$\frac{\partial p}{\partial x} \propto \frac{\omega}{c} e^{-j\beta x} . \tag{4.24}$$

This means that for $x_\Delta \ll \lambda/2$, the driving force and, thus, the transducer sensitivity are proportional to the frequency, that is, $\underline{F} \propto \omega$.

Figure 4.13 illustrates this fact. It shows two sound-pressure waves of different frequencies, $\omega_2 > \omega_1$, frozen at an instant $t = t_1$. Clearly, the pressure difference, $\underline{p}_\Delta$, between two points x_Δ apart is higher for the higher frequency. Note what would happen if the microphones were not small compared to the wavelength, λ. For a finite x_Δ, the pressure difference, $\underline{p}_\Delta$, vanishes for $x_\Delta = n\,\lambda/2$. For real microphones this happens above about 4–10 kHz.

For a diverging spherical wave of zero order, which we introduce in more detail in Sect. 9.1, a simplified equation for the sound pressure is

---

[6] The monopolar and the bipolar directional characteristics represent horizontal-plane sections of first-order and second-order spherical harmonics—compare Fig. 9.5.

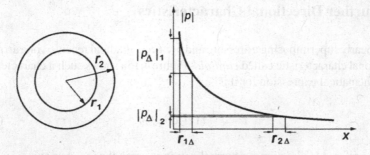

Fig. 4.14 Illustrating the near-field component of the sound-pressure gradient

$$\underline{p}(r) \propto \frac{1}{r}\, e^{-j\beta r} . \tag{4.25}$$

The factor $1/r$ is due to lossless spherical expansion, causing that the same active power passes through all spherical areas (shells) around the sound source. The sound-pressure gradient, then, results as

$$\frac{\partial \underline{p}}{\partial r} \propto \left(\frac{1}{r} + \frac{j\omega}{c}\right) \frac{1}{r}\, e^{-j\beta r} \quad \text{with} \quad \beta = \omega/c . \tag{4.26}$$

When approaching the sound source, the frequency-independent sum term, $1/r$, increases relative to the frequency dependent one, $j\omega/c$. This leads to a relative gain of the low frequencies compared to the higher ones.

The first term in the sum is sometimes called the *near-field gradient*, and the second one the *far-field gradient*. While the far-field gradient originates from phase differences—explained in Fig. 4.13—the near-field gradient stems from the distance-related amplitude decrease with on side of the gradient receiver being closer to the source. This effect is independent of frequency. Figure 4.14 provides an explanation.

Pressure-gradient receivers are usually much less sensitive than pressure receivers because of $\partial \underline{p}/\partial r \ll p(r)$. The relative increase of the low frequencies in the near field—the so-called *proximity effect*—is exploited to construct microphones for acoustically adverse conditions like very noisy or reverberant situations. These microphones are less sensitive for distant sources than they are, for instance, for a speaker's voice when held close to the mouth. We find them used by bus-drivers or announcers at fairs, for example. In particular, all directional stage microphones depend on this effect. The low-frequency-sensitivity increase can partly be compensated by adequate mechanical tuning of the microphone—compare Sect. 5.2.

4.7   Further Directional Characteristics

When linearly superimposing a pressure and pressure-gradient receiver, one arrives at a directional characteristic called *cardioid*. Figure 4.15a shows such a characteristic. The mathematical expression for it is

$$\Gamma = \frac{1}{2}\left(1 + \cos\theta\right),\tag{4.27}$$

where the reference direction for normalization is again the receiver axis.

There are two ways of realizing such a receiver. The first one is to arrange both receivers in practically the same position—*coincident microphones*—and add their output signals in phase and with the same amplitude at frontal incidence. The second one—illustrated in Fig. 4.15b—uses an acoustic delay line to guide sound from the rear side of the receiver to the back of the membrane.

Receivers that select higher-order pressure gradients from the sound field than gradient receivers with a figure-of-eight characteristic achieve even sharper spatial selectivity. The higher orders are determined according to $\partial^n p/\partial r^n$ and lead to directional characteristics as depicted in Fig. 4.16. These receivers are, however, even less sensitive and very frequency dependent, according to $\underline{F} \propto \omega^n$. The analytical expression for their directional characteristic is

$$\Gamma = \cos^n\theta.\tag{4.28}$$

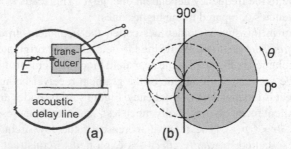

Fig. 4.15 Cardioid microphone. **(a)** Construction. **(b)** Directional characteristic

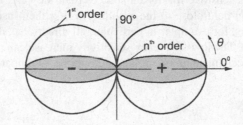

Fig. 4.16 Higher-order figure-of-eight directional characteristic

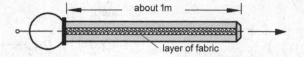

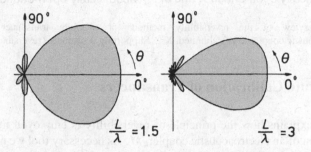

Fig. 4.17 Line microphone

Fig. 4.18 Directional characteristics of line microphones of different length

The receivers we dealt with so far were all small compared to the wavelengths considered, so that no interference takes place. We now discuss an example of a receiver that is distinctly larger than typical wavelengths. This so-called *line microphone*—schematically depicted in Fig. 4.17—deliberately exploits interference of incoming waves.

A line microphone is a receiver with an extremely narrow directionality, feasible for a fairly broad frequency band. It consists of a tube that is typically open at one end and fitted with a microphone at the other one. Along the tube there is a slit through which sound can enter.

Now consider the following two extreme cases.

– Sound impinges laterally, meaning that all points on the slit are excited in phase, prompting waves that all propagate to the microphone. They cancel each other out upon arrival because of their different travel distances, making the receiver insensitive to this direction.
– Sound impinges frontally. Now all waves propagate in phase to the microphone and add up there upon arrival. The receiver has its maximum sensitivity in this direction.

Figure 4.18 illustrates the complete directional characteristics for two receivers of different length. We discuss the computation of such diagrams in Sect. 9.4. Note that the slit is often covered with fabric with a flow resistance that varies along the tube. This feature compensates for losses during wave propagation in the tube.

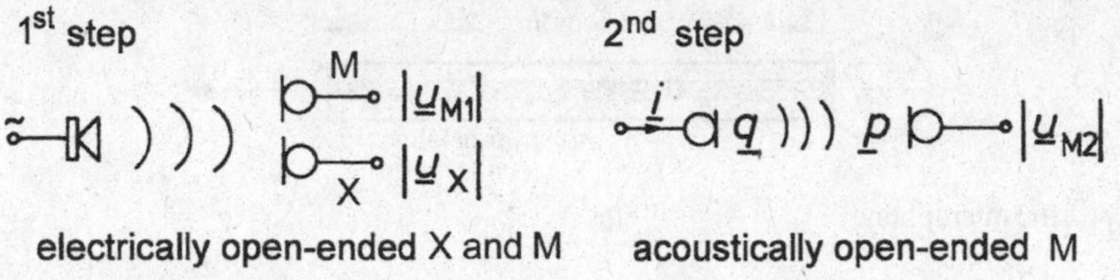

1st step   **2nd step**

electrically open-ended X and M   acoustically open-ended M

Fig. 4.19 Overview of the reversibility method for absolute transducer calibration. **M** … Electroacoustic coupler to be calibrated. **X** … Supporting transducer (reversible)

4.8 Absolute Calibration of Transducers

This section explains how the principle of reversibility is employed for the absolute calibration of an electroacoustic coupler, M. As necessary tool we need a small supporting transducer, X, which is reversible and has a spherical directional characteristic. A constant sound source is also required. At the supporting transducer we have, due to reversibility as stated in (4.16),

$$\left|\frac{u}{F}\right|_{i=0} = \left|\frac{v}{i}\right|_{F=0} \quad \text{and, thus,} \quad \left|\underline{T}_{\text{up}}\right|_X = \left|\frac{u}{p}\right|_{i=0} = \left|\frac{q}{i}\right|_{F=0}. \qquad (4.29)$$

Figure 4.19 denotes the two steps to be taken. Firstly, both the supporting transducer and the microphone to be measured are positioned closely together (spatially coincident) in a free sound field. The output voltages of both components are measured with their electric output ports at idle. The ratio of the two voltages measured corresponds to the ratio of the sensitivities of the two transducers and is expressed as

$$\left|\frac{u_X}{u_{M_1}}\right| = \frac{|\underline{T}_{\text{up}}|_X}{|\underline{T}_{\text{up}}|_M}. \qquad (4.30)$$

Secondly, we use the supporting transducer as a spherical sound source. This source is excited with a current, $\underline{i}$, and generates a volume velocity of $\underline{q}$. Due to the negligible power efficiency of such a source, we assume that its acoustic port is mechanically idling, that is, $F = 0$. This means with (4.29) that find

$$|\underline{q}| = |\underline{T}_{\text{up}}|_X \, |\underline{i}|. \qquad (4.31)$$

With the acoustic impedance, $|\underline{p}/\underline{q}| = |\underline{Z}_a|$, which is computable concerning the spherical sound field with the source distance known, we arrive at

$$|\underline{u}|_{M_2} = |\underline{p}| \, |\underline{T}_{\text{up}}|_M = |\underline{q}| \, |\underline{Z}_a| \, |\underline{T}_{\text{up}}|_M. \qquad (4.32)$$

Inserting (4.30) into (4.31) produces

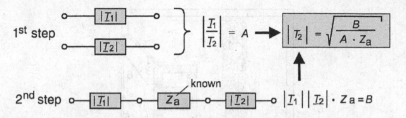

Fig. 4.20 Conceptual summary of the absolute-calibration method

$$|\underline{q}| = |\underline{T}_{up}|_X |\underline{i}| = \left|\frac{u_X}{u_{M_1}}\right| \left|\underline{T}_{up}\right|_M |\underline{i}|. \tag{4.33}$$

Filling this expression into (4.32) results in

$$|\underline{u}|_{M_2} = |\underline{Z}_a| (|\underline{T}_{up}|_M)^2 \left|\frac{u_X}{u_{M_1}}\right| |\underline{i}|, \text{ or} \tag{4.34}$$

$$|\underline{T}_{up}|_M = \sqrt{\frac{|u_{M_1}| \, |u_{M_2}|}{|\underline{Z}_a| \, |u_X| \, |\underline{i}|}}. \tag{4.35}$$

Note that only electric measurements are necessary to determine the sensitivity coefficient, $|T_{up}|_M$, of the electroacoustic coupler in calibration. The coupler to be calibrated needs not be reversible.

An overall accuracy of 1% can be achieved with this calibration method because electric measurements are very accurate. Therefore, certified calibration laboratories apply it. The conceptual essence of the method is summarized in Fig. 4.20. The point is that by knowing the ratio and the product of two quantities, both of them are determined.

4.9 Exercises

Electromechanic and Electroacoustic Transduction

Problem 4.1 For the mechanic system shown in Fig. 4.21, construct an analog electric circuit using analogy #2. The oscillating coil is excited by a constant force, $F =$ constant. Neglect the volume behind the coil and the radiation impedance.

Problem 4.2 Derive the directional characteristic for an arbitrary linear combination of two collocated sound sources, one with a spherical and the other one with a figure-of-eight characteristic. Sketch the following special cases: sphere, hypocardioid, cardioid, hypercardioid, figure-of-eight.

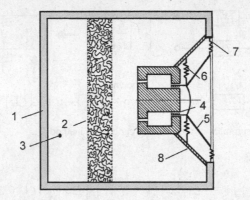

Fig. 4.21 Mechanic-acoustic circuit of a loudspeaker box to be converted into an equivalent electric one by analogy #2

Problem 4.3 Sketch the directional characteristic of a line microphone, that is, a microphone at the end of a narrow tube with a slit opening. Assume plane-wave sound incidence.

Problem 4.4 Pressure-gradient receivers are implemented in practice as pressure-difference receivers.

Determine the magnitude of the pressure difference as a function of frequency, given the distance, d, to the measurement point.

Problem 4.5 In a set-up for determining the sensitivity coefficient of a transducer of magnetic-field type, dedicated for measurements of structure-born sound, the magnetic-field transducer under test is rigidly connected to two other transducers, one of which is reversible. Mass and compliance of the rigid-connection elements are given.

Determine the sensitivity coefficient of the transducer under test, based solely on the electric measurements where the reversible transducer is first used as a transmitter, and then as a receiver.

Problem 4.6 Transmission systems are often mathematically represented by chain matrices.

(a) What advantages does the chain-matrix representation have?

(b) What do the matrix elements mean?

(c) Pure acoustic or electric networks for real transducers can be presented in a form as shown in Fig. 4.22. Find corresponding matrix elements.

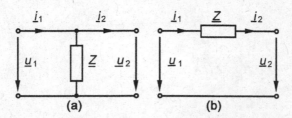

Fig. 4.22 Two-port representation of acoustic and/or electric elements

Problem 4.7 Determine the directional characteristic of a microphone as schematically depicted in Fig. 4.23.

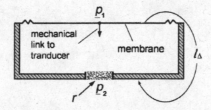

Fig. 4.23 Schematic sketch of a microphone with rear opening $\underline{p}_1$ and $\underline{p}_2$ ...Sound-pressure signals at the frontside and at the backside

Chapter 5
Magnetic-Field Transducers

While dealing with magnetic-field transducers in this chapter and electric-field transducers in the next, we demonstrate that the force-law relationship between the mechanic force, F, and the coupled electric quantity, u or I, is either linear or quadratic. A linear force law, characterized by $F \propto u$ or $F \propto i$, is observed when the energy content of the magnetic or electric field does not vary when the movable component changes position. A quadratic force law, characterized by $F \propto u^2$ or $F \propto i^2$, appears when the movable component meets the electric or magnetic field at a boundary, implying that the energy of the field varies when the movable component changes position. The force in the boundary area is given by a relationships that we obtain by imagining a small *virtual shift* of the border area, namely,

$$F(x) = -\frac{\mathrm{d}}{\mathrm{d}x}\left[\frac{1}{2} L(x)\, i^2\right],$$

(5.1)

for the magnetic field, and

$$F(x) = -\frac{\mathrm{d}}{\mathrm{d}x}\left[\frac{1}{2} C(x)\, u^2\right],$$

(5.2)

for the electric field, having recalled from electrodynamics that the energy, $W = \int F(x)\,\mathrm{d}x$, stored in an inductance is $W_{\mathrm{L}} = (L\, i^2)/2$, and the energy stored in a capacitance is $W_{\mathrm{C}} = (C\, u^2)/2$.

Such quadratic force laws require linearization before applying them to transducers. This linearization is performed by adding a constant offset quantity (bias) to the alternating quantity under consideration. Therefore, we produce a magnetization bias for magnetic-field transducers with a permanent magnet or a constant magnetization current. Similarly, for electric-field transducers, we achieve a polarization bias with an electret or a constant polarization voltage.

© Springer-Verlag GmbH Germany, part of Springer Nature 2021
N. Xiang and J. Blauert, *Acoustics for Engineers*,
https://doi.org/10.1007/978-3-662-63342-7_5

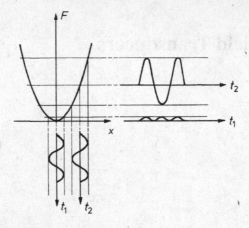

Fig. 5.1 Linearization of the force characteristic

Figure 5.1 illustrates the basic idea of this kind of linearization. Adding bias shifts the alternating quantity to a less curved part of the force plot. In mathematical form we get for $x_- \gg x_\sim$,

$$F \propto x^2 = (x_- + x_\sim)^2 = \underbrace{x_-^2}_{1} + \underbrace{2\,x_-\,x_\sim}_{2} + \underbrace{x_\sim^2}_{3}\,. \tag{5.3}$$

The right side of equation (5.3) has three parts,

- Part 1 denotes a constant quantity. We filter it out with an appropriate high-pass filter.
- Part 2 shows is linearly related with the force. We are primarily interested in the alternating force, $x_\sim$, but it is important to note that the amount of bias, x_-, controls the amplitude.
- Part 3 becomes more and more irrelevant with increasing bias, that is, for $x_\sim \ll x_-$.

Part 3 describes an alternating quantity with a doubled frequency. For sinusoidal excitation it behaves according to

$$\sin^2 \omega t = \frac{1}{2}\left[1 - \cos(2\,\omega t)\right]. \tag{5.4}$$

Energy-converting elements with an intrinsically quadratic force law are no transducers in the strict sense unless they have been linearized. It is also only after linearization that they become reversible. Yet, even non-linearized, they are usable as sound emitters, but they then emit sounds with twice the frequency of the electric excitation signal.

Magnetic-field transducers are often called *velocity transducers* because the output voltage of the innner transducer is proportional to the particle velocity, that is $u_i \propto v_i$. In comparison, electric-field transducers are often called *displacement transducers* because the output voltage of the inner transducer is proportional to the particle displacement, namely, $u_i \propto \xi_i$. The distinction between velocity and displacement transducers is relevant for an optimal mechanic tuning—an item that we discuss later.

5.1 The Magnetodynamic Transduction Principle

The Inner Transducer

Consider a rod-shaped conductor exposed to a stationary magnetic field, $\vec{B}$ —shown in Fig. 5.2. When an electric current, i, passes through this conductor, a force, $\vec{F}$, acts on it according to the expression

$$\vec{F} = i\,(\vec{l} \times \vec{B}).\tag{5.5}$$

Moving the rod within the stationary field induces an electric voltage, u, of

$$u = \vec{l}\,(\vec{v} \times \vec{B}).\tag{5.6}$$

These equations are vector equations. We simplify them here by only considering movements within one spatial dimension. Thus, for movements of the rod perpendicular to the direction of the magnetic induction, $\vec{B}$, the equations for the inner transducer in complex notation reduce to (5.7) and (5.8). The first transducer equation then becomes

$$\underline{F}_i = B\,l\,\underline{i}_i,\tag{5.7}$$

and the second one, showing that this is a velocity transducer, results as

$$\underline{u}_i = B\,l\,\underline{v}_i.\tag{5.8}$$

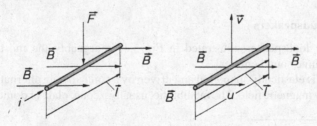

Fig. 5.2 Rod conductor in a stationary magnetic field

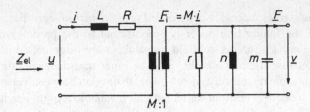

Fig. 5.3 Equivalent circuit for a magnetodynamic transducer

In these equations, the term $M = B\,l$ is referred to as the *transducer coefficient*. It is also possible to derive (5.8) from (5.7) as shown below, by defining the power at both ports to be equal, which means

$$\underline{u}_i\, \underline{i}_i^* = \underline{F}_i\, \underline{v}_i^*.$$

(5.9)

The Real Transducer

Real magnetodynamic transducers[1] have elements besides the inner transducer. A simple equivalent circuit for a real magnetodynamic transducer is given in Fig. 5.3. At the electrical end, we see an inductance, L, and a resistance, R. On the mechanical side, there are the mass of the movable component, m, the necessary spring, n, and some fluid damping, r.

From such an equivalent circuit, it is possible to derive the electric input impedance of the device, $\underline{Z}_{el}$, when used as a sound emitter. The efficiency of electroacoustic sound sources is normally very low, usually only a few percent. For this reason, we assume that the mechanic port runs idle, meaning that it has no load connected to it. For this case we get

$$\underline{Z}_{el}\big|_{\underline{F}=0} = j\,\omega\,L + R + \left[\frac{M^2}{j\,\omega\,m + \frac{1}{j\,\omega\,n} + r} \right].$$

(5.10)

5.2 Magnetodynamic Sound Emitters and Receivers

Dynamic Loudspeakers

The dynamic loudspeaker, illustrated in Fig. 5.4, is arguably the most important magnetodynamic sound emitter.
A membrane is elastically supported and driven by a coil carrying alternating current in a stationary magnetic field. The membrane, usually a cone, plate or dome, transmits

---

[1] The name for this kind of transducer, *magnetodynamic* or just *dynamic*, is of historic origin. In the mechanics terminology, transducers of all kinds are dynamic devices.

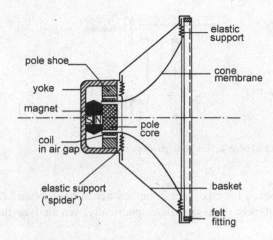

Fig. 5.4 Section view of a dynamic cone loudspeaker

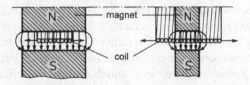

Fig. 5.5 Coil and air-gap of a dynamic loudspeaker

sound into the surrounding air. It should oscillate as a whole, without any bending waves. Membranes are typically of a *non-deconvolveable* form, that is, they cannot be unfolded to a plane. They are manufactured from stiff, sandwich-like layered foils or light foam.

Figure 5.5 illustrates how short coils in a long air-gap or long coils in a short air-gap are used to make sure that the coil does not suffer from field nonlinearities when moving inside the air-gap. The long-coil solution allows for smaller permanent magnets, making it more economical for high magnetic fields. The resistive part of the coil impedance is normally 4–$8\,\Omega$. However, the reactive part is often much larger but reducible by a built-in copper ring. This ring also decreases non-linearities and increases mechanic damping. The damping, in turn, decreases the power efficiency of the device.

The power efficiency of a loudspeaker is proportional to the square of the magnetic-flux density, B, in the air-gap. With a common B of 1–2 Tesla, which is 1–$2\,\mathrm{Vs/m^2}$, the power efficiency is only a few percent. Horn loudspeakers—as dealt with in Sect. 8.3—may achieve up to $15\,\%$.

Loudspeakers are normally built into cabinets (boxes) or baffles. This provision prohibits acoustic shortening, which is a balancing airflow between the front and back sides of the loudspeaker. Absorptive materials like rock wool or porous foam, dampen undesired cavity resonances of the housing—refer to Sect. 11.2. For the resonance frequency of loudspeakers, the compliance of the enclosed air is of concern when they are housed in sealed cabinets.

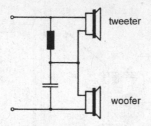

Fig. 5.6 Tweeters and woofers driven through a cross-over network

At low frequencies, loudspeakers in baffles act as hemispheric radiators while those in closed cabinets (boxes) radiate spherically. We analyze this behavior in detail in Sect. 9.3.

Increasing the size of loudspeakers increases the linear dimensions of the membranes until they are on the order of the sound wavelengths in air, thus causing previously omnidirectional radiation to become increasingly directional. We discuss the theory behind this phenomenon in Sect. 10.3. The membrane may also enter into bending movements, which contradicts the otherwise increasing directionality. Making the center of the membrane cone a little stiffer encourages the high frequencies to be emitted from a smaller center area, which reduces the effective mass of the system as frequency increases.

Because the cone center moves with the whole membrane, high frequencies are radiated from a moving source, leading to audible *Doppler* shifts. To avoid this, the high and low frequencies are often radiated by different loudspeakers. The arrangement illustrated in Fig. 5.6 consists of a tweeter for high frequencies and a woofer for low frequencies. The cross-over networks require a careful design.

Magnetodynamic tweeters[2] (dome tweeters) look very much like the moving-coil microphone shown in Fig. 5.10. To avoid focusing of their directional characteristics at high frequencies, more than one tweeter are employed. However, this leads to a problem experienced in all multi-path loudspeaker arrays, namely, interference of signals originating from loudspeakers at different locations. This issue needs careful treatment.

Balancing all the different parameters of loudspeaker systems makes loudspeaker design an *art*, which requires extensive experience and knowledge of materials. A membrane, for example, must not be too stiff, which would cause resonance peaks, but also not too compliant, which would reduce efficiency. The inner membrane support is normally fitted with some damping, the outer support provides an impedance match that prevents the bending waves of the membrane from being reflected from the rim.

The loudspeakers' mechanical tuning usually shifts the main resonance to the low end of the transmitted frequency range. In this frequency range, spherical waves

---

[2] For piezoelectric tweeters see Sect. 6.4.

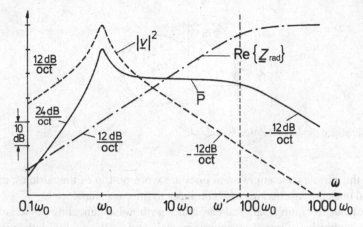

Fig. 5.7 Tuning the frequence response for the case that F and, consequently, i, are constant. $\overline{P}$ denotes the active power

are predominately emitted. Thus, below a limiting frequency, ω', the real part of the radiation impedance, $\underline{Z}_{rad}$, increases proportional to ω^2. The radiated-power situation—depicted in Fig. 5.7—accounts for both of these effects. In Fig. 5.7, we assume that the system is excited with a constant force, $\underline{F}$, which is equivalent to constant-current excitation.

Tuning the mechanic resonance to lower frequencies extends the range. The acceleration, a, of the membrane should be constant to keep the radiated power independent of frequency since from $|\underline{v}^2| \propto 1/\omega^2$ follows $|\underline{a}| = \omega |\underline{v}| = $ const.

The decrease of power radiation at high frequencies is partly counterbalanced by increased focusing toward receiving points inside the range of the radiation beam. Properly designed higher-order resonances further improve the efficiency of the system.

Possible audible *booming*, that is, low-frequency ringing, is avoided by reducing the inner impedance of the electric source until it becomes a matched-power source. This reduces the peak of the resonance. Further reduction of the inner impedance and the current excitation causes the constant-force excitation to convert into a constant-voltage excitation and, thus, into a constant-velocity excitation. Modern power amplifiers have inner impedances of only a few mΩ.

Electric equalization using the techniques discussed here are in use to overcome the mechanic deficiencies of the loudspeaker, but be aware that they require mechanic forces which may considerable. Providing these forces reduces the power efficiency.

Monitoring the movement of the membrane with sensors and controlling the driving force accordingly is called *motional feedback* and has been successful at reducing distortions—particularly at the lower end of the frequency range.

Dynamic Headphones

Headphones/earphones are commonly grouped according to the way that they are worn, namely, as *circum-aural*, *supra-aural*, and *intra-aural*. The circum-aural ones

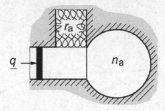

Fig. 5.8 Simplified model of a closed headphone

surround the auricles, the supra-aural ones are worn on top of the auricles, and the intra-aural ones are inserted into the entrances to the ear canals.

It is popular to equip head- and earphones with noise-canceling devices. These reduce the audibility of intruding environmental sounds. We deal with the principle of active noise control in Sect. 14.5 Sound-canceling requires an electrical power supply. This also holds for head- and earphones that are controlled via a wireless link. Yet, our following discussion solely concerns acoustic-design criteria.

Acoustically, the specific way of coupling of head- and earphones to the sound-field is the relevant feature. There we discriminate between open and closed phones.

Open (open-back) headphones are mechanically tuned like loudspeakers (low-end tuning), although these headphones are not mounted in a baffle or box. This situation results in acoustic shorting, but the effect is not problematic in this instance as the listeners' ears are very close to the transducers.

Closed headphones work on a closed air volume, n_a, which may have some degree of parallel leakage. According to an extremely simplified model that disregards leakage-mass effects, sound passing through this leakage experiences some fluid damping, r_a. The situation is depicted in Fig. 5.8.

Resulting from

$$\frac{\underline{q}}{\underline{p}} = j\,\omega\,n_a + \frac{1}{r_a}, \quad \text{namely} \quad \underline{p} = \frac{\cdot\,\underline{q}}{j\,\omega\,n_a + \frac{1}{r_a}}. \tag{5.11}$$

to achieve constant sound pressure, $\underline{p}$, over the entire frequency range, $\underline{q}_0$ and $\underline{v}$ must increase proportionally with ω from the middle frequencies on. This goal is typically accomplished with high-end tuning and aperiodic damping, where ξ is held constant and $v \propto \omega$. At low frequencies, an additional resonance helps creating a "fuller-sounding" auditory event.

Dynamic Microphones

Dynamic microphones have been the most widely used consumer microphone type for a long time. Recently, they are outnumbered by electret and dielectric silicon-chip microphones—in particular, *MEMS microphones*[3]—see Sect. 6.6 for details.

---

[3] MEMS ... Micro-electric-mechanical system.

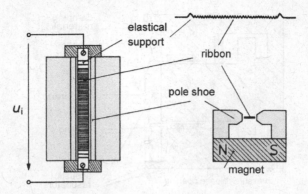

Fig. 5.9 Ribbon microphone

The *ribbon microphone* represents the fundamental principle of dynamic microphones in the clearest form. A softly supported ribbon of light conductive material like aluminum moves between the poles of a permanent magnet—illustrated in Fig. 5.9.

As the sound field reaches the ribbon from both sides, the system acts as a pressure-gradient receiver. It is mechanically low-end tuned. This kind of tuning causes that it works in the region above its characteristic frequency, that is, where the ribbon responds as a mass, and we have $v \propto 1/\omega$. Therewith the typical decrease of the sensitity with ω is compensated. As a direct result of this tuning, the ribbon microphone is unfortunately mechanically delicate and very sensitive to structure-borne sound transmitted from the floor, and to mechanic impact in general. Properly adjusted elastically support reduces these effects—compare Sect. 13.6.

The sensitivity and inner impedance of the ribbon microphone are low since there is only one ribbon in the magnetic field. With a transformer in chain (cascade), which transforms the inner impedance to about 200 Ω, a typical sensitivity would be $T_{up} \approx 1\,mV/Pa$.

Ribbon microphones come as spherical or cardioid receivers. The latter kind is accomplished by positioning an acoustic sink to one side of the ribbon. Ribbon microphones are rare today but sometimes irreplaceable. Trumpet sounds in studios, for instance, are often picked up with these microphones since they are difficult to overload, meaning that they do not produce nonlinear distortions even at very high sound pressures like $> 130\,dB$.

Moving-coil microphones are velocity transducer. Their sensitivity increases due to the usage of a coil instead of a ribbon—depicted in Fig. 5.10. As a result, the induced voltage increases with the number of turns, v. Unfortunately, the mass of the movable component also increases, which may cause audible transient effects due to arrival-time distortions. Typical sensitivities are $T_{up} > 1.5\,mV/Pa$. There are many different, sometimes very sophisticated constructions of dynamic microphones with sizes as small as about 20 mm in diameter. Since moving-coil microphones have a low inner impedance, they do not require an amplifier close to the transducer.

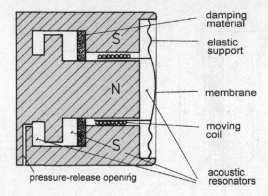

Fig. 5.10 Moving-coil microphone

However, the coil makes them susceptible to magnetic interference, a susceptibility
that is reducible by a fixed compensation coil in series with the moving coil. The
mechanic tuning depends on whether the receiver is a pressure or pressure-gradient
one.

The input impedance of pressure receivers like the one shown in Fig. 5.10 should
be purely resistive to achieve a constant ratio of p/v and a constant resultant output
voltage. To this end, the system is usually heavily damped, and the main resonance of
the system is tuned to middle frequencies—illustrated in Fig. 5.11. Further resonances
are applied at the upper and lower end of the frequency region for equalization. Such
resonances are created by additional cavities behind the membrane in Fig. 5.10. The
figure also shows the felt rings that produce the damping.

Pressure-gradient microphones, that is, figure-of-eight microphones as well as
cardioid microphones are characterized by a second sound inlet on their backside,
leading to a carefully specified acoustic delay-line. The system is mechanically tuned
to the low end so that the system acts as a mass in its main operational frequency

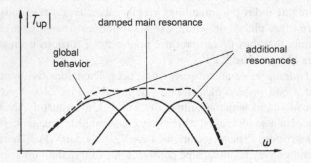

Fig. 5.11 Mechanic tuning of an omnidirectional moving-coil microphone

range. The following two construction features distinguish cardioid microphones. Both of them decrease the *proximity effect*—compare Sect. 4.7.

- The variable-distance principle uses delay lines with different effective lengths for different frequencies to increase the driving force at low frequencies.
- The double-path principle uses two transducers with different delay lines, one for the high frequencies and one for the low. The two transducers are connected by an appropriate cross-over network.

5.3   The Electromagnetic Transduction Principle

The Inner Transducer

Figure 5.12 illustrates the fundamental arrangement for the inner-transducer principle. To derive the transducer equation, we compute the force on the movable armature by imagining a small *virtual shift* of the armature, dx.

We start with a fundamental relation from electrodynamics, *Ampere*'s law, which states that in an arrangement as shown in Fig. 5.12, the electric current, i, and the magnetic-flux density, B, are proportional as follows,

$$i\, v = (B/\mu_0)\, 2\, x\,, \quad v \ldots \text{number of turns in the coil} \qquad (5.12)$$

whereby it is assumed that the iron cores of the yoke and the movable armature are highly permeable, so that the energy of the magnetic field is concentrated in the air-gap, $2\, x$. Multiplication with the cross-sectional area of the air-gap, A, renders

$$A \cdot i \cdot v = (A \cdot B/\mu_0)\, 2\, x \; = (\Phi/\mu_0)\, 2\, x\,, \qquad (5.13)$$

Φ is the magnetic flux. By inserting the definition of the inductance, $L = v\, \Phi/i$, into (5.13), we get the inductance of our arrangement in the form of

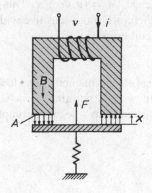

Fig. 5.12 Electromagnet with a movable armature

$$L = v^2 \frac{\mu_0 A}{2x}. \tag{5.14}$$

Referring to (5.1) and (5.14), we now compute the contracting force as

$$F(x) = -\frac{d}{dx}\left[\frac{1}{2}L(x)i^2\right] = \underbrace{\frac{1}{2}\frac{i^2 v^2}{2x^2}}_{(B/\mu_0)^2}\mu_0 A = \frac{B^2 A}{\mu_0} = \frac{\Phi^2}{\mu_0 A}. \tag{5.15}$$

This is clearly a *quadratic* power law. To linearize it, we add a permanent magnetic flux, Φ_-, as a magnetic bias—either by applying a permanent magnet or due to a constant current, i_-. This flux is large as compared to the alternating flux, $\Phi_\sim$, and the two result in $\Phi = \Phi_- + \Phi_\sim$, leading to

$$F(x)_\sim \approx \frac{2\,\Phi_-\,\Phi_\sim}{\mu_0\,A} = \frac{2\,\Phi_-}{\mu_0\,A}\,\frac{v\,\mu_0 A}{2x}, \tag{5.16}$$

with

$$\Phi_\sim = \frac{L\,i_\sim}{v} = v\,\frac{\mu_0\,A}{2x}\,i_\sim. \tag{5.17}$$

In this way, we obtain the first transducer equation,

$$\underline{F}_i \approx \left(\frac{v\,\Phi_-}{x}\right)\underline{i}_i. \tag{5.18}$$

The second equation is easily derived by considering the equality of the in- and output power in (5.9), resulting in

$$\underline{u}_i \approx \left(\frac{v\,\Phi_-}{x}\right)\underline{v}_i. \tag{5.19}$$

Note that the permanent flux, Φ_-, and the number of turns, v, appear in the relationship for the transducer coefficient, $M = (v\,\Phi_-)/x$. This means that the transducer becomes more efficient and, therefore, more sensitive with increasing magnetic bias and an increasing number of coil turns.

The Real Transducer

The equivalent circuit of the real electromagnetic transducer corresponds to the magnetodynamic transducer circuit, with the addition of one very interesting feature. The mechanic compliance is supplemented by a negative compliance called the field compliance, $n_f = dx/dF|_{i_-}$, which results from decreasing the air-gap[4]. The decreasing gap increases the force attracting the armature and opposing the reversing mechanic force. For large displacements this may cause the membrane to bounce to one of the magnet poles and cling there. This is a very undesired effect that becomes more likely with increasing permanent magnetic bias—compare the solutions to problem 5.2.

5.4 Electromagnetic Sound Emitters and Receivers

Electromagnetic transducers come in very small and efficient form. Yet, nonlinear distortions are harder to manage than with magnetodynamic transducers because of the intrinsically quadratic force law. Examples of traditional applications include telephone-receiver capsules, miniature microphones for hearing aids, pick-up transducers for record players, and free-swinging loudspeakers. An electromagnetic telephone-receiver capsule and a hearing-aid microphone are shown in Figs. 5.13 and 5.14 for historical reasons.

In the traditional telephone capsule, a *Helmholtz* resonator tuned to the middle of the speech spectrum is put on top of the membrane to improve the capsules sensitivity for speech signals. Note that low-frequency tuning supports membrane clinging— undesired as explained above.

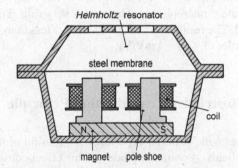

Fig. 5.13 Telephone-receiver capsule

---

[4] The field compliance is derivable by a virtual shift as follows,

$$\frac{1}{n_f}\Big|_{i_-} = \frac{dF}{dx} = \frac{2\Phi_-}{\mu_0 A}\frac{d\Phi_-}{dx} = \frac{2\Phi_-}{\mu_0 A}\frac{-vi_-\,\mu_0 A}{2x^2}$$
$$= \frac{2\Phi_-}{\mu_0 A}\frac{-\Phi_-}{x} = \left(\frac{v\Phi_-}{x}\right)^2\frac{-2x}{\mu_0 A v^2} = -\frac{M^2}{L}.$$

$$(5.20)$$

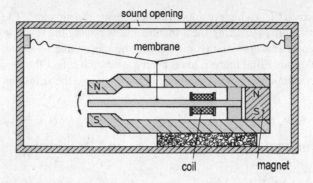

sound opening

membrane

coil   magnet

Fig. 5.14 Hearing-aid microphone

Table 5.1 Magnetization coefficients of different magnetostrictive materials

Material	Magnetization Coefficient
Iron	$l_\Delta/l = -8 \cdot 10^{-6}$
Cobalt	$l_\Delta/l = -55 \cdot 10^{-6}$
Nickel	$l_\Delta/l = -35 \cdot 10^{-6}$
Ferrite	$l_\Delta/l = -100 \text{ to } +40 \cdot 10^{-6}$

   A special construction trick—shown in Fig. 5.14—has long been used to manu-
facture efficient miniature microphones. The reed is very thin. This is possible as it
is not pre-magnetized. The result is a microphone that is less than 1 cm large and has
a sensitivity on the order of $\underline{T}_{\text{up}} \approx 1\,\text{mV/Pa}$.

5.5   The Magnetostrictive Transduction Principle

Rods made of ferromagnetic material experience a variation of their lengths when
exposed to magnetic fields. A common model of this effect considers the distances
between the molecules as forming a *fictive air-gap*. Such a model leads to the same
transducer equations as those of the electromagnetic transducer. The force law is
intrinsically quadratic and requires linearization for transducer use.

   Table 5.1 lists one-dimensional magnetization coefficients, l_Δ/l, for a number
of magnetostrictive materials. A negative sign means that the rod length decreases
when the material is magnetized. Yet, increases do also occur, for example, in ferrite
rods. More sophisticated models than the simple air-gap model are able to explain
this effect.

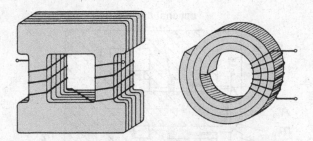

Fig. 5.15 Magnetostrictive transducers

5.6 Magnetostrictive Sound Transmitters and Receivers

Magnetostrictive transducers have a very high mechanic impedance, which makes them well-suited to underwater and/or ultrasound applications. This transducer type achieve efficiencies of more than 90 % when used in water at ultrasound frequencies. Sometimes pre-magnetization is not applied when this principle is used to build narrow-band emitters. In such cases, the transducer emits sound signals with twice the frequency of the exciting electric signals. It is noteworthy in this context that electric transformers for 60 Hz emit sound at 120 Hz, making them electrostrictive devices.

Figure 5.15 shows two exemplary realizations of magnetostrictive transducers. The left one is a two-mass resonator with the bridging bars acting as springs. The different layers of ferromagnetic material are divided by thin isolating foils to avoid eddy currents, which would cause losses. Typical applications include emitters and receivers for echo sounders, and ultrasound emitters for drilling, cleansing, and melding purposes.

5.7 Exercises

Magnetic-Field Transducers

Problem 5.1 Establish an equivalent circuit for an electromagnetic sound transducer that considers the loss, the inductance of the arrangement as well as the field compliance.

How does the transducer behave at electrical short cut—ignoring the loss?

Problem 5.2 Given a moving-coil microphone as depicted in Fig. 5.16.

The microphone has the following parameters.

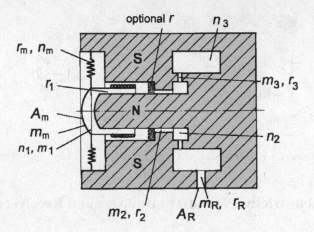

Fig. 5.16 Schematic plot of a moving-coil microphone

m_m ...Mass of membrane and coil
n_n ...Compliance of the support
r_m ...Friction of the support
m_1 ...Mass of co-vibrating air
r_1 ...Friction in the air, particularly in the air gap
m_2 ...Acoustic mass of the air in duct 2
r_2 ...Friction due to porous material in duct 2
n_2 ...Compliance of the air in chamber 2

(a) Establish the equivalent circuit for the moving-coil microphone.

(b) Discuss the frequency course (curve) of the membrane velocity given a frequency-independent sound pressure on the membrane.

Chapter 6
Electric-Field Transducers

In electric-field transducers, electric fields cause mechanic forces or, in the reverse effect, mechanic forces create electric polarization. Even more than magnetic-field transducers, electric-field transducers exist in a wide variety of forms and shapes. Some transducers are intrinsically linear while others have a quadratic force law and must be linearized. Also irreversible controlled couplers exist. In this chapter, we concentrate on the basic principles of electric-field transducers and discuss some illustrative examples.

6.1 The Piezoelectric Transduction Principle

Certain crystals are exploitable as transducers because they have the following properties.

(a) The crystal's physical dimensions change when an electric field is applied to it.

(b) Deformations caused by mechanic forces cause electric polarization on the surfaces of the crystal. These effects are subsumed under the term *piezoelectricity* (pressure electricity).

For piezoelectric effects to occur, the crystals must lack a center of symmetry—illustrated in Fig. 6.1b.

Two characteristic material parameters are used to characterize piezoelectricity, the piezoelectric coefficient, e, and the piezoelectric module, d. These parameters are defined by the following equations,

$$\sigma = e\,s, \quad \text{and} \quad \sigma = d\,\theta, \tag{6.1}$$

© Springer-Verlag GmbH Germany, part of Springer Nature 2021
N. Xiang and J. Blauert, *Acoustics for Engineers*,
https://doi.org/10.1007/978-3-662-63342-7_6

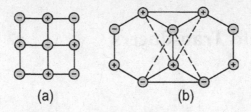

Fig. 6.1 Crystal lattices. **(a)** With … **(b)** Without center of symmetry

Thereby σ is the electric polarization, $s = \xi/x$ is the strain (amount of stretching) and θ is the stress (mechanic tension). The piezoelectric module, d, is often preferred over e by application experts.[1] In the current chapter, we work with e for clearer presentation.

There are generally six different strains that may exist in solid bodies. Three of these exist as normal strains in the directions of the spatial coordinates, x, y, z, and the other three are shear strains along the planes, x/y, y/z, and x/z. As an example, the dielectric flux density, D, for just the x–direction is expressed as

$$D_x = \sum_i e_{x,i}\, s_i + \sum_j \varepsilon_{x,j}\, E_j\,, \qquad \varepsilon \dots \text{dielectric permittivity} \qquad (6.2)$$

with $i \dots x$, y, z, x/y, y/z, x/z, and $j \dots x$, y, z. In many materials, however, most of the piezoelectric coefficients are zero. In the following examples, we only deal with one coefficient at a time. Due to linearity, the total result is obtained by superposition.

Important materials with *inherent piezoelectricity* are the natural crystals *quartz* and *turmaline*, and the artificial crystals *potassium sodium tartrate* (known as *Rochelle* or *Seignette* salt), *lithium niobate, lithium-sulfate hydrate* and *cadmium sulfide*. Table 6.1 lists material characteristics exemplary, where ε_0 is the permittivity of the vacuum. Quartz is particularly stable with regard to temperature and has minimal internal losses. *Rochelle* salt has a high piezoelectric effect but is very sensitive to temperature and humidity.

In addition to the materials noted above, many electrostrictive substances become piezoelectric by particular treatment. We deal with this so-called *influenced piezoelectricity* in Sect. 6.3.

The Inner Transducer

We now derive the piezoelectric-transducer equation for the longitudinal piezoelectric effect as employed in the quartz thickness vibrator shown in Fig. 6.2. The other five dimensions ensue similarly.

As stated earlier, we have $\sigma = e \cdot s$ and $\xi = s \cdot x$, from which we get $\sigma = (e\,\xi)/x$. This allows to write

$$A\,\sigma = Q_{el} = A\,\frac{e}{x}\,\xi\,, \qquad (6.3)$$

---

[1] It is possible to derive d from a given e through the modulus of elasticity, *Young*'s modulus—assuming linearity according to *Hooke*'s law.

Table 6.1 Typical characteristics of inherently piezoelectric materials*

Name	Material	Piezoelectric coefficient, e	Piezoelectric module, d	Dielectric permittivity, ε	Remark
Quartz, SiO_2	Natural crystal	$170\,mC/m^2$ (normal strain)	$2.3\,pC/N$	$4.6\,\varepsilon_0$	*Curie* temp. 570 °C
Rochelle salt	Synthetic crystal	$4.7\,C/m^2$ (shear strain)	$2300\,pC/N$	$200\,\varepsilon_0$ to $1300\,\varepsilon_0$	*Curie* temp. 24–35 °C

*$[C/N] = [m/V]$, $\varepsilon_0 = 8.854188$ pF/m

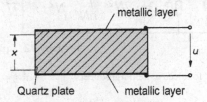

Fig. 6.2 Arrangement for the longitudinal piezoelectric effect

where $Q_{el} = \sigma\,A$ is the electric charge. By making use of $i(t) = dQ_{el}/dt$ and $v(t) = d\xi/dt$, and by assuming that the power is equal at the two ports, that is, $\underline{F}\,\underline{v}^* = \underline{u}\,\underline{i}^*$, we arrive at the two transducer equations,

$$\underline{F} = \left(\frac{e\,A}{x}\right)\underline{u}, \quad \text{and} \quad \underline{i} = \left(\frac{e\,A}{x}\right)\underline{v}. \tag{6.4}$$

The transducer coefficient, N, thus, comes out as

$$N = \frac{e\,A}{x}. \tag{6.5}$$

Piezoelectric transducers, like all electric-field transducers, are elongation (displacement) transducers. With the capacitance being $C = (A\varepsilon)/x$, we get

$$\underline{Q}_{el} = A\,\frac{e}{x}\,\xi = C\,\underline{u}, \tag{6.6}$$

which results in

$$\underline{u} = \frac{e}{\varepsilon}\,\underline{\xi}, \quad \text{that is} \quad \underline{u} \propto \underline{\xi}. \tag{6.7}$$

Note that the ratio of e/ε is the relevant quantity for characterizing the distinction of the piezoelectric effect.

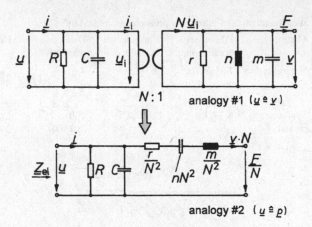

Fig. 6.3 Equivalent circuits for electric-field transducers

As stated above, we consider the inner transducers lossless. Yet, in ferroelectric materials we have to expect losses due to hysteresis. The equivalent circuit of the real transducer in Fig. 6.3 accounts for such losses.

The Real Transducer

There is a gyrator in the equivalent circuit based on electromechanic analogy #2. To avoid this, we replace the mechanic of the circuit with its dual. This action, which is equivalent to switching to analogy #1. Figure 6.3, renderes a simplified equivalent circuit in both forms. In the literature, we usually find the bottom one.

The electric input impedance is most easily derived from analogy #1 and is equal to

$$\left. \underline{Z}_{\mathrm{el}} \right|_{\underline{F}=0} = \frac{1}{G + \mathrm{j}\,\omega C + \left[\dfrac{N^2}{r + \mathrm{j}\,\omega\,m + \frac{1}{\mathrm{j}\omega n}} \right]} \, . \tag{6.8}$$

The equivalent circuits depicted in Fig. 6.3 is also applicable to the other kinds of electric-field transducers.

6.2   Piezoelectric Sound Emitters and Receivers

Piezoelectric sound emitters and receivers have a wide range of use case, involving airborne, waterborne and solid-borne sound. Their realizations depend heavily on the particular application. Figure 6.4 provides an overview of more important vibration forms.

The bending oscillator is particularly interesting. Two piezoelectric layers are glued back-to-back with a conducting foil in between—shown in Fig. 6.5. The high

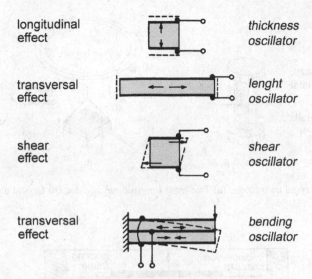

Fig. 6.4 Various vibration forms induced by piezoelectricity

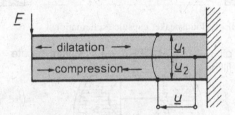

Fig. 6.5 Piezoelectric layers as a bending vibrator

capacitance, C, results in a low inner electric impedance. This provides for a proper impedance match to allow for long connecting wires. The mechanic impedance is low as well, which is advantageous, particularly in airborne-sound fields. Since the two layers act electrically in parallel, they provide good electrical shielding.

A traditional domain for piezoelectric sound emitters is underwater applications, especially utilizing ultrasound. Figure 6.6a shows a two-mass longitudinal vibrator that, for instance, are suitable as echo sounders. The transducer is reversible, thus allowing for its usage as an underwater sound receiver, that is, as a *hydrophone*.

Piezoelectric sound receivers exist in a variety of forms. A relatively high capacitance of about 1–3 μF is valuable from a practical standpoint because it allows for a few meters of cable before amplification becomes necessary.

Figure 6.6b shows a traditional microphone construction that has been known as the *crystal microphone*. The interesting construction trick, worth mentioning here, is that saddle oscillators help provide low compliance.

This kind of microphones features high-frequency tuning when used as a pressure receiver and mid-frequency tuning with high damping when used as a pressure-

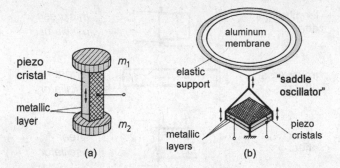

Fig. 6.6   Piezo-crystal transducers. (**a**) Two-mass longitudinal vibrator. (**b**) Crystal microphone

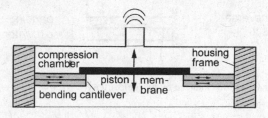

Fig. 6.7   Piezoelectric MEMS-loudspeaker capsule (schematic)

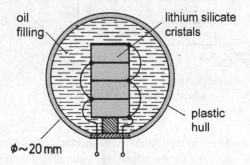

Fig. 6.8   Piezoelectric hydrophone

gradient receiver. To understand these rules, recall that they are elongation receivers, that is, $\underline{\xi} = \underline{v}/j\omega$ holds.

Figure 6.7 shows a section of a piezoelectric miniature loudspeaker schematically. The moving part of it is realized on a silicon chip—so-called "MEMS" loudspeakers.[2] MEMS loudspeakers, and also MEMS microphones—compare Fig. 6.20—come in a variety of constructions. They are, for example, used in hearing aids and mobile phones.

Figure 6.8 depicts a hydrophone where several piezoelements are mechanically stacked (cascaded). Stacking is a widely used method to increase the sensitivity of

---

[2] MEMS ... Micro-electric-mechanical system.

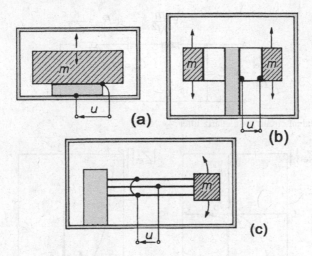

Fig. 6.9 Acceleration/velocity sensors

piezoelectric devices. Piezoelectric transducers are also used as acceleration and/or velocity sensors in vibration meters. Three different construction principles are depicted in Fig. 6.9.

Piezoelectric resonators are used in large quantities as frequency-determining elements in electric circuits, for example, quartz filters in electric watches. Quartz is a preferred material here because of its low internal damping and high thermal stability. The quartz crystal operates in a vacuum to achieve outstanding quality factors on the order of $Q \approx 5 \cdot 10^5$.

Figure 6.10 illustrates an equivalent circuit for a quartz filter. There are two resonances, one parallel and one serial, that lie very close together so that $\omega_p = 1.01 \, \omega_s$. The formulae for the resonance frequencies are

$$\omega_s = \frac{1}{\sqrt{L_m \, C_m}}, \quad \text{and} \quad \omega_p = \frac{1}{\sqrt{L_m \left(\frac{C \, C_m}{C + C_m}\right)}}. \tag{6.9}$$

The curve of the impedance and admittances of the quartz filter over frequency is roughly estimated by applying *Foster*'s reactance rules—which are well known to network specialists. Figure 6.11 shows the resulting plots of the reactance and the magnitude of the input-impedance as a function of frequency schematically.[3]

---

[3] When plotting such curves, it is often advantageous to start with the lossless case and then introduce small losses afterward.

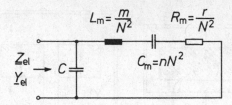

Fig. 6.10 Equivalent circuit for a quartz filter

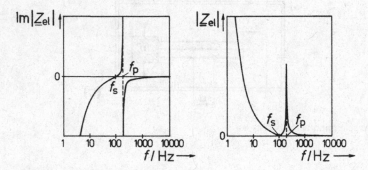

Fig. 6.11 Reactance and magnitude of the input impedance of a quartz filter

6.3   The Electrostrictive Transduction Principle

All dielectric materials experience mechanic deformations when exposed to an electric field. Basically, there is a quadratic relationship between the strength of the field and the stress of the material. However, linearization is accomplishable with a permanent-electric-field bias. In this way, the electrostrictive effect becomes reversible and, thus, applicable for transducer use.

A particularly powerful electrostrictive effect is observed in ferroelectric materials. Some of these materials are polarizable permanently by heating them up beyond their *Curie* temperature and letting them cool down while being exposed to a strong electric field. The ferroelectric *Weiss* domains orient themselves according to the direction of the electric field and keep this orientation when cooled down.

The behavior of polarized electrostrictive materials is called *influenced piezoelectricity*. It acts equivalently to inherent piezoelectricity. Literature often does not even distinguish between inherent and influenced piezoelectricity. Important materials for technological applications are,

(a) Ceramic (polycrystalline) materials, such as *barium titanate* and *lead zircone titanate*

(b) Amorphous *piezopolymeres* (polycrystalline or high-polymers semicrystaline artificial materials), such as *polyvinyliden fluoride* and *polyvinyl chloride*.

Table 6.2 Characteristics of materials with influenced piezoelectricity (line 1 and 3). The material in line 3, shown for comparison, is a piezoelectret*

Name	Material	Piezolelectric coefficient, e	Piezoelectric module, d	Dielectric permittivity, ε	Remark
Lead-zircone titanat, PZT	Ceramic	$\approx 20\,\text{C/m}^2$ (shear strain)	$300\,\text{pC/N}$	$1400\,\varepsilon_0$ to $1700\,\varepsilon_0$	*Curie* temp. $\approx 350\,°\text{C}$
Polyvinyliden fluoride, PVF$_2$	Polymer	$50\,\text{mC/m}^2$ (transversal strain)	$25\,\text{pC/N}$	$13\,\varepsilon_0$	*Curie* temp. $> 100\,°\text{C}$
Polypropylene cellular film, PP	Polymer	$0.50\,\text{mC/m}^2$ (normal str.)	$500\,\text{pC/N}$	$1.5\,\varepsilon_0$	temp. limit $\approx 50\,°\text{C}$

*$[\text{C/N}] = [\text{m/V}]$, $\varepsilon_0 = 8.854188\,\text{pF/m}$

Among the piezoceramics, there are several proven alloys with different characteristics, including very high piezoelectric coefficients. Piezopolymers obtain their piezoelectric features by stretching them in one direction first and, subsequently, polarize them permanently. The transducer equations and simple equivalent circuits are identical for inherent and influenced piezoelectricity.

Some example material characteristics are compiled in Table 6.2. Note that the material in the third line is not piezoelectric but an electret. Sect. 6.6 provides more details on electrets.

6.4 Electrostrictive Sound Emitters and Receivers

Among the electrostrictive sound emitters and receivers there are many shapes in addition to those used in piezoelectric transducers. It is indeed the great advantage of influenced piezoelectricity that many relevant materials are freely formable. For instance, piezopolymeres are manufactured into foils (films) as thin as a few mm. Such foils are elastic and stretchable across curved surfaces. Figures 6.12 and 6.13 present a collection of realized forms.

6.5 The Dielectric Transduction Principle

Historically, this principle was called the *electrostatic* principle. The result of the arrangement shown in Fig. 6.14 is a quadratic force law. Two conducting plates, one fixed and one movable, are exposed to an electric voltage, u. The plates become electrically charged, resulting in an electrostatic pulling force.

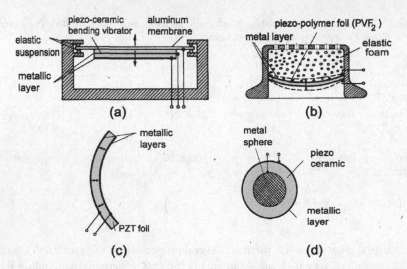

Fig. 6.12 Piezoelectric transducers. (**a**) Piezoceramic telephone receiver. (**b**) Headphone with piezopolymer foil. (**c**) Ultrasound beamer with piezopolymer foil. (**d**) Piezoceramic hydrophone for ultrasound (diameter ≤ 0.5 mm)

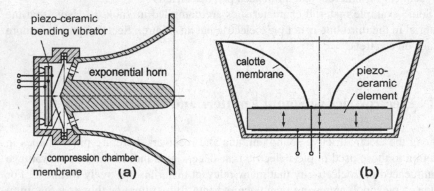

Fig. 6.13 Further piezoelectric transducers. (**a**) Piezoceramic horn tweeter. (**b**) Piezoceramic calotte tweeter. Due to their high impedance, piezoelectric tweeters may be operated in parallel with dynamic loudspeakers without cross-over networks

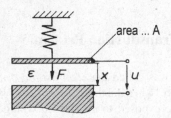

Fig. 6.14 Schematic sketch illustrating the dielectric transducer principle

The Inner Transducer

The transducer equations are again derived by assuming a virtual shift, dx, of the movable plate after the electric source has been disconnected while the plates hold a constant electric charge, Q_{el}. For the energy of the field between the plates, we obtain

$$W_C = \frac{1}{2} C u^2 = \frac{1}{2} \frac{\varepsilon A}{x} u^2 = \frac{1}{2} \frac{Q_{el}^2}{C}, \qquad (6.10)$$

with

$$Q_{el} = C u, \quad \text{and} \quad C = \varepsilon A/x. \qquad (6.11)$$

With the electric charge, Q_{el}, held constant, a virtual shift of dx reveals a quadratic power law as follows.

$$F(x) = -\frac{d}{dx} \left[\frac{1}{2} C(x) u^2 \right] = \frac{Q_{el}^2}{2\varepsilon A} = \frac{C^2}{2\varepsilon A} u^2 = \frac{A\varepsilon}{2x^2} u^2 = \frac{1}{2} \frac{C}{x} u^2. \qquad (6.12)$$

After linearizing with a DC-voltage, so that $u = u_- + u_\sim$ with $u_- \gg u_\sim$, and assuming equality of the port powers, that is $\underline{F}\, \underline{v}^* = \underline{u}\, \underline{i}^*$, the following transducer equations result,

$$\underline{F} \approx \left(\frac{C u_-}{x} \right) \underline{u}, \quad \text{and} \quad \underline{i} \approx \left(\frac{C u_-}{x} \right) \underline{v}. \qquad (6.13)$$

The dielectric transducer coefficient then becomes

$$N = \left(\frac{C u_-}{x} \right). \qquad (6.14)$$

The dielectric transducer is an elongation (displacement) transducer since, with $u = Q_{el}/C$, it follows that $du \propto dx$ and $\underline{u} \propto \underline{\xi}$.

The Real Transducer

The equivalent circuit for the real dielectric, linearized transducers is identical to the one shown in Fig. 6.3 for piezoelectric transducers. However, electrical losses are negligible in practice. The mass of the membrane and, at least for pressure receivers, the mechanic damping are very small compared to the compliance of the air cushion. In other words, the electric input impedance becomes practically a capacitance, which results in

$$\underline{Z}_{el}\Big|_{\underline{F}=0} \approx \frac{1}{j\omega(C + n N^2)}. \qquad (6.15)$$

As with electromagnetic transducers, a detailed analysis shows that a negative field-compliance appears in parallel with the compliance of the air volume. As a result, we again face some danger of the membrane jumping to one plate and clinging to it, especially when the polarization voltage is too high.

6.6   Dielectric Sound Emitters and Receivers

Dielectric sound emitters are used as loudspeakers, especially tweeters, and as ultra-sound emitters—see Fig. 6.15. When light, tightly stretched membranes are used, mechanic tuning to the high-frequency end is unavoidable but maybe compensated for with equalization in the electric circuit.

Since the membrane experiences only tiny displacements, large areas that move in phase are necessary for efficient sound emission. The efficiency increases proportional to polarization-voltage square, but high voltages carry the danger of electric burn-throughs. Fortunately, there are self-healing membrane materials. Various shapes are possible, including large planes and spheres, whereby wire meshes are in use for the back electrodes. Loudspeakers have been built with a frequency range down to 50 Hz using this principle. Figure 6.16 shows two realized push-pull arrangement. There are also dielectric (electrostatic) headphones, which are known for their particularly "clear" presentation. This auditory clearness, which also holds for electrostatic loudspeakers, is attributed to their tiny moving masses, which do not cause any severe phase distortions.

Dielectric sound receivers are known as *condenser microphones*. They have a broad frequency range and only minimal distortions, making a high-quality condenser microphone a good choice for studio and measuring applications. Figure 6.17 presents a schematic section of such a microphone. The air-gap behind the membrane is typically 5–100 μm, and capsule capacitances are 20–100 pF for studio and measuring microphones. The membranes are usually gold-coated plastic foils. Polarization voltages are 40–200 V. The losses of the capsule capacitance are minimal, represented by a so-called charge resistance as high as 0.5–300 GΩ. Due to the high inner impedance, the connecting wires are prone to induce noise, such as humming. Further, they add a parasitic capacitance and may cause mechanical

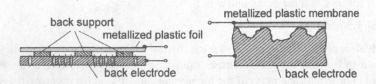

Fig. 6.15 Examples of dielectric sound emitter constructions

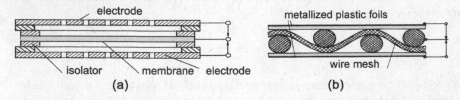

Fig. 6.16 Electrostatic loudspeaker with push/pull driving forces. (a) Arrangement as used in high-quality loudspeaker. (b) Arrangement as, for example, used in spherical loudspeakers

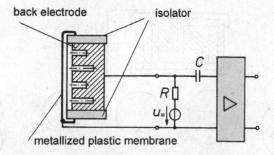

Fig. 6.17 Condenser microphone with accompanying electric circuit

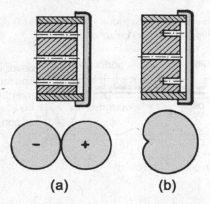

Fig. 6.18 Figure-of-eight or cardioid microphone

problems. To avoid such issues, there is often an impedance-converting amplifier positioned back-to-back with the capsule. After this stage, typical sensitivities are about $T_{up} \approx 10\text{--}30\,\text{mV/Pa}$,

Condenser microphones, when driven in a so-called *low-frequency circuit*— shown in Fig. 6.17—have, on principle, a low-end roll-off that is determined by the first-order R/C high-pass formed by the capsule capacitance and the charge resistance.

An alternative method of operating dielectric microphones, the so-called *high-frequency circuit*, avoids this high-pass behavior. Here, due to their varying capacitance, condenser microphones are integrated into electric-oscillator circuits as frequency-determining elements. However, in this mode, the condenser microphones no longer act as reversible transducers but as controlled couplers. The principle is applied for measurement microphones because it is extremely resistant to induced noise. Also, as it does not have a lower cut-off frequency, you can even measure the static air pressure with it!

Condenser microphones are elongation transducers. As pressure receivers they are high-end tuned, and as figure-of-eight or cardioid microphones—such as shown in Fig. 6.18—they are low-end tuned with prominent damping.

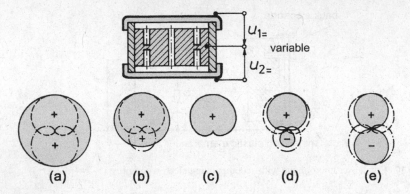

Fig. 6.19 Microphone with a steerable directional characteristic. (**a**) Spherical. (**b**) Wide-angle cardioid. (**c**) Cardioid. (**d**) Super cardiod. (**e**) Figure-of-eight

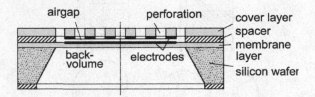

Fig. 6.20 Silicon-chip microphone (schematic)

High symmetry of the figure-of-eight directional characteristic is achieved by placing a counter electrode on the other side of the capsule. This detail also allows the directional characteristic to be steered electrically since the sensitivities of the membranes are controlled by the applied polarization voltages. Figure 6.19 provides an overview.

Miniature dielectric microphones, MEMS, are manufactured very much like electronic chips, for instance, by etching little membranes with an air-gap into the substrate of silicon wafers. This process leads to silicon-chip microphones—schematically shown in Fig. 6.20. Due to their very small dimensions, MEMS microphones are easily arranged in arrays. Both during the production process and the actual use, these microphones withstand fairly high temperatures.

In a similar way as *magnetization* is achieved with magnets, *polarization* is achieved with so-called *electrets*. These are materials that show a permanent polarization after appropriate treatment, such as exposure of heated material to strong electric fields and subsequent down-cooling, corona discharge, or other forms of electron bombardment. If such a material, for example, *teflon* or *fluorocarbon*, is positioned behind the membrane, an external polarization voltage becomes superfluous. These electrets easily mimic polarization voltages of 100 V. Figure 6.21 illustrates the principle.

Electret microphones, including an integrated impedance converter, are manufactured very economically and have become a widespread microphone type worldwide.

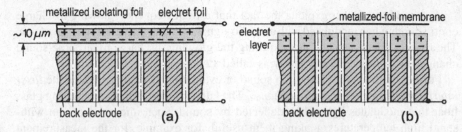

Fig. 6.21 Electret microphones. (**a**) Version with electret membrane. (**b**) Version with electret back electrode

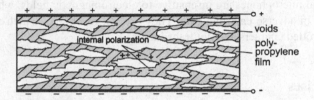

Fig. 6.22 Piezoelectret film

Recently, *elastic electrets* have become available. An example would be piezoelectric polymer films, made from *polyethylene*. They are extruded and treated mechanically afterward in such a way that they develop a cellular structure with ample void bubbles in it—sketched in Fig. 6.22. Polarization is performed by corona discharge with the effect that surface polarization develops around the voids. These so-called *piezoelectrets* behave very much like piezoelectric material—see Table 6.1 for comparison. Piezoelectret films, coated with conducting layers, act as microphones without additional attachments. Stacked $\approx$50 μm layers of these films achieve sensitivities of $\geq$20 mV/Pa.

6.7 Further Transducer and Coupler Principles

In addition to the transducer principles that we already discussed in the current and prior sections, there are some further techniques, even without magnetic and electric fields, that we like to mention briefly.

The first bases on the fact that a wire with a varying current passing through it heats up, causes variations of the air pressure around it, and radiates sound. This is called a *thermophone*. The wire resistance also varies in response to the varying velocity of airflow, making it useful as a *velocity receiver*. Such a device is called a *hot-wire anemometer*. Modern constructions employ micro-machined sensors with two very thin, heated wires. These cover a frequency range from 0 Hz up to more than 20 kHz.

A second transducer exploits the fact that a high-frequency glowing discharge emits sounds when modulated with an audio signal, creating the basis for a *ionophone*. The reverse effect, namely, that exposing the glowing discharge to airborne-sound changes its resistance in synchrony, is called a *cathodophone*.

Further methods for picking up sound or evaluating vibrations include the *laser interferometer* and *optical microphones*. The latter contains a light-conducting glass fiber that modulates light when deflected by sound. Such microphones can withstand high temperatures, making them useful, for example, for the measurement of sounds inside combustion engines. Further, silicon-chip microphones are built in such a way that the membrane movements are monitored by optic rather than electric means. Optical microphones are insensitive to electromagnetic fields, which can be an advantage in adverse environments. They are not reversible, which means that they are controlled couplers and not transducers.

6.8 Exercises

Electric-Field Transducers

Problem 6.1 The following data are given for a dielectric microphone.

Membrane area	...$A = 9.1 \cdot 10^{-4}\,\text{m}^2$
Membrane thickness	...$h = 10^{-5}\,\text{m}$
Membrane distance	...$x_0 = 2 \cdot 10^{-5}\,\text{m}$
Membrane-material density	...$\rho_m = 8 \cdot 10^3\,\text{kg/m}^3$
Membrane-resonance frequency	...$f_0 = 12\,\text{kHz}$
Polarization voltage	...$u_0 = 150\,\text{V}$
Resistance	...$R_r = 10\,\text{M}\Omega$

It is assumed that only half of the membrane mass actually moves. Dielectric loss and membrane loss should not be considered.

(a) Determine the elements of the simplified electric equivalent circuit, consider the field-compliance (...*field-spring*).

(b) Determine the frequency course (curve) of the microphone.

(c) What is the transducer coefficient, $\underline{u}/\underline{p}$, within its effective frequency range?

Problem 6.2 Given an electret microphone as plotted in Fig. 6.23, whereby

1 ...thin metal layer with a conductivity $\rightarrow \infty$

2 ...isolating layer

3 ...electret with surface polarization, σ

4 ...air-gap,

5 ...metallic back plate with a conductivity $\rightarrow \infty$

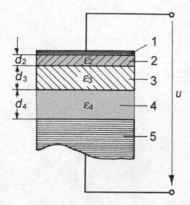

Fig. 6.23 Schematic plot of an electret microphone

In this microphone, the air-gap, d_4, varies as

$$d_4 = d_{40} + d_{41} \sin(\omega t) \quad \text{with } d_{41} \ll d_{40}. \tag{6.16}$$

(a) Calculate approximately the AC-component, $u_\sim$, of the voltage, u, on an electret microphone.

(b) What is the AC-component of the voltage, u, on a dielectric microphone of the same construction, driven by a DC-voltage, u_{0-}, but without electret? Find the equivalent DC-voltage, u_{0-}, for the electret microphone with $\sigma = 10^{-8}\,\mathrm{C/cm^2}$, $d_3 = 1.3 \cdot 10^{-3}\,\mathrm{cm}$ and $\varepsilon_2 = \varepsilon_3 = 3\,\varepsilon_0$.

(c) What is the transducer coefficient, $\underline{u}/\underline{p}$, of an electret microphone below its resonance frequency?

Problem 6.3 A quartz employed in an oscillator—see Fig. 6.24—is specified as follows.

$R_1 = 60\,\Omega$, $L_1 = 1.5\,\mathrm{H}$, $C_2 = 0.016\,\mathrm{pF}$, and $C_3 = 60\,\mathrm{pF}$

(a) Determine both the parallel and the series resonance, the frequency interval between the two resonances, and the quality factor. Sketch the admittance in the complex plane.

(b) How does an additional capacitor, C, parallel to the quartz influence these quantities?

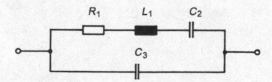

Fig. 6.24 Equivalent circuit of a quartz oscillator

Chapter 7
The Wave Equation in Fluids

So far in this book, we dealt with *vibrations*. These are processes that vary as functions of time. We were able to describe relevant types of vibrations with *common* differential equations. The current chapter now focuses on *waves*. These are processes that vary with both time and space. Their mathematical description requires *partial* differential equations.

Motivated by the electroacoustic analogies, we start with an excursion into electromagnetic waves. The discussion brings us back to sound waves in the next section.[1]

Figure 7.1 illustrates the equivalent circuit for an elementary section of a homogeneous, lossless electrical transmission line. For this section, the following loop and node equations hold,

$$u - \left(u + \frac{\partial u}{\partial x}\, dx\right) = L'\, dx\, \frac{\partial i}{\partial t}, \quad \text{and} \tag{7.1}$$

$$i - \left(i + \frac{\partial i}{\partial x}\, dx\right) = C'\, dx\, \frac{\partial u}{\partial t}. \tag{7.2}$$

The following two linear differential equations are obtained by neglecting the higher-order differentials, thus we get

$$-\frac{\partial u}{\partial x} = L'\, \frac{\partial i}{\partial t} \quad \text{and} \quad -\frac{\partial i}{\partial x} = C'\, \frac{\partial u}{\partial t}. \tag{7.3}$$

This set of equations shows that a temporal variation of the slope of one of the variables results in a proportional spatial variation of the slope of the other one. This

---

[1] An advice for readers who have not yet heard of electromagnetic waves: Start with reading Sects. 7.1–7.2.

© Springer-Verlag GmbH Germany, part of Springer Nature 2021
N. Xiang and J. Blauert, *Acoustics for Engineers*,
https://doi.org/10.1007/978-3-662-63342-7_7

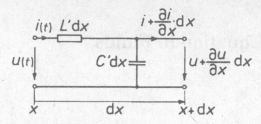

Fig. 7.1 Equivalent circuit for a differential section of a homogeneous, lossless electrical transmission line

causes the total energy on the line to swing between two types of complementary energies, namely,

– Magnetic energy per length, $W' = \frac{1}{2}L'\,i^2$
– Electric energy per length, $W' = \frac{1}{2}C'\,u^2$

We now combine the two linear differential equations (7.3), to a differential equation of the second order, which is

$$\frac{\partial^2 u}{\partial x^2} = \frac{1}{c_{\text{line, el}}^2}\,\frac{\partial^2 u}{\partial t^2}\,.\tag{7.4}$$

This is the so-called *wave equation*, here in the formulation for electromagnetic waves on an electrical transmission line. Hereby $c_{\text{line, el}}$ is the propagation speed of electromagnetic waves on the transmission line, namely,

$$c_{\text{line, el}} = \frac{1}{\sqrt{L'\,C'}}\,.\tag{7.5}$$

In acoustics, we also have waves that propagate along with one coordinate, for example, in gas-filled tubes with diameters that are small compared to the wavelength. Such *longitudinal* compression waves are schematically sketched in Fig. 7.2.

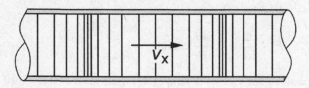

Fig. 7.2 One-dimensional longitudinal acoustic wave. Zones of compression and rarefaction are schematically indicated

Two complementary forms of energies are again required for these types of waves to occur, but in this case we choose energies per volume. In this way, we stay compatible with the use of p and v as characteristic sound-field quantities. We thus get,

- Potential energy per volume, $W'' = \frac{1}{2} \kappa p^2$ κ ...volume compressibility
- Kinetic energy per volume, $W'' = \frac{1}{2} \varrho v^2$ ϱ ...mass density

Mimicking the wave equation for the electromagnetic waves, we consequently suggest the following wave equations for one-dimensional acoustic waves,

$$-\frac{\partial p}{\partial x} = \varrho \frac{\partial v}{\partial t} \quad \text{and} \quad -\frac{\partial v}{\partial x} = \kappa \frac{\partial p}{\partial t}, \quad \text{or} \quad -\frac{\partial v}{\partial x} = \frac{1}{\varrho c^2} \frac{\partial p}{\partial t}, \qquad (7.6)$$

and the combined-second order expression as

$$\frac{\partial^2 p}{\partial x^2} = \frac{1}{c^2} \frac{\partial^2 p}{\partial t^2}, \qquad (7.7)$$

with the propagation speed of sound waves to be

$$c = \frac{1}{\sqrt{\kappa \varrho}}, \qquad (7.8)$$

and the volume compressibility to be

$$\kappa = \frac{1}{\varrho c^2}. \qquad (7.9)$$

The relevant question is now whether these supposed equations, which are so neatly analog to electric expressions, are actually in compliance with the physical reality. The answer is *yes*, at least approximately. Yet, there are several features of the media where sound exists that must be taken as idealized. A total of five linearizations is necessary. The following section elaborate on these.

7.1 Derivation of the One-Dimensional Wave Equation

We assume an idealized medium as a model of the real physical medium. Only compression and expansion but no shear stress are allowed. This restriction limits us to a media category called *fluids*. Many gases and liquids are treatable treated as fluids. Fluids only experience longitudinal waves. The following features of the idealized medium are assumed.

- The medium is homogeneous and does not crack, meaning that there are no inclusions of vacuum as might be caused by cavitation.
- The thermal conductivity is zero. This assumes adiabatic compression.

– The inner friction is zero, meaning that there are no energy losses and no viscosity.

– The medium has a defined mass and elasticity.

We now assume that the medium is not flowing and that there is no drift, meaning that $v_- \approx 0$. The alternating part of the pressure be small compared to the static pressure, that is, $p_- \gg p_\sim$. The alternating part of the density is also small in comparison to the static one, in other words, $\varrho_- \gg \varrho_\sim$. This results in a so-called *small-signal operation* of the medium.

To arrive at the wave equation, we recollect three fundamental physical relationships. These are listed in the following.

- *Hooke*'s law, applied to fluids
- *Newton*'s mass law
- The mass-conservation law

We now formulate these three relationships in their differential forms as required for the rest of our discussion. In differential form, they are known as

- State equation
- *Euler*'s equation
- Continuity equation

The State Equation

The relationship $p = f(\varrho)$, as conceptually depicted in Fig. 7.3, applies to the medium because it has mass and elasticity. For small-signal operation, $\varrho_\sim \ll \varrho_-$, we substitute this function with its tangent at the operating point, which is

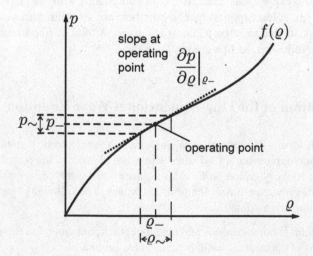

Fig. 7.3 Conceptual relationship between the sound pressure and the medium density

$$p_\sim = \left.\frac{\partial p}{\partial \varrho}\right|_{\varrho_-} \varrho_\sim \quad \text{or} \quad p_\sim = c^2 \varrho_\sim . \tag{7.10}$$

This leads us to the *state equation* in the form

$$\partial \varrho = \frac{1}{c^2} \left.\partial p\right|_{\varrho_-} \quad \text{or} \quad c = \sqrt{\left.\frac{\partial p}{\partial \varrho}\right|_{\varrho_-}} . \tag{7.11}$$

In this way, we make our *first linearization* by assuming a linear spring characteristic for the fluid. We prove later that c is the speed of sound, which is specific for a material but depends on its temperature, humidity, and static pressure.

The proportionality coefficient in the state equation, c^2, is estimated theoretically for most single-atom gases by assuming a *perfect gas*. If such a gas is compressed in the absence of heat conduction, we have so-called *adiabatic compression*. The then applicable adiabatic law of thermodynamics states that

$$p\, V^\eta = \text{const} = p_- \, V_-^\eta , \tag{7.12}$$

where V is the volume of a mass element concerned, and $\eta = c_p/c_v$ (usually known as γ) is the ratio of the specific heat capacities,[2] with which we derive

$$p = \left(\frac{V_-}{V}\right)^\eta p_- = \left(\frac{\varrho}{\varrho_-}\right)^\eta p_- . \tag{7.13}$$

Differentiation of this expression at the position $\varrho = \varrho_-$ renders the following expression,

$$\left.\frac{\partial p}{\partial \varrho}\right|_{\varrho_-} = \eta \underbrace{\left(\frac{\varrho}{\varrho_-}\right)^{\eta-1}}_{1 \text{ for } \varrho = \varrho_-} \frac{1}{\varrho_-} p_- , \tag{7.14}$$

and then, consequently,

$$\left.\frac{\partial p}{\partial \varrho}\right|_{\varrho_-} = c^2 = \frac{\eta\, p_-}{\varrho_-} . \tag{7.15}$$

Finally we get for the speed of sound in the perfect gas,

$$c = \sqrt{\frac{\eta\, p_-}{\varrho_-}} = \sqrt{\frac{1}{\varrho_-\, \kappa_-}} , \tag{7.16}$$

---

[2] These quantities are taken from thermodynamics, where c_p is the specific heat capacity at constant pressure and c_v the specific heat capacity at constant volume.

Table 7.1 Values of typical materials

Material	Density, ϱ_-	Sound speed, c	Characteristic impedance, Z_w
Air at 1000 hPa and 20 °C	1.2 kg/m^3	343 m/s	412 Ns/m^3
Water at 10 °C	1000 kg/m^3	1440 m/s	$1.44 \cdot 10^6$ Ns/m^3
Steel (longitudinal wave)	≈ 7500 kg/m^3	≈ 6000 m/s	$\approx 45 \cdot 10^6$ Ns/m^3

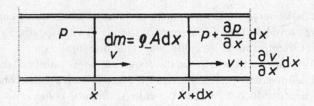

Fig. 7.4 An accelerated differential mass, dm, of fluid in a tube

where

$$\kappa_- = \kappa \mid_{\varrho_-} = 1/(\eta \; p_-) = \frac{1}{\varrho_- c^2} \, . \tag{7.17}$$

This expression is very similar to the one for liquids, but there the reciprocal value of the volume compressibility, or bulk compressibility, κ, the so-called the compression modulus, which is $K = 1/\kappa$, is more often used. Some specific material values are given in Table 7.1.[3]

Euler's Equation

Figure 7.4 depicts a differential mass of gas in a tube with rigid walls and a diameter small compared to the wavelength. The mass element is accelerated due to a pressure difference. The total differential of the particle velocity, v, being a function of time and space, is

$$dv(x, t) = \frac{\partial v}{\partial t} dt + \frac{\partial v}{\partial x} dx \, . \tag{7.18}$$

From this we arrive at the acceleration, a,

$$a = \frac{dv}{dt} = \frac{\partial v}{\partial t} + \frac{\partial v}{\partial x} \frac{dx}{dt} \, , \quad \text{where} \quad \frac{dx}{dt} = v \, . \tag{7.19}$$

---

[3] The characteristic field impedance, Z_w, also known as wave impedance, is introduced in Sect. 7.4.

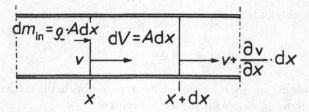

Fig. 7.5 A spatially-fixed differential volume, dV, of fluid in a tube

According to *Newton*'s mass law, $F = m \cdot a$, we now write

$$A \left[p - \left(p + \frac{\partial p}{\partial x} dx \right) \right] = (\varrho \, A \, dx) \left(\frac{\partial v}{\partial t} + \frac{\partial v}{\partial x} v \right), \qquad (7.20)$$

which is *Euler*'s equation.

Note that a *second linearization* is implemented by the replacement of ϱ with its static value, ϱ_-. Further, the second term in the sum, $v \cdot (\partial v / \partial x)$, is usually irrelevant in acoustics since v is zero in resting media, and also because the variation of v over space is usually small. This means that this term is a differential of the second order. Neglecting it, constitutes our *third linearization*.[4] We thus arrive at our first linear wave equation as follows,

$$-\frac{\partial p}{\partial x} = \varrho_- \frac{\partial v}{\partial t}. \qquad (7.21)$$

The Continuity Equation

Figure 7.5 depicts a fixed volume in a tube where mass flows in and out. This situation follows the mass-conservation law, which states that what flows in and does not come out again remains inside the volume since mass does not vanish.

The mass flowing in during a time interval, dt, is

$$dm_{\text{in}} = A \, \varrho_- \, \overbrace{v \, dt}^{dx}, \qquad (7.22)$$

and the mass flowing out during the same time interval is

$$dm_{\text{out}} = A \, \varrho_- \, v \, dt + A \, \varrho_- \frac{\partial v}{\partial x} dx \, dt. \qquad (7.23)$$

---

[4] Consider whether this linearization is still justified when the medium is flowing fast as it does in an exhaust system, and/or if there are rapid changes of v over x as occur at area steps in a tube. The second term is the more significant one in *flow dynamics* while the first term is typically neglected in that scientific field.

The difference, also known as the mass surplus, is

$$\mathrm{d}m_\Delta = -A\,\varrho_-\,\frac{\partial v}{\partial x}\,\mathrm{d}x\,\mathrm{d}t\,. \tag{7.24}$$

If the mass inside the volume has changed, the mass density in the volume must have also changed, allowing us to write the mass surplus as

$$\mathrm{d}m_\Delta = A\,\mathrm{d}x\,\frac{\partial \varrho}{\partial t}\mathrm{d}t\,. \tag{7.25}$$

Combining the two surplus expressions yields

$$-\varrho_-\,\frac{\partial v}{\partial x} = \frac{\partial \varrho}{\partial t}\,, \tag{7.26}$$

which is the *continuity equation*.

Since we want to use the field quantities p and v, we now substitute p by v by applying the state equation (7.11) as follows,

$$\partial \varrho_- = \frac{\partial p}{c^2} = \kappa_-\varrho_-\,\partial p\,. \tag{7.27}$$

Combining equations (7.26) and (7.27), finally renders the second linear differential wave equation as

$$-\frac{\partial v}{\partial x} = \kappa_-\,\frac{\partial p}{\partial t}\,, \tag{7.28}$$

with $\kappa_- = 1/(\varrho_-\,c^2)$. Note that κ has been replaced by κ_-, which is our *fourth linearization*.[5]

7.2   Three-Dimensional Wave Equation in *Cartesian* Coordinates

The two wave equations that we just derived are valid for a one-dimensional axial wave in a tube. The walls of the tube do not play any role since we have a purely longitudinal wave that propagates parallel to the walls.

---

[5] Because of $\kappa_- \approx \kappa$ and $\varrho_- \approx \varrho$, we omit the subscripts of κ_- and ϱ_- from now on to simplify the notation—such as in $\kappa = 1/(\varrho c^2)$.

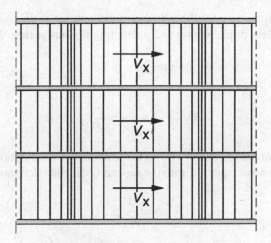

Fig. 7.6 Bundle of tubes with one-dimensional waves in them

Our equations are also valid for a bundle of tubes—shown in Fig. 7.6—and nothing would change concerning the wave if we took the tubes away. Consequently, the equations still hold for one-dimensional longitudinal waves in infinitely extended media.

As there is no shear stress in fluids, waves from different orthogonal directions in the medium just superimpose without influencing each other. This property allows us to formulate the three-dimensional wave equations in *Cartesian* coordinates by superimposing the wave components of the x-, y-, and z-direction. Denoting $\overrightarrow{x}$, $\overrightarrow{y}$, and $\overrightarrow{z}$ for the *unit vector* along the x-, y-, and z-directions, we utilize the three operators taken from vector analysis,

Gradient

$$\overrightarrow{\operatorname{grad} p} = \overrightarrow{\nabla} p = \left\{ \frac{\partial p_x}{\partial x}; \frac{\partial p_y}{\partial y}; \frac{\partial p_z}{\partial z} \right\}$$

$$= \left\{ \frac{\partial p_x}{\partial x} \overrightarrow{x} + \frac{\partial p_y}{\partial y} \overrightarrow{y} + \frac{\partial p_z}{\partial z} \overrightarrow{z} \right\} \dots \text{a vector} \quad (7.29)$$

Divergence

$$\operatorname{div} \cdot \overrightarrow{v} = \nabla \cdot \overrightarrow{v} = \left\{ \frac{\partial v_x}{\partial x} + \frac{\partial v_y}{\partial y} + \frac{\partial v_z}{\partial z} \right\}, \dots \text{a scalar} \quad (7.30)$$

Laplacian

$$\operatorname{div} \cdot \overrightarrow{\operatorname{grad} p} = \nabla^2 p = \left\{ \frac{\partial^2 p_x}{\partial x^2} + \frac{\partial^2 p_y}{\partial y^2} + \frac{\partial^2 p_z}{\partial z^2} \right\}, \dots \text{a scalar} \quad (7.31)$$

Using these operators, we write the following handsome set of linear acoustical wave equations for lossless fluids, with $\dot{v}$ and $\dot{p}$ denoting first-order time derivatives,

$$- \operatorname{grad} p = \varrho \, \dot{v} \,, \tag{7.32}$$

and

$$- \operatorname{div} \cdot v = \kappa \, \dot{p} = \frac{1}{\varrho \, c^2} \, \dot{p} \,, \tag{7.33}$$

where we have omitted the vector arrows for simplicity. Combination of (7.32) and (7.33) renders the wave equation with the second-order time derivative,

$$\nabla^2 p = \frac{1}{c^2} \, \ddot{p} \,. \tag{7.34}$$

The corresponding equations for the particle velocity, v, shown below, looks formally identical,[6] namely,

$$\nabla^2 v = \frac{1}{c^2} \, \ddot{v} \,. \tag{7.35}$$

Note that both the first-order (7.32)–(7.33) and the second-order wave equations (7.34)–(7.35) are of general form and independent of the actual coordinate systems.

7.3   Solutions of the Wave Equation

We now present some prominent solutions of the wave equation. To keep the discussion simple, we restrict ourselves to one-dimensional cases, specifically the wave equation in its following the formulation[7] as predicted at the beginning of this chapter.

$$\frac{\partial^2 p}{\partial x^2} = \frac{1}{c^2} \frac{\partial^2 p}{\partial t^2} \,. \tag{7.36}$$

Two or three-dimensional solutions evolve from superposition of one-dimensional solutions.

General Solution

The following general solution is known as the *d'Alembert* solution. *d'Alembert* assumes two waves propagating in opposite directions, x and $-x$. Both the forward-progressing and returning waves have the same speed, c. The solution is written as

---

[6] In theoretical acoustics, the velocity potential, Φ, is frequently used. It is defined as $p = \varrho \, \dot{\Phi}$, $\vec{v} = -\overrightarrow{\operatorname{grad} \Phi}$. Its wave equation is $\nabla^2 \Phi = (1/c^2) \, \ddot{\Phi}$.

[7] This form follows also from combining the differential of x in (7.21) with that of t in (7.28).

$$p(x,t) = p_\rightarrow \left(t - \frac{x}{c}\right) + p_\leftarrow \left(t + \frac{x}{c}\right). \tag{7.37}$$

Note that, since this equation is a differential equation of the second order, it is satisfied by any function $f(t \pm x/c)$ that is differentiable twice with respect to both time and space. In the above equation, $p_\rightarrow(\cdot)$ and $p_\leftarrow(\cdot)$ denote two functions with opposite directions of propagation along the x-axis.

Solution for Harmonic Functions

This solution, called the *Bernoulli* solution, is a special case of *d'Alembert*'s solution applied to harmonic functions (sinusoids). Dealing with harmonic functions allows us to take advantage of complex notation by rewriting the wave equation as follows,

$$\frac{\partial^2 p}{\partial x^2} - (j\beta)^2 \underline{p} = 0. \tag{7.38}$$

This is known as the *Helmholtz* form, in which

$$\beta = \frac{\omega}{c} = \omega \sqrt{\varrho \kappa}, \tag{7.39}$$

is called *phase coefficient*.[8]

The solution unfolds from trying $\underline{p} = e^{\gamma x}$. This approach leads to the characteristic equation, $\gamma^2 = (j\beta)^2$, and its solutions,[9] $\underline{\gamma}_{1,2} = \pm(j\beta)$. These solutions again allow for two waves in opposite directions, namely,

$$\underline{p}(x) = \underline{p}_\rightarrow e^{-j\beta x} + \underline{p}_\leftarrow e^{+j\beta x}. \tag{7.40}$$

Similarly, the solutions for the particle velocity is expressible as

$$\underline{v}(x) = \underline{v}_\rightarrow e^{-j\beta x} + \underline{v}_\leftarrow e^{+j\beta x}. \tag{7.41}$$

Both the forward progressing and the returning waves go through a complete period for $\beta \lambda = 2\pi$, with λ being the *wavelength*, we obtain the well-known formulas

$$\lambda f = c \quad \text{and} \quad \beta = \frac{\omega}{c} = \frac{2\pi}{\lambda}. \tag{7.42}$$

---

[8] In physics this quantity is often called *wave number* or, more precisely, *angular wave number* and denoted by k. In fact, it is not a "number" as it has the dimension [1/length].

[9] γ is called *complex propagation coefficient*, also denoted complex wave number, $\underline{k}$. We make use of it later, particularly, in Sects. 8.3 and 11.2.

7.4   Field Impedance and Power Transport in Plane Waves

First we look at the forward progressing wave, expressed as

$$\underline{p}(x) = \underline{p}_{\rightarrow} e^{-j\beta x} \quad \text{and} \quad \underline{v}(x) = \underline{v}_{\rightarrow} e^{-j\beta x} . \tag{7.43}$$

Following *Euler*'s equation (7.21), we obtain

$$\varrho \frac{\partial v}{\partial t} = -\frac{\partial p}{\partial x} \quad \text{and} \quad j\omega \underline{v}_{\rightarrow} e^{-j\beta x} \varrho = -(-j\beta) \underline{p}_{\rightarrow} e^{-j\beta x} \tag{7.44}$$

and, finally, considering the returning way likewise,

$$Z_{\mathrm{w}} = \frac{\underline{p}_{\rightarrow} e^{-j\beta x}}{\underline{v}_{\rightarrow} e^{-j\beta x}} = -\frac{\underline{p}_{\leftarrow} e^{+j\beta x}}{\underline{v}_{\leftarrow} e^{+j\beta x}} = \varrho c = \sqrt{\frac{\varrho}{\kappa}} . \tag{7.45}$$

The real quantity, Z_{w}, is known as *characteristic field impedance* or *wave impedance*.[10] Since Z_{w} is real in our case, sound pressure and particle velocity are in phase, which generally holds for plane waves in lossless fluids.

The thus purely resistive (active) intensity, $\overline{I} = \mathrm{Re}\{\underline{I}\}$, transported by the forward progressing wave is

$$\overline{I} = \frac{1}{2} \mathrm{Re}\left[\underline{p}_{\rightarrow} e^{-j\beta x} \left(\underline{v}_{\rightarrow} e^{-j\beta x} \right)^{*} \right] = \frac{1}{2} |\underline{p}_{\rightarrow}| |\underline{v}_{\rightarrow}| = \frac{1}{2} Z_{\mathrm{w}} |\underline{v}_{\rightarrow}|^{2} , \tag{7.46}$$

which also holds for the returning wave. The transported active power, $\overline{P}$, results from multiplying the intensity with the area which it crosses perpendicularly. This multiplication resulting in the following scalar product of two vectors, $\overline{P} = \overrightarrow{I} \cdot \overrightarrow{A_{\perp}}$.

7.5   Transmission-Line Equations and Reflectance

As previously mentioned, there is only axial wave propagation in tubes with much smaller diameters than the wavelengths, $d \ll \lambda$.

Since the ratio of $\underline{p}$ and $\underline{v}$ inside the tube is only dependent on the terminating impedance of the tube, $\underline{Z}_0 = \underline{p}_0 / \underline{v}_0$, it is convenient to formulate the solution of the wave equation in terms of the distance from the position of $\underline{Z}_0$. To this end, we substitute the coordinate x with $-l$ as shown in Fig. 7.7.

---

[10] When considering losses in the medium, $\underline{Z}_{\mathrm{w}}$ becomes complex due to the addition of an imaginary component.

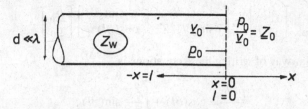

Fig. 7.7 One dimensional wave in a tube with a termination

Now, because of $e^{\pm j\beta l}|_{l=0} = 1$ at position $l = 0$, we have $\underline{p}_0 = \underline{p}_\rightarrow + \underline{p}_\leftarrow$ and $\underline{v}_0 = \underline{v}_\rightarrow + \underline{v}_\leftarrow$. Keeping (7.45) in mind, we further have $Z_w = \underline{p}_\rightarrow/\underline{v}_\rightarrow = -\underline{p}_\leftarrow/\underline{v}_\leftarrow$, which leads to the expressions

$$Z_w \, \underline{v}_0 = \underline{p}_\rightarrow - \underline{p}_\leftarrow \quad \text{and} \quad \underline{p}_0/Z_w = \underline{v}_\rightarrow - \underline{v}_\leftarrow. \tag{7.47}$$

Consequently it follows that

$$\underline{p}_\rightarrow = \frac{1}{2}\left(\underline{p}_0 + Z_w \underline{v}_0\right) \qquad \underline{p}_\leftarrow = \frac{1}{2}\left(\underline{p}_0 - Z_w \underline{v}_0\right) \tag{7.48}$$

$$\underline{v}_\rightarrow = \frac{1}{2}\left(\frac{\underline{p}_0}{Z_w} + \underline{v}_0\right) \qquad \underline{v}_\leftarrow = -\frac{1}{2}\left(\frac{\underline{p}_0}{Z_w} - \underline{v}_0\right). \tag{7.49}$$

For an arbitrary position, $l = -x$, the following expressions result,

$$\underline{p}(l) = \frac{1}{2}\left(\underline{p}_0 + Z_w \underline{v}_0\right) e^{j\beta l} + \frac{1}{2}\left(\underline{p}_0 - Z_w \underline{v}_0\right) e^{-j\beta l} \quad \text{and} \tag{7.50}$$

$$\underline{v}(l) = \frac{1}{2}\left(\frac{\underline{p}_0}{Z_w} + \underline{v}_0\right) e^{j\beta l} - \frac{1}{2}\left(\frac{\underline{p}_0}{Z_w} - \underline{v}_0\right) e^{-j\beta l}. \tag{7.51}$$

Complex decomposition with $e^{-j\alpha} = \cos\alpha - j\sin\alpha$ leads to the following two equations that are known as *transmission-line equations*,[11] namely,

$$\underline{p}(l) = \underline{p}_0 \, \cos(\beta l) + j Z_w \underline{v}_0 \sin(\beta l), \tag{7.52}$$

$$\underline{v}(l) = j\frac{\underline{p}_0}{Z_w} \, \sin(\beta l) + \underline{v}_0 \cos(\beta l), \tag{7.53}$$

or in matrix form

---

[11] This name originates from electrical engineering where the same equations hold but with p being replaced by u and v by i.

$$\begin{bmatrix} \underline{p}(l) \\ \underline{v}(l) \end{bmatrix} = \begin{bmatrix} \cos(\beta l) & \mathrm{j}Z_{\mathrm{w}} \, \sin(\beta l) \\ \mathrm{j} \, \sin(\beta l)/Z_{\mathrm{w}} & \cos(\beta l) \end{bmatrix} \begin{bmatrix} \underline{p}_0 \\ \underline{v}_0 \end{bmatrix}. \tag{7.54}$$

Another popular way of writing these equations is

$$\frac{\underline{p}(l)}{\underline{p}_0} = \cos(\beta l) + \mathrm{j}\frac{Z_{\mathrm{w}}}{\underline{Z}_0} \, \sin(\beta l), \tag{7.55}$$

$$\frac{\underline{v}(l)}{\underline{v}_0} = \cos(\beta l) + \mathrm{j}\frac{\underline{Z}_0}{Z_{\mathrm{w}}} \, \sin(\beta l). \tag{7.56}$$

We now discuss different terminations of the tube. The following three cases are of particular interesting in this context.

(1) $\underline{Z}_0 = \underline{p}_0/\underline{v}_0 = Z_{\mathrm{w}}$. This means that there is no reflection at the terminal, and all power moves on. This case is called *wave match* and results in no returning wave. Consequently, (7.52) becomes

$$\underline{p}(l) = \underline{p}_0 \, \mathrm{e}^{\mathrm{j}\beta l}. \tag{7.57}$$

(2) $\underline{Z}_0 = \infty$ and, hence, $\underline{v}_0 = 0$. This case is called *hard termination* and is achieved by closing the tube with a rigid surface. In this case, the forward progressing wave is fully reflected, and zero power leaves through the terminal. Equations (7.52) and (7.53) render

$$\underline{p}(l) = \underline{p}_0 \, \cos(\beta l) \quad \text{and} \quad \underline{v}(l) = \mathrm{j}\frac{\underline{p}_0}{Z_{\mathrm{w}}} \, \sin(\beta l). \tag{7.58}$$

These two formulas describe a so-called *standing wave*. The sound pressure varies at all positions sinusoidally as a function of time. Sound pressure and particle velocity are 90° out of phase.

(3) $\underline{Z}_0 = 0$ and, thus, $\underline{p}_0 = 0$. Now the end of the tube is open, resulting in *soft termination*

$$\underline{p}(l) = \mathrm{j}Z_{\mathrm{w}} \, \underline{v}_0 \, \sin(\beta l) \quad \text{and} \quad \underline{v}(l) = \underline{v}_0 \cos(\beta l) \tag{7.59}$$

Neglecting any radiation out of the open end—which is of course idealizing—the sound pressure is zero on the terminating plane. The forward progressing and the returning wave superimpose again to a standing wave, but they have a different phase angle than the one in the hard-termination case. Again, there is no power leaving through the terminal.

The reflectance, $\underline{R}$, is a useful quantity for describing the reflection at the terminal plane of the tube. Using (7.48), it is defined as

$$\underline{R} = \frac{\underline{p}_{\leftarrow}}{\underline{p}_{\rightarrow}} = \frac{(\underline{p}_0 - Z_{\mathrm{w}} \, \underline{v}_0)/2}{(\underline{p}_0 + Z_{\mathrm{w}} \, \underline{v}_0)/2} = \frac{\underline{Z}_0 - Z_{\mathrm{w}}}{\underline{Z}_0 + Z_{\mathrm{w}}}. \tag{7.60}$$

We have $\underline{R} = 0$ for wave match, $\underline{R} = +1$ for hard/rigid termination and $\underline{R} = -1$ for soft termination. If intensity or power are under consideration, the following definitions[12] are often used.

$|R|^2$ ...Degree of reflection

$1 - |R|^2$ ...Degree of absorption, α.

7.6 The Acoustic Measuring Tube

The acoustic measuring tube, also known as *Kundt*'s tube, is applied to measure field impedances and reflectances in the terminal plane of a tube. Such a tube is schematically shown in Fig. 7.8. The tube is excited by a sound source from one end—a sinusoidal sound in our case. The other end is terminated by the impedance to be measured. There is an arrangement of absorbing material in front of the source to avoid back reflection.

Classical Method

A microphone moves along the axis of the tube and measures the sound pressure as a function of its position. The sound pressure inside the tube follows the following equation, whereby $\underline{R} = |\underline{R}| e^{j\phi_R}$,

$$\underline{p}(l) = \underline{p}_{\rightarrow} (e^{+j\beta l} + \underline{R} e^{-j\beta l}) = \underline{p}_{\rightarrow} \left[e^{+j\beta l} + |\underline{R}| e^{-j(\beta l - \phi_R)} \right]. \qquad (7.61)$$

For any $|\underline{R}| \neq 0$, this function shows maxima and minima along l. In other words, a standing-wave behavior shows up—depicted in the lower panel of Fig. 7.8. The maxima are located at all positions of l where the phase angles of the two terms in (7.61) are equal, the minima accordingly, where the phase angles are opposite. The positions of the extremes thus are,

$$\beta l = -(\beta l - \phi_R) \pm n\pi, \qquad (7.62)$$

with minima when n is odd, and maxima when n is even.

This results in the following *standing-wave ratio*, S,

$$S = \left| \frac{p_{max}}{p_{min}} \right| = \frac{1 + |\underline{R}|}{1 - |\underline{R}|}, \qquad (7.63)$$

---

[12] In the literature, these terms are often called reflection coefficient and absorption coefficient. However, their dimension is [1] and their value range is 0–1, or 0%–100%, respectively. This is why we prefer the term *degree* here .

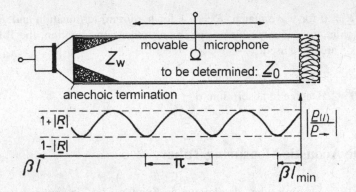

Fig. 7.8 *Kundt*'s tube for impedance measurement

from which we obtain the magnitude of the reflectance[13] as

$$|\underline{R}|(l = 0) = \frac{S - 1}{S + 1}. \tag{7.64}$$

The phase angle of the reflectance, ϕ_R, is given by the position of the first minimum, l_{min}. It results from

$$\phi_R(l = 0) = \beta\, l_{min}. \tag{7.65}$$

Transfer-Function Method

In an alternative approach, the so-called *transfer-function method*, the complex reflectance, $\underline{R}\,(l = 0)$, and, consequently, the terminating impedance, $\underline{Z}_0 = \underline{p}_0/\underline{v}_0$, is measured without a microphone to be shifted mechanically. Instead, two microphones at a mutual distance of $l_\Delta = l_2 - l_1$ are employed at the positions 2 and 1— depicted in Fig. 7.9. Starting from (7.52), we write[14]

$$\underline{p}_2 = \underline{p}_1 \cos(\beta\, l_\Delta) + \mathrm{j} Z_w \underline{v}_1 \sin(\beta\, l_\Delta) \quad \text{and, thus,} \tag{7.66}$$

$$\frac{\underline{p}_2}{\underline{p}_1} = \cos(\beta\, l_\Delta) + \mathrm{j}\frac{Z_w}{\underline{Z}_1} \sin(\beta\, l_\Delta), \quad \text{or} \tag{7.67}$$

$$\underline{Z}_1 = \frac{\mathrm{j} Z_w \sin(\beta\, l_\Delta)}{(\underline{p}_2/\underline{p}_1) - \cos(\beta\, l_\Delta)}. \tag{7.68}$$

---

[13] The theory of *Kundt*'s tube is homomorphic to the theory of the measuring line for electromagnetic waves. To convert reflectance into impedance, and vice versa, a graphical tool called *Smith* chart is useful. It transforms the plot of a function in the complex $\underline{Z}$-plane into the corresponding plot in the complex $\underline{R}$-plane.

[14] Note that all involved distances, that is, l_1, l_2 and $l_\Delta = l_2 - l_1$, have positive values.

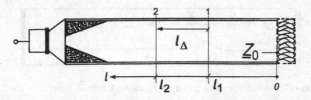

Fig. 7.9 Impedance measurement with the transfer-function method

With a straight-forward transformation along the distance $-l_2$, which follows from the transmission-line equations (7.55) and (7.56), $\underline{Z}_2$ transformes into the impedance at the surface of the probe to be tested, $\underline{Z}_0$. The relevant formula is

$$\underline{Z}_0 = \frac{\underline{Z}_1 \cos(\beta l_1) + j Z_w \sin(\beta l_1)}{\cos(\beta l_1) + j(\underline{Z}_1/Z_w)\sin(\beta l_1)}, \tag{7.69}$$

In a similar way, we obtain a widely used formula for the complex reflectance, namely

$$\underline{R} = \frac{(\underline{p}_1/\underline{p}_2) - e^{-j\beta l_\Delta}}{e^{+j\beta l_\Delta} - (\underline{p}_1/\underline{p}_2)} \, e^{+j 2\beta l_2}. \tag{7.70}$$

The ratio of $\underline{p}_1/\underline{p}_2 = \underline{T}_{pp}$ for a given frequency, ω_1, is a transfer factor. Its frequency function, $\underline{H}_{pp}(\omega)$, is called a *transfer function*. This fact lends the method its name.

The transfer-function method fails for frequencies where the nominator or denominator in (7.69) or (7.70) is equal to zero. This is the case whenever $l_\Delta = \lambda/2$. To cover a wide frequency range, the condition $l_\Delta < \pi c/\omega$ must be fulfilled. To this end, more than two microphones at different positions are employed—or one microphone that measures sequentially at different positions.

7.7 Exercises

Wave Equation in Fluids, Solutions, *Kundt*'s Tube

Problem 7.1 An impulsive disturbance in a fluid medium propagates along a thin tube as illustrated in Fig. 7.10. The wavelength is much longer than the tube diameter. To ease the solution, the sawtooth-formed impulses can be simplified by straight dotted lines.

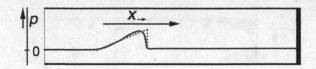

Fig. 7.10 A form-of-sawtooth impulse, propagating along a narrow tube

(a) The sawtooth-formed pressure impulse propagates along a tube—see Fig. 7.10. Sketch the reflected pressure impulse, and the superposition between the forward-propagating and reflected pressure impulses with the tube being rigidly terminated. Sketch at least more than four refined steps as indicated in Fig. 7.10.

(b) Consider now that a sawtooth-formed velocity impulse propagates along a tube. Sketch the reflected velocity impulse and the superposition of the forward-propagating and reflected impulses with the tube being rigidly terminated. Sketch at least more than four refined steps as indicated in Fig. 7.10.

Problem 7.2 A thin tube—with a diameter of $d \ll \lambda$, that is, much smaller than the wavelength—is terminated by an impedance, $\underline{Z}_0$, see Fig. 7.7. $\underline{p}_0$ and $\underline{v}_0$ are the sound pressure and the velocity at the termination.

Prove that the following holds,

$$Z_w \underline{v}_0 = \underline{p}_\rightarrow - \underline{p}_\leftarrow \quad \text{and} \quad \underline{p}_0/Z_w = \underline{v}_\rightarrow - \underline{v}_\leftarrow \tag{7.71}$$

$$\underline{p}_\rightarrow = \frac{1}{2}\left(\underline{p}_0 + Z_w \underline{v}_0\right) \quad \text{and} \quad \underline{p}_\leftarrow = \frac{1}{2}\left(\underline{p}_0 - Z_w \underline{v}_0\right) \tag{7.72}$$

$$\underline{v}_\rightarrow = \frac{1}{2}\left(\frac{\underline{p}_0}{Z_w} + \underline{v}_0\right) \quad \text{and} \quad \underline{v}_\leftarrow = -\frac{1}{2}\left(\frac{\underline{p}_0}{Z_w} - \underline{v}_0\right), \tag{7.73}$$

as stated in (7.47), (7.48), and (7.49).

Problem 7.3 A thin tube with a length, L, and a wave resistance, Z_L, is terminated by an impedance, Z_0. At location L, which is the input port to the tube, a plane wave enters the tube and travels toward its termination, Z_L.

(a) Determine, in general form, the input impedance of the tube.

(b) Consider the special cases of $Z_0 = 0$, $Z_0 = \infty$, and $Z_0 = Z_L$. How is the sound pressure distributed in these particular cases?

Problem 7.4 Similar to the sound pressure inside the tube in (7.57) with a termination expressed by a reflectance, $\underline{R} = |\underline{R}|\, e^{j\phi_r}$, and the complex sound pressure expressed as (7.61),

$$\underline{p}(l) = \underline{p}_\rightarrow \left[e^{+j\beta l} + |\underline{R}|\, e^{-j(\beta l - \phi_R)}\right] \tag{7.74}$$

(a) Calculate the magnitude (envelope) of $\underline{p}\,(l)$, that is, $|\underline{p}(l)|$, of the complex sound pressure.

(b) Express the conditions for the envelope, $|\underline{p}(l)|$, to assume minima (nodes) or the maxima (anti-nodes).

(c) Determine the ratio of the maxima and the minima of the sound-pressure envelope.

(d) Derive the relations of the locations of the standing-wave nodes and anti-nodes and the termination impedance.

Problem 7.5 Derive the wave equation of an oscillating string under the small-displacement assumption, in other words, that displacements induced by the string oscillations are much smaller than the length of the string.

Chapter 8
Horns and Stepped Ducts

The wave equations derived in the preceding chapter allow to calculate arbitrary sound fields with any possible, physically meaningful boundary conditions. Yet, we only dealt with one-dimensional waves so far. Such one-dimensional waves exist, for instance, in tubes with diameters which are small compared to the wavelength, that is, $d \ll \lambda$. This condition guarantees that no waveforms other than axial ones propagates in the tube. One-dimensional propagation also means that all wave planes perpendicular to the axial direction are planes of constant phase.

In the following, we consider ducts where the diameter varies with x. In other words, the area function, $A = f(x)$, is no longer constant. Nevertheless, the condition $d \ll \lambda$ still holds.

This section discusses the two cases depicted in Fig. 8.1.

- *Continuous variation* of the cross area As long as this variation is only gradual compared to the wavelength, it is still justified to assume one-dimensional, axial propagation. Radial propagation is then insignificant.
- *Step-like variations* of the cross area as a function of x—so-called *stepped ducts* Very close to the position where the step occurs, radial components of the particle velocity do exist, but they do not propagate away from the step. Therefore, there are only plane waves left. The radial wave components are negligible already at small distances away from the the cross area. We thus assume for our calculations that just x_Δ in front of the step and, consequently, x_Δ behind it, the axial component of the volume velocity, $q = A \underline{v}$ is the same in both cross-sections. Accordingly, we disregard any *modal dispersion* in the immediate vicinity of the position of the step by setting the volume velocity at both sides of the step to equal.

© Springer-Verlag GmbH Germany, part of Springer Nature 2021
N. Xiang and J. Blauert, *Acoustics for Engineers*,
https://doi.org/10.1007/978-3-662-63342-7_8

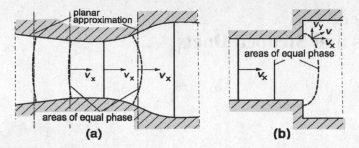

Fig. 8.1 Two types of ducts with non-constant area functions (**a**) Continuous variation of the cross-sectional area. (**b**) Stepped variations

8.1 *Webster*'s Differential Equation—The Horn Equation

We now deal with the cases where the area function varies only gradually, and perpendicular areas are areas of approximately constant phase. This condition is captured by *Webster*'s equation—also called *Horn equation*.

Figure 8.2 illustrates the derivation of this differential equation. Please consider the elementary volume between the cross areas at x and $x + dx$.

We apply the state equation and *Euler*'s equation in their original form, that is,

$$\partial\varrho = \frac{1}{c^2}\partial p \quad \text{and} \quad -\frac{\partial p}{\partial x} = \varrho\frac{\partial v}{\partial t}, \tag{8.1}$$

whereby, to be sure, $p = p(x, t)$ and $v = v(x, t)$ are functions of both space and time. In the continuity equation, the non-constant area function, A(x), is considered as follows.

The in-flowing mass is

$$dm_{\text{in}} = \varrho A(x)\, v\, dt. \tag{8.2}$$

The out-flowing mass is

$$dm_{\text{out}} = \varrho\left[A(x) + \frac{dA}{dx}\, dx\right]\left(v + \frac{\partial v}{\partial x}\, dx\right) dt$$

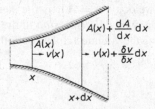

Fig. 8.2 Cross-section of a *horn*—a duct with gradually increasing cross area

$$= \varrho \left[A(x)\, v + A(x)\, \frac{\partial v}{\partial x}\, dx + v\, \frac{dA}{dx}\, dx + \overbrace{(\cdots\cdots)}^{\text{2}^{\text{nd}}\text{ order differentials}} \right] dt. \quad (8.3)$$

Neglecting the second-order differentials in the sum, we get the mass surplus as follows,

$$dm_\Delta = -\varrho\, A(x) \left[\frac{\partial v}{\partial x} + \frac{1}{A(x)}\, \frac{dA}{dx}\, v \right] dt\, dx$$

$$= \overbrace{\left[A(x) + \frac{dA(x)}{2} \right]}^{\text{average area}} dx\, \frac{\partial \varrho}{\partial t}\, dt . \quad (8.4)$$

Neglecting second-order differentials again yields

$$-\varrho \left[\frac{\partial v}{\partial x} + \frac{1}{A(x)}\, \frac{dA}{dx}\, v \right] = \frac{\partial \varrho}{\partial t} , \quad (8.5)$$

which is the *modified continuity equation*.

To employ p as the second field quantity instead of ϱ, the state equation (8.1) is used, and we arrive at

$$-\left[\frac{\partial v}{\partial x} + \frac{1}{A(x)}\, \frac{dA}{dx}\, v \right] = \frac{1}{\varrho c^2}\, \frac{\partial p}{\partial t} . \quad (8.6)$$

Combining (8.1) and (8.6) leads to *Webster's* equation, which is

$$\frac{\partial^2 p}{\partial x^2} + \left[\frac{1}{A(x)}\, \frac{dA}{dx} \right] \frac{\partial p}{\partial x} = \frac{1}{c^2}\, \frac{\partial^2 p}{\partial t^2} . \quad (8.7)$$

The term $[1/A(x)]\,(dA/dx)$ is identical to $d[\ln A(x)]/dx$. For $A(x) = $ const, *Webster's* equation reduces to the normal one-dimensional wave equation.[1]

For some analytically defined area functions, *Webster's* equation is solvable in closed form. In the following sections, we take conical and exponential horns as examples.

---

[1] Note that *Webster's* equation for v differs from the one derived above for p.

8.2 Conical Horns

For the conical horn—sketched in Fig. 8.3—the area function is

$$A(x) = A_0 \left(\frac{x}{x_0} \right)^2 . \tag{8.8}$$

Inserting this area function into *Webster*'s equation results in

$$\frac{\partial^2 p}{\partial x^2} + \frac{2}{x} \frac{\partial p}{\partial x} = \frac{1}{c^2} \frac{\partial^2 p}{\partial t^2} , \tag{8.9}$$

whereby we have used

$$\frac{1}{A(x)} \frac{\mathrm{d}A}{\mathrm{d}x} = \frac{A_0}{A(x)} 2 \left(\frac{x}{x_0} \right) \frac{1}{x_0} = \frac{2}{x} . \tag{8.10}$$

Equation (8.9) is rewritable in the following form—as is proven by differentiating,

$$\frac{\partial^2 (p\, x)}{\partial x^2} = \frac{1}{c^2} \frac{\partial^2 (p\, x)}{\partial t^2} . \tag{8.11}$$

Inspecting this formula, we realize that its form corresponds to the one-dimensional wave equation, yet, instead of p, we now have a product $p \cdot x = g$.
We approach the solution of this expression with the trial

$$\underline{g}(x) = \underline{p}(x)\, x = \underline{g}_\rightarrow\, \mathrm{e}^{-\mathrm{j}\beta x} + \underline{g}_\leftarrow\, \mathrm{e}^{+\mathrm{j}\beta x} . \tag{8.12}$$

For the forward-progressing (outbound) wave, we obtain the following results for p and v,

$$\underline{p}_\rightarrow(x) = \frac{\underline{g}_\rightarrow}{x}\, \mathrm{e}^{-\mathrm{j}\beta x} \quad \text{and} \tag{8.13}$$

$$\underline{v}_\rightarrow(x) = \underline{g}_\rightarrow \left(\frac{1}{\varrho c x} + \frac{1}{\mathrm{j}\omega \varrho x^2} \right) \mathrm{e}^{-\mathrm{j}\beta x} , \tag{8.14}$$

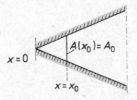

Fig. 8.3 Cross-section of a conical horn

where the solution for the particle velocity, v, has been found via the solution for p by applying *Euler*'s equation (8.1) with $\beta = \omega/c$ as follows,

$$- \underline{g}_{\rightarrow} \left(-\frac{1}{x^2}e^{-j\beta x} - j\beta e^{-j\beta x}\frac{1}{x} \right) = j\omega\varrho\underline{v}. \tag{8.15}$$

The sound-field from the conical horn is separable in a *near-field* and a *far-field*. The boundary between these two regions is defined as the position where the magnitudes of the real and imaginary parts of the velocity are equal, which is at

$$\left| \frac{1}{\varrho c\, x_{\mathrm{ff}}} \right| = \left| \frac{1}{j\omega\varrho\, x_{\mathrm{ff}}^2} \right|, \tag{8.16}$$

resulting in a far-field distance of

$$x_{\mathrm{ff}} = \frac{\lambda}{2\pi} = \frac{1}{\beta_{\mathrm{ff}}} = \frac{c}{\omega_{\mathrm{ff}}}, \tag{8.17}$$

with $x < \lambda/2\pi$ defining the *near-field* and $x > \lambda/2\pi$ the *far-field*.

The sound pressure, p, decreases to half with a doubling of the distance, x, which is a decrease of 6 dB per distance doubling. For v the situation is more complicated. In the near field, v decreases with $1/x^2$ per distance doubling, which is a 12 dB decrease. However, in the far-field, v behaves like p, that is, with a 6 dB decrease per distance doubling.

The field-impedance, $\underline{Z}_{\mathrm{f}}$, of the conical sound field results from dividing (8.13) by (8.14) as

$$\underline{Z}_{\mathrm{f}}(x) = \frac{\underline{p}_{\rightarrow}(x)}{\underline{v}_{\rightarrow}(x)} = \frac{1}{\frac{1}{\varrho c} + \frac{1}{j\omega\varrho x}} = \varrho c\, \frac{j\frac{2\pi x}{\lambda}}{1 + j\frac{2\pi x}{\lambda}}. \tag{8.18}$$

As a substitute for this field-impedance we draw a long tube with a short branch where a concentrated mass is positioned—depicted in Fig. 8.4. The right panel of the figure shows an electro-acoustic analogy.

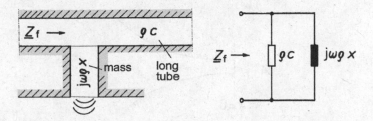

Fig. 8.4 Equivalent circuits for conical horns

The reactive (imaginary) component, $j\omega\varrho x$, is the so-called *co-vibrating medium mass*. This component swings about without transporting active power. The active (real) component, ϱc, becomes relatively (not absolutely!) stronger with increasing distance. For $x \gg \lambda/2\pi$, $\underline{Z}_f$ approaches ϱc. Note that ϱc is the field-impedance in a tube with a constant diameter and, thus, the specific field-impedance of the medium, Z_w.

8.3   Exponential Horns

The area function of the exponential horn—see Fig. 8.5—is given by

$$A(x) = A_0\, e^{2\epsilon x}, \tag{8.19}$$

with $\epsilon > 0$ being the so-called *flare coefficient*.
Now differentiation deliveres

$$\frac{1}{A(x)}\frac{dA}{dx} = \frac{d\,[\ln A(x)]}{dx} = 2\epsilon \quad \text{and, thus,} \tag{8.20}$$

$$\frac{\partial^2 p}{\partial x^2} + 2\epsilon\frac{\partial p}{\partial x} = \frac{1}{c^2}\frac{\partial^2 p}{\partial t^2}. \tag{8.21}$$

The structure of this equation is easily understood by applying complex notation, which leads to

$$\frac{\partial^2 \underline{p}}{\partial x^2} + 2\epsilon\frac{\partial \underline{p}}{\partial x} + \frac{\omega^2}{c^2}\,\underline{p} = 0, \tag{8.22}$$

an equation which recalls the equation of the damped oscillator—see Sect. 2.3.

We confine to the forward progressing wave again and, consequently, try the approach $\underline{p}(x) = e^{\underline{\gamma} x}$. This trial leads to the characteristic quadratic equation

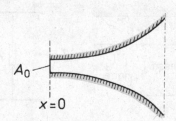

Fig. 8.5 Cross-section of an exponential horn

$$\underline{\gamma}^2 + 2\,\epsilon\,\underline{\gamma} + \frac{\omega^2}{c^2} = 0 \tag{8.23}$$

with its two solutions

$$\underline{\gamma}_{1,2} = -\epsilon \pm \sqrt{\epsilon^2 - \frac{\omega^2}{c^2}} = -\epsilon \pm j\sqrt{\frac{\omega^2}{c^2} - \epsilon^2}. \tag{8.24}$$

The complex quantity, $\underline{\gamma}$, is termed *propagation coefficient*, whereby

$$\underline{\gamma} = \alpha + j\,\beta, \tag{8.25}$$

where α is the *damping coefficient* and β the *phase coefficient*.

Hence, the solution of the wave equation is an exponential function decreasing with x. This kind of decreasing is called *spatial damping*. The general solutions for p and v in the forward-progressing wave results as

$$\underline{p}_\rightarrow(x) = \underline{p}_\rightarrow(0)\,e^{-\epsilon x}e^{-j\left(\sqrt{\frac{\omega^2}{c^2} - \epsilon^2}\right)x} = \underline{p}_\rightarrow(0)\,e^{\alpha x}e^{-j\beta x} \quad \text{, and} \tag{8.26}$$

$$\underline{v}_\rightarrow(x) = \frac{\epsilon + j\sqrt{\frac{\omega^2}{c^2} - \epsilon^2}}{j\,\omega\,\varrho}\,\underline{p}_\rightarrow(x). \tag{8.27}$$

Again, the solution for v has been derived from the one for p by applying *Euler*'s equation (8.1).

A prerequisite for wave propagation is that the expression under the square root is positive and, thus, results in a phase coefficient, β. This is the case when $\omega^2/c^2 > \epsilon^2$ and, accordingly, $2\pi/\lambda > \epsilon$ holds. The condition is fulfilled above a limiting angular frequency

$$\omega_l = \epsilon\,c. \tag{8.28}$$

Below ω_l, there is an exponential fade-out as the expression under the root becomes negative, and we end up with pure damping without wave propagation. This condition means physically that mass is shifted about, but no energy is transported because no sufficient compression takes place. ω_l decreases with decreasing flare coefficient, ϵ. In other words, the slimmer the horn, the lower the limiting frequency.

Note that the phase speed, $c_{\rm ph}$, in the exponential horn, is different from that in a free plane wave, c, viz,

$$c_{\rm ph} = \frac{\omega}{\beta} = \frac{\omega}{\sqrt{\left(\frac{\omega}{c}\right)^2 - \epsilon^2}}. \tag{8.29}$$

Furthermore, c_{ph} is frequency-dependent. This effect is called *dispersion* since different frequency components travel with different speed and, thus, the different wave components arrive at the end of the horn at different instances.[2]

The so-called *group-delay distortions*, which describe the frequency-dependent delay of the envelope of a transmitted signal, are highest close to the limiting frequency. The group delay, τ_{gr}, over a wave-traveling distance of l is in our case

$$\tau_{gr} = \frac{d\beta}{d\omega} = \frac{l}{c\sqrt{1 - \left(\frac{\omega_l}{\omega}\right)^2}}, \tag{8.30}$$

The field-impedance in the exponential horn, $\underline{Z}_f$, is given by

$$\underline{Z}_f = \frac{\underline{p}_\rightarrow}{\underline{v}_\rightarrow} = \frac{j\omega\varrho}{\epsilon + j\sqrt{\frac{\omega^2}{c^2} - \epsilon^2}} = \varrho c \left[\sqrt{1 - \left(\frac{\omega_l}{\omega}\right)^2} + j\left(\frac{\omega_l}{\omega}\right)\right]. \tag{8.31}$$

As with the conical horn, $\underline{Z}_f$ approaches $\varrho c = Z_w$ with increasing frequency because of $\underline{Z}_f \Rightarrow \varrho c$ for $\omega \gg \omega_l$.

8.4 Radiation Impedances and Sound Radiation

The acoustic power that is sent out by an electro-acoustic transducer or any other sound source is proportional to the real part of the impedance, $r_{rad} = \text{Re}\{\underline{Z}_{rad}\}$, that terminates the source at its acoustic output port. Since this impedance is formed by coupling the sound field with the source, we call it *radiation impedance*, $\underline{Z}_{rad}$, and its real part *radiation resistance*, r_{rad}. The radiation impedance is a mechanic impedance—refer to Sect. 4.5—namely,

$$\underline{Z}_{rad} = \frac{\underline{F}}{\underline{v}}. \tag{8.32}$$

The radiated power, then, is

$$\overline{P} = \frac{1}{2} \text{Re}\{\underline{Z}_{rad}\} |\underline{v}|^2 = \frac{1}{2} r_{rad} |\underline{v}|^2. \tag{8.33}$$

The following relation holds between the field-impedance, $\underline{Z}_f$, and the radiation impedance, $\underline{Z}_{rad}$,

$$\underline{Z}_{rad} = \int_A \underline{Z}_f \, dA, \tag{8.34}$$

---

[2] This fact contributes to the characteristic sound of horn loudspeakers.

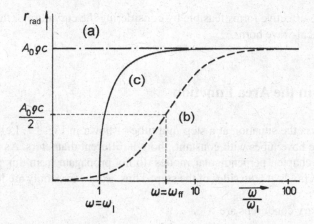

Fig. 8.6 Schematic plot of the radiation resistance. Frequencies normalized to the limiting frequency, ω_l, of the exponential horn **(a)** Tube with constant cross-section. **(b)** Conical horn. **(c)** Exponential horn

with A being the effective radiation area.

For transducers that radiate into a horn, the effective area is equal to the area of the horn's mouth in the optimal case. In the synopsis shown in Fig. 8.6, we assume that the tube/horn is so long that no waveforms are reflected from the opening, but that the diameter is still small compared to the wavelength, namely, $d \ll \lambda$. This is an idealizing assumption of course.

In summing up, we get for the tube with a constant cross-section,

$$r_{rad} = Z_w \, A_0 = \varrho c \, A_0 \, , \tag{8.35}$$

for the conical horn

$$r_{rad}\,(A_0) = A_0 \, \text{Re}\{\underline{Z}_f\} = A_0 \, \varrho c \, \frac{(\frac{\omega}{c} x_0)^2}{1 + (\frac{\omega}{c} x_0)^2} \, , \tag{8.36}$$

and for the exponential horn

$$r_{rad}\,(A_0) = A_0 \, \text{Re}\{\underline{Z}_f\} = A_0 \, \varrho c \, \sqrt{1 - \left(\frac{\omega_l}{\omega}\right)^2} \, . \tag{8.37}$$

For the conical horn, ω_{ff}, which forms the near-field/far-field division at a given distance from the mouth, x_1, is independent of the opening angle of the horn. For the exponential horn, however, the limiting angular frequency, ω_l depends on the flare coefficient, ϵ.

The exponential horn, among all horns that *Webster*'s equation covers, is the one with the steepest increase of $\text{Re}\{\underline{Z}_{rad}\} = r_{rad}$ as a function of frequency. However, there are more-effective forms feasible by considering the curvature of the waves—such as spherical-wave horns.

8.5 Steps in the Area Function

We now discuss the situation at a step in a tube—shown in Fig. 8.7. Left and right of the step, we have tubes with constant, though different diameters. As mentioned earlier in this chapter, perpendicular modes do not propagate from this position as long as $d \ll \lambda$ holds at both sides of the step. Thus, axial waves only are left slightly off the step.

The boundary conditions are

$$\underline{p}_1 = \underline{p}_2, \quad \text{and} \tag{8.38}$$

$$A_1 \underline{v}_1 = A_2 \underline{v}_2, \quad \text{which is} \quad \underline{q}_1 = \underline{q}_2. \tag{8.39}$$

Therefore, we take both quantities as continuous at the step. At the step a reflected wave is created. Accordingly, by combining

$$\underline{p}_{1\rightarrow} + \underline{p}_{1\leftarrow} = \underline{p}_{2\rightarrow} \quad \text{with} \quad A_1 \left(\frac{\underline{p}_{1\rightarrow}}{Z_{\mathrm{w}}} - \frac{\underline{p}_{1\leftarrow}}{Z_{\mathrm{w}}} \right) = \frac{\underline{p}_{2\rightarrow}}{Z_{\mathrm{w}}} A_2, \tag{8.40}$$

we get a reflectance of

$$\underline{R} = \frac{\underline{p}_{1\leftarrow}}{\underline{p}_{1\rightarrow}} = \frac{A_1 - A_2}{A_1 + A_2}. \tag{8.41}$$

As q is continuous at the step, it makes sense to introduce this quantity to deal with stepped-duct problems rather than the particle velocity, $\underline{v}$. Thus, we rewrite the transmission-line equations (7.52) and (7.53) with q instead of $\underline{v}$ as

$$\underline{p}(l) = \underline{p}_0 \cos(\beta l) + \mathrm{j} Z_{\mathrm{L}} \underline{q}_0 \sin(\beta l), \quad \text{and} \tag{8.42}$$

$$\underline{q}(l) = \mathrm{j} \frac{\underline{p}_0}{Z_{\mathrm{L}}} \sin(\beta l) + \underline{q}_0 \cos(\beta l), \tag{8.43}$$

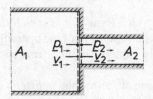

Fig. 8.7 Tube with steps in the area function

where

$$Z_L = \frac{Z_w}{A} = \frac{1}{A}\sqrt{\frac{\varrho}{\kappa}} = \sqrt{\frac{m'_a}{n'_a}}, \tag{8.44}$$

is the specific acoustic impedance of the respective tube. m'_a is the *mass load*, that is, the acoustic mass/length. n'_a is the *compliance load*, that is, the acoustic compliance/length.

Thereupon, the two relevant energies are length-related quantities as well, namely, the kinetic-energy/length,

$$W' = \frac{1}{2} m'_a q^2, \tag{8.45}$$

and the potential-energy/length,

$$W' = \frac{1}{2} n'_a p^2. \tag{8.46}$$

The reflectance at the step between two tubes, each with constant cross-section, results as

$$\underline{R} = \frac{Z_{L_2} - Z_{L_1}}{Z_{L_2} + Z_{L_1}}. \tag{8.47}$$

Please note that by taking p as analog to $\underline{u}$, and q as analog to $\underline{i}$, we experience a complete analogy to the electric transmission line where we observe $\underline{i}_1 = \underline{i}_2$ and $\underline{u}_1 = \underline{u}_2$ at steps.

8.6 Stepped Ducts

Taking $\underline{q}$ as the second wave quantity, helps to deal with stepped ducts via electric analogies—see Fig. 8.8. Furthermore, this allows to include acoustic concentrated elements into our consideration within the same analog circuits—refer to Sect. 2.6. So the theories of analysis and synthesis of electric networks, including transmission lines with and without losses, are directly applicable to acoustical problems. For example, it is feasible to design acoustic filters with specified transfer functions in this way—including high-pass, low-pass, and band-pass filters. This possibility is, for instance, exploited in the design of mufflers for car-exhaust systems to achieve brand-specific auditory imprints.

For the application of this method, it is worthy of recalling the relevant paragraph on transmission lines in Sect. 7.5. Consequently, acoustic tubes, represent T-elements as given in the following matrix equation—see Fig. 8.8.

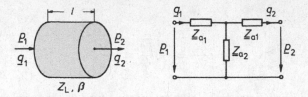

Fig. 8.8 Equivalent circuit for an acoustic tube segment

$$\begin{bmatrix} \underline{p}_1 \\ \underline{q}_1 \end{bmatrix} = \begin{bmatrix} \cos(\beta l) & jZ_L \sin(\beta l) \\ j\frac{1}{Z_L} \sin(\beta l) & \cos(\beta l) \end{bmatrix} \begin{bmatrix} \underline{p}_2 \\ \underline{q}_2 \end{bmatrix}. \tag{8.48}$$

These are the so-called two-port equations of a tube section, formulated in wave-parameter form. Two-port theory says that the following relations holds,

$$1 + \frac{\underline{Z}_{a_1}}{\underline{Z}_{a_2}} = \cos(\beta l) \quad \text{and} \quad \frac{1}{\underline{Z}_{a_2}} = j\frac{1}{Z_L} \sin(\beta l), \tag{8.49}$$

further,

$$\underline{Z}_{a_1} = jZ_L \tan\left(\frac{\beta l}{2}\right) \quad \text{and} \quad \underline{Z}_{a_2} = -jZ_L \frac{1}{\sin(\beta l)}. \tag{8.50}$$

Note that, because transcendental functions (tan, sin) are involved, the elements are, on principle, not realizable as concentrated acoustic elements. Yet, for sections of small lengths, that is $l \ll \lambda$, the following approximations apply,

$$\tan\left(\frac{\beta l}{2}\right) \approx \frac{\beta l}{2} \quad \text{and} \quad \frac{1}{\sin(\beta l)} \approx \frac{1}{\beta l}. \tag{8.51}$$

Hence, with

$$Z_L = \frac{1}{A}\sqrt{\frac{\varrho}{\kappa}} \quad \text{and} \quad \beta = \omega\sqrt{\varrho\kappa}, \tag{8.52}$$

we get

$$\underline{Z}_{a_1} \approx \frac{1}{2}j\omega\frac{\varrho}{A}l = \frac{1}{2}j\omega m'_a l \quad \text{and} \quad \underline{Z}_{a_2} \approx \frac{1}{j\omega\kappa A l} = \frac{1}{j\omega n'_a l}. \tag{8.53}$$

Figure 8.9 shows an equivalent circuit for a short section of a tube. For instance, this equivalent circuit is applied for calculating the transfer function of the human vocal tract or the ear canal. Figure 8.10 depicts the principle. The higher the attempted accuracy of the calculation, the more sections have to be assumed.

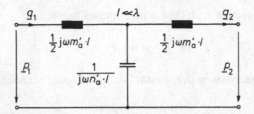

Fig. 8.9 Equivalent circuit for a tube segment (details)

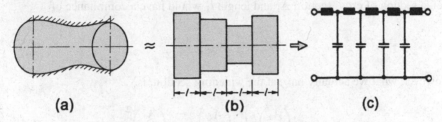

Fig. 8.10 Approximation of a tube with varied cross-sectional area

Finally, in this section, we treat the case of a very short narrowing or widening in a tube with an otherwise constant cross-section. The widening acts as a concentrated, branching spring, n_Δ, the narrowing as a concentrated serial mass, m_Δ. Figure 8.11a conceptualizes this situation by taking a widening section with the length of $l_2 \ll \lambda$, as the example.

From (8.53) we learn that

$$m'_{a_1} = \frac{A_2}{A_1} m'_{a_2} \quad \text{and} \quad n'_{a_1} = \frac{A_1}{A_2} n'_{a_2}. \tag{8.54}$$

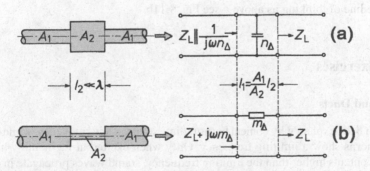

Fig. 8.11 Equivalent circuit of a short tube segments. (a) Short widening in a long constant-diameter tube. (b) Short narrowing

Now, the widening section with the length contains the mass

$$m_a = m'_{a_2}\, l_2\,.\tag{8.55}$$

If this mass were loaded upon the constant-diameter tube, we needed the length

$$l_1 = \frac{A_1}{A_2}\, l_2\,.\tag{8.56}$$

Yet, a section of cross section A_1 and length l_1 would have a compliance of

$$n_a = \frac{A_2}{A_1}\, n'_{a_2}\, l_1\,.\tag{8.57}$$

However, what we actually have at the widening section, is

$$n_a = n'_{a_2}\, l_2 = \left(\frac{A_1}{A_2}\right) n'_{a_1} \left(\frac{A_1}{A_2}\right) l_1 = \left(\frac{A_1}{A_2}\right)^2 n'_{a_1} l_1\,.\tag{8.58}$$

In other words, the widening section acts like a section of the constant-diameter tube with a cross-section of A_1 and a length of

$$l_1 = \frac{A_1}{A_2} l_2\,,\tag{8.59}$$

plus an additional parallel spring of

$$n_\Delta = \left[\left(\frac{A_2}{A_1}\right)^2 - 1\right] n_a\,.\tag{8.60}$$

At the narrowing section, we have an additional serial mass, m_Δ, derived along the same line of thinking as above—see Fig. 8.11b.

8.7   Exercises

Horns and Ducts

Problem 8.1 Explain why conical horns radiate sound at any frequency, while exponential horns show a limiting frequency. Only when excited at its mouth with frequency contents higher than the limiting frequency, sound waves propagate in exponential horns.

Problem 8.2 A horn radiator consists of an electrodynamic driver, a pressure chamber, and an exponential horn—see Fig. 8.12. Assuming the horn to be long enough,

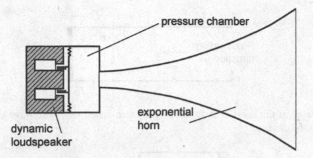

Fig. 8.12 Horn loudspeaker with pressure chamber

backward reflected waves from the open end are insignificant. Furthermore, the pressure chamber is tiny compared to the shortest wavelengths concerned.

(a) Determine the mechanical input impedance of the horn for $\omega \ll \omega_l$.

(b) Assuming that the horn is excited by a constant velocity, find the horn's flare coefficient, ε, such that the radiated power, above a frequency of $f_u = 500\,\text{Hz}$, does not vary by more than 3 dB.

(c) Develop an equivalent electric circuit for the pressure chamber.

(d) Develop an equivalent electric circuit for the complete horn-loudspeaker system.

(e) Determine the transformation coefficient (area ratio) of the pressure chamber such that the horn radiates the maximum sound power, given the following specifications.

Specific impedance of air	$\ldots \varrho c = 406\,\text{Ns/m}^3$
Input resistance of the loudspeaker	$\ldots R_l = 24\,\Omega$
Inner resistance of the signal generator	$\ldots R_g = 24\,\Omega$
Square value of the transducer coefficient	$\ldots B^2 l^2 = 44\,\text{Vs}^2/\text{m}^2$
Mechanical resistance	$\ldots r_m = 1.2\,\text{Ns/m}$
Diameter of the speaker membrane	$\ldots d = 6\,\text{cm}$

Problem 8.3

(a) Draw the radiation resistance, r_{rad}, of the following three configurations,[3] all of which are covered by *Webster*'s equation, namely,

- Tubes with constant diameter
- Conical horns

---

[3] The use of a suitable programming tool, such as MATLAB or PYTHON, is recommended. Even a spread-sheet can do.

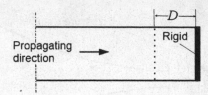

Fig. 8.13 Air layer of thickness D in front of rigidly terminated tube. The wave length is much longer than the tube diameter

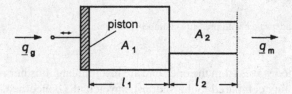

Fig. 8.14 Stepped tube composed of two sections and with a pressure chamber

– Exponential horns

Plot the graph as a function of the frequency, ω, normalized by the limiting frequency of the exponential horn, ω_1, or, alternatively, normalized by the near-field/far-field boundary, ω_{ff}.

(b) Discuss the resulting graphs.

Problem 8.4 A (plane) wave is traveling towards a rigidly terminated tube as sketched in Fig. 8.13. Apply (8.48) to determine the field impedance, $\underline{Z}_D$, at the surface that is located at a distance, D, away from the rigid termination.

Problem 8.5 Given a stepped tube composed of two sections and with a pressure chamber—as shown in Fig. 8.14,

(a) Determine the function $\underline{H}(\omega) = \underline{q}_m(\omega) / \underline{q}_g(\omega)$.

(b) Find the poles of this function. Note that the termination impedance is assumed to be very small compared to the pipe resistance.

Chapter 9
Spherical Waves, Harmonics, and Line Arrays

The wave equation, $\nabla^2 p = \ddot{p}/c^2$, as derived in Sects. 7.1 and 7.2, theoretically covers all possible sound fields in idealized fluids, that is, gases and liquids. However, the task of computing specific sound fields requires solutions of the wave equation for the particular boundary conditions in each case. In general, this task is be mathematically expensive, but there are helpful computer programs available, some of which are based on numerical methods like the finite-element method, FEM, or the boundary-element method, BEM. Yet, In practice, approximations are often sufficient to understand the structure of a problem.

Closed solutions of the wave equation only exist for a limited number of special cases. We already introduced the plane wave as one-dimensional solution in *Cartesian* coordinates. A few further one-, two- and three-dimensional cases are solvable in closed form, especially when symmetries allow simplified formulations using appropriate coordinate systems as is the case for spherical or cylindrical coordinates.

In the current chapter, we discuss basic solutions of the wave equation in spherical coordinates. In the same way that periodical time signals are decomposed into *Fourier* harmonics, spherical sound waves are decomposable into *spherical harmonics*.[1]

To start with the essential basics, we focus on the spherical harmonics of 0th and 1st order and the sound sources that emit them. This also makes sense from the engineering standpoint since 0th- and 1st-order sound sources are of significant practical relevance, mainly for the following two reasons.

1. At low frequencies many sound emitters act approximately like sources of 0th- or 1st-order spherical waves.

2. According to *Huygen*'s principle, each point on a wavefront is the origin of a spherical wave. Many sound fields are conceivable in a comparatively simple way by employing this principle. With spherical sound waves, the

---

[1] Spherical harmonics are *eigen-functions* of the wave equation in spherical coordinates.

© Springer-Verlag GmbH Germany, part of Springer Nature 2021
N. Xiang and J. Blauert, *Acoustics for Engineers*,
https://doi.org/10.1007/978-3-662-63342-7_9

synthesis of sound radiators with arbitrary directional characteristics is possible.

9.1 The Spherical Wave Equation

The wave equation allows for a one-dimensional, point-symmetric solution. This is a sound wave where all parameters only depend on the distance, r, from the origin. The solution does not depend on the direction of propagation, which is always radial and directed either outward or toward the origin. This type of wave is called a spherical wave of the 0th order, and a sound source that emits such a wave is called a spherical source of 0th order.

To derive the appropriate wave equation, it is helpful to transform the wave equation from *Cartesian* coordinates, x, y, z, into spherical coordinates, ϕ, θ, r. This is accomplished with the following well known operator,[2]

$$\Delta = \nabla^2 = \left[\frac{\partial^2}{\partial x^2} + \frac{\partial^2}{\partial y^2} + \frac{\partial^2}{\partial z^2} \right]$$
$$= \frac{1}{r^2} \left[\frac{\partial}{\partial r} \left(r^2 \frac{\partial}{\partial r} \right) + \frac{1}{\sin \theta} \frac{\partial}{\partial \theta} \left(\sin \theta \frac{\partial}{\partial \theta} \right) + \frac{1}{\sin^2 \theta} \frac{\partial^2}{\partial \varphi^2} \right]. \tag{9.1}$$

Because the assumed sound field is point symmetric and only changes in the radial direction, we state that

$$\frac{\partial}{\partial \theta} = \frac{\partial}{\partial \varphi} \equiv 0. \tag{9.2}$$

This leads to the wave equation for the 0th-order spherical wave,

$$\frac{1}{r^2} \frac{\partial}{\partial r} \left(r^2 \frac{\partial}{\partial r} \right) p = \frac{\partial^2 p}{\partial r^2} + \frac{2}{r} \frac{\partial p}{\partial r} = \frac{1}{c^2} \frac{\partial^2 p}{\partial t^2}. \tag{9.3}$$

Note that this equation is identical to the wave equation for conical horns—which was derived in Sect. 8.2. The only difference is that x has been replaced by r. This congruence is intuitively plausible when we think of the spherical wave as a sound field composed of an infinite number of adjacent, very slim conical horns. Figure 9.1 illustrates this concept. When removing the "walls" between these conical horns, the sound field nevertheless remains the same because there is radial propagation only.

The solutions for the outward-progressing wave in the 0th-order spherical sound field are

$$\underline{p}_\rightarrow (r) = \frac{\underline{g}_\rightarrow}{r} e^{-j\beta r}, \quad \text{and} \tag{9.4}$$

$$\underline{v}_\rightarrow (r) = \underline{g}_\rightarrow \left(\frac{1}{\varrho c r} + \frac{1}{j \omega \varrho r^2} \right) e^{-j\beta r}. \tag{9.5}$$

---

[2] For a derivation of this expression see the solution to Problem 9.6.

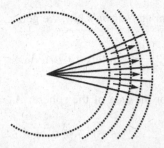

Fig. 9.1 Spherical waves of 0th order as a composition of conical waves

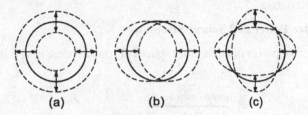

Fig. 9.2 Sound sources for spherical waves (**a**) 0th-order source, also called *breathing* sphere. (**b**) Example of a 1st-order source, also called rigid *oscillating sphere*. (**c**) Example of a 2nd-order source. Note that there are $2n + 1$ possible modes per order, n, with $n = 0, 1, 2 \ldots$ Compare Fig. 9.5

The field impedance of the diverging wave is

$$\underline{Z}_f = \varrho c \frac{j\frac{2\pi r}{\lambda}}{1 + j\frac{2\pi r}{\lambda}} = \frac{1}{\frac{1}{\varrho c} + \frac{1}{j\omega \varrho r}} . \tag{9.6}$$

The region of $r < \lambda/2\pi$ is called the *near-field*, the one of $r > \lambda/2\pi$ the *far-field*, as discussed in Sect. 8.2.

Spherical sound fields of the 0th order are radiated by spherical sound sources of 0th order, also called *breathing spheres*—shown in Fig. 9.2a.

The Co-vibrating Medium Mass

It is an interesting exercise to calculate which part of the near field medium mass moves back and forth without being compressed. Because this part does not transmit active power, it is sometimes called the *Watt*-less-vibrating mass. The equivalent circuit in Fig. 8.4 illustrates this situation. The diameter of the breathing sphere is r_0.

In the far-field, the real term outweighs the imaginary one. As a result, there is no reactive power and no *Watt*-less vibrating mass. In the near-field the particle velocity flows through the mass so that

$$\left| \frac{\underline{p}}{\underline{v}} \right| = \omega \varrho r_0, \tag{9.7}$$

and, therefore, by implementing *Newton*'s law, we get

$$\left| \frac{\underline{F}}{\underline{v}} \right| = \omega m = \omega \varrho r_0 A_0 . \tag{9.8}$$

By inserting the formula for the area of the sphere, $A_0 = 4 \pi r_0^2$, the co-vibrating mass is found to be

$$m_{co} = 4 \pi r_0^3 \varrho . \tag{9.9}$$

This is three times the mass of the medium inside the sphere if the medium is the same inside and outside.

Radiated Active Power and Source Strength

The radiated active power of a 0th-order spherical sound source is as follows—refer to Sect. 8.4,

$$\overline{P} = \frac{1}{2} \overbrace{A(r) \operatorname{Re}\{\underline{Z}_f(r)\}}^{\text{radiation resistance, } r_{\text{rad}}} |\underline{v}(r)|^2$$

$$= \frac{1}{2} \varrho c \frac{\left(\frac{\omega r}{c}\right)^2}{1 + \left(\frac{\omega r}{c}\right)^2} 4 \pi r^2 |\underline{v}|^2$$

$$= \frac{1}{2} \varrho c \frac{\left(\frac{\omega}{c}\right)^2}{4\pi \left[1 + \left(\frac{\omega r}{c}\right)^2\right]} \underbrace{(4 \pi r^2 |\underline{v}|)^2}_{\text{vol. velocity } q} . \tag{9.10}$$

In the near-field, we have $2\pi r/\lambda = \omega r/c \ll 1$, allowing us to write

$$\overline{P} = \frac{1}{2} \varrho c \frac{\left(\frac{\omega}{c}\right)^2}{4 \pi} (4 \pi r^2 |\underline{v}|)^2 = \frac{1}{2} \frac{\varrho \omega^2}{4 \pi c} |\underline{q}_0|^2 . \tag{9.11}$$

With $|v| \sim 1/r^2$, the following also holds,

$$\left(4 \pi r^2 |\underline{v}|\right)^2 = |\underline{q}_0|^2 \approx \text{const} , \tag{9.12}$$

which means that in the near-field the volume velocity, $\underline{q}$, is fairly independent of the distance, r, and converges to $\underline{q}_0$. This primary volume velocity, $\underline{q}_0$, is called the *source strength* of spherical radiators.

The active power transmitted, $\overline{P}$, does not depend on the distance, r, given that the medium is lossless. As a result of this and the fact that active power flows through all spherical shells, we write

$$\overline{P} = 4 \pi r^2 \operatorname{Re}\{\underline{I}\} \neq f(r) . \tag{9.13}$$

The term 4π in the denominator of (9.11) denotes the full spherical angle, $\Omega_\Sigma = 4\pi$. If a 0th-order spherical source with the source strength $\underline{q}_0$ radiates into a smaller spherical angle, Ω_1, that is only a section of the available volume, then the radiated power increases by a ratio of $4\pi/\Omega_1$. Since this power is only radiated into the smaller angle, the intensity, $\mathrm{Re}\{\underline{I}\}$, in this section increases by $(4\pi/\Omega_1)^2$ or $20\lg(4\pi/\Omega_1)\,\mathrm{dB}$.

This relationship is of practical relevance, for instance, for horn loudspeakers, further for all 0th-order spherical sound sources when placed in front of a wall or in a corner or edge of a room. The following level increases result from such placements.[3]

- Placement in front of a wall (hemisphere) $\implies$ +6 dB
- Placement in a room edge (quarter sphere) $\implies$ +12 dB
- Placement in a corner (1/8th sphere) $\implies$ +18 dB

Point Sources of the 0th Order (Monopoles)

In the 0th-order spherical sound field we have

$$\frac{\underline{p}(r)}{\underline{v}(r)} = \frac{1}{\frac{1}{\varrho c} + \frac{1}{j\omega\varrho r}}, \quad \text{or} \quad \underline{g}_\rightarrow \frac{e^{-j\beta r}}{r} = \frac{\underline{v}(r)}{\frac{1}{\varrho c} + \frac{1}{j\omega\varrho r}}, \tag{9.14}$$

from which follows

$$\underline{g}_\rightarrow = \frac{4\pi\,\underline{v}(r)\,r^2}{4\pi\left(\frac{r}{\varrho c} + \frac{1}{j\omega\varrho}\right)}\,e^{+j\beta r}. \tag{9.15}$$

We now let the radius of the sphere go to zero while keeping $\underline{g}_\rightarrow$ constant. In this way we obtain

$$\lim_{r\to 0}\,[4\pi r^2\,\underline{v}(r)] = \underline{q}_0, \tag{9.16}$$

from which follows

$$\lim_{r\to 0}\underline{g}_\rightarrow = \frac{j\omega\varrho\,\underline{q}_0}{4\pi}. \tag{9.17}$$

Finally, we arrive at the sound field of the *point source* of 0th-order, which is also known as *monopole*,

$$\underline{p}_\rightarrow(r) = j\omega\varrho\underline{q}_0\,\frac{e^{-j\beta r}}{4\pi r}. \tag{9.18}$$

Any 0th-order spherical sound source, that is, any breathing sphere, is representable by an equivalent monopole with the same source strength, $\underline{q}_0$.

---

[3] Note that loudspeakers in closed boxes become spherical radiators at low frequencies—refer to Sect. 9.4. Adjustment of their-frequency response is thus possible by appropriate placement in the space.

9.2 Spherical Sound Sources of the First Order

A rigid sphere may oscillate according to the sketch in Fig. 9.2b, and a sound field created in this way is called a 1st-order spherical sound field. Such a sound field is no longer point-symmetric, which means that the shells around the sphere do not represent areas of equal phase. This may also be expressed as $\partial/\partial\theta \neq 0$ and/or $\partial/\partial\varphi \neq 0$.

Since the problem is axial-symmetric it is sufficient to deal with one section through the sphere. Here we consider a vertical section along the x–axis. The following boundary condition is valid for the radial component on the surface of the sphere,

$$\underline{v}(\theta) = \underline{v}(0)\cos\theta. \tag{9.19}$$

Point Sources of the 1st Order (Dipoles)

The fields of two complementary monopoles with opposite phase are combinable for creating a sound field like that of an oscillating sphere. This allow to derive the wave equation for 1st-order spherical sound fields in a relatively easily way.

Two point sources with equal strength but of opposite phase, that is, $\underline{q}_1 = -\underline{q}_0$ and $\underline{q}_2 = +\underline{q}_0$, are positioned a distance of $2\,d$ apart, forming a so-called *dipole*. Due to the linearity of the wave equation, the sound field of this arrangement is given by superposition of the two individual sound fields, namely,

$$\underline{p}_\rightarrow = \frac{j\omega\varrho}{4\pi}\,\underline{q}_0\left(\frac{e^{-j\beta r_2}}{r_2} - \frac{e^{-j\beta r_1}}{r_1}\right). \tag{9.20}$$

Figure 9.3a illustrates this situation. Since the two 0th-order point sources have zero radius, possible reflection or diffraction caused by their presence need not be considered.

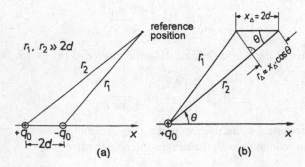

Fig. 9.3 Derivation of the dipole sound field, (a) Two monopoles of opposite phase at a distance of $2\,d$, (b) Equivalent situation with only one monopole

The next step is to perform a limit-operation in such a way that $2\,d$ goes to zero, while, by definition, the *dipole strength*, $\underline{\mu}_d = 2\,d\,\underline{q}_0$, is kept constant. This condition prohibits the two monopoles from canceling each other, and we write

$$\underline{p}_\rightarrow = \frac{j\omega\varrho}{4\pi}\,\underline{\mu}_d\,\lim_{2d\to 0}\left[\frac{1}{2\,d}\left(\frac{e^{-j\beta r_2}}{r_2} - \frac{e^{-j\beta r_1}}{r_1}\right)\right]. \tag{9.21}$$

Figure 9.3b illustrates that the previous equation is interpretable as the result of the differentiation of a monopole sound field in the x–direction. By taking $\partial x \approx \partial r/\cos\theta$ we get

$$\frac{\partial}{\partial x}\,f(x) = \lim_{x_\Delta\to 0}\left[\frac{f(x+x_\Delta) - f(x)}{x_\Delta}\right], \quad \text{where} \quad x_\Delta = 2\,d\,. \tag{9.22}$$

The resulting solutions of the wave equation for the dipole field are as follows— outward-progressing waves only,

$$\begin{aligned}
\underline{p}_\rightarrow(r,\theta) &= \frac{j\omega\varrho\underline{\mu}_d}{4\pi}\,\cos\theta\,\frac{\partial}{\partial r}\left(\frac{e^{-j\beta r}}{r}\right) \\
&= -j\omega\varrho\underline{\mu}_d\left(\frac{1}{r} + j\beta\right)\cos\theta\,\frac{e^{-j\beta r}}{4\pi r} \\
&= \varrho c\,\underline{\mu}_d\left(\beta^2 - j\frac{\beta}{r}\right)\cos\theta\,\frac{e^{-j\beta r}}{4\pi r},
\end{aligned} \tag{9.23}$$

$$\underline{v}_\rightarrow(r,\theta) = \underline{\mu}_d\left(\beta^2 - \frac{2}{r^2} - j\frac{2\beta}{r}\right)\cos\theta\,\frac{e^{-j\beta r}}{4\pi r}\,. \tag{9.24}$$

The solution for $\underline{v}$ has again been derived via *Euler*'s equation. Note that the sound pressure possesses a $1/r^2$-component, which means that it has a near-field. From (9.23) and (9.24). the field impedance is, with $\beta = 2\pi/\lambda = \omega/c$,

$$\underline{Z}_f = \frac{\underline{p}_\rightarrow}{\underline{v}_\rightarrow} = \varrho c\,\frac{(\beta r)^2 - j\beta r}{(\beta r)^2 - 2 - j2\beta r} = \varrho c\,\frac{\left(\frac{2\pi r}{\lambda}\right)^2 - j\frac{2\pi r}{\lambda}}{\left(\frac{2\pi r}{\lambda}\right)^2 - 2 - j2\left(\frac{2\pi r}{\lambda}\right)}\,. \tag{9.25}$$

The real part thereof is

$$\mathrm{Re}\{\underline{Z}_f\} = \varrho c\,\frac{(\beta r)^4}{\left[2 + (\beta r)^2\right]^2} = \varrho c\,\frac{\left(\frac{2\pi r}{\lambda}\right)^4}{\left[2 + \left(\frac{2\pi r}{\lambda}\right)^2\right]^2}\,. \tag{9.26}$$

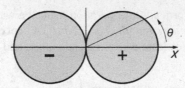

Fig. 9.4 Directional characteristic, Γ, of a dipole sound source

In the near-field, we thus have an approximate proportionality of $\mathrm{Re}\{\underline{Z}_f\} \propto \omega^4$. Recall that for the monopóle we found $\mathrm{Re}\{\underline{Z}_f\} \propto \omega^2$ for the near-field, that is, a profoundly less dependence on the frequency.

The dipole sound field shows the following directional characteristic for both the sound pressure and the particle velocity,

$$\Gamma = \frac{\underline{p}_\rightarrow(r, \theta)}{\underline{p}_\rightarrow(r, 0)} = \frac{\underline{v}_\rightarrow(r, \theta)}{\underline{v}_\rightarrow(r, 0)} = \cos\theta\,, \tag{9.27}$$

as plotted in Fig. 9.4. Note that the plot only shows the vertical plane, but the directional characteristics are axial-symmetric around the x-axis.

We see a figure-of-eight characteristic that complies with the boundary conditions of the rigid oscillating sphere. It follows that sound field of oscillating spheres is represented by 1st-order spherical point sources (dipoles).

9.3 Spherical Harmonics

It is possible to consider any sound-field as being composed of a series of orthogonal spherical harmonics of different orders. These spherical harmonic waves are eigenfunctions of the wave equation in spherical coordinates. Taking the spherical-wave equation in the *Helmholtz* form using (9.1) with a trial solution

$$\underline{p}(r, \theta, \varphi) = \underline{R}(r)\,\underline{\Theta}(\theta)\,\underline{\Phi}(\varphi)\,\mathrm{e}^{\mathrm{j}\omega t}, \tag{9.28}$$

yields

$$\frac{1}{r^2}\left[\frac{1}{\underline{R}}\frac{\partial}{\partial r}\left(r^2\frac{\partial\underline{R}}{\partial r}\right) + \frac{1}{\underline{\Theta}\,\sin\theta}\frac{\partial}{\partial\theta}\left(\sin\theta\,\frac{\partial\underline{\Theta}}{\partial\theta}\right)\right.$$

$$\left. + \frac{1}{\underline{\Phi}\,\sin^2\theta}\frac{\partial^2\underline{\Phi}}{\partial\varphi^2}\right] + \beta^2 = 0. \tag{9.29}$$

We separate this equation[4] in a radial, r, an elevation, θ, and an azimuth, φ, component, respectively,

$$\frac{1}{r^2}\frac{\partial}{\partial r}\left(r^2\frac{\partial \underline{R}}{\partial r}\right) + \beta^2\,\underline{R} - \frac{n(n+1)}{r^2}\,\underline{R} = 0, \tag{9.30}$$

$$\frac{1}{\sin\theta}\frac{\partial}{\partial\theta}\left(\sin\theta\,\frac{\partial \Theta}{\partial\theta}\right) + \left[n(n+1) - \frac{m^2}{\sin^2\theta}\right]\Theta = 0, \tag{9.31}$$

and

$$\frac{\partial^2 \Phi}{\partial\varphi^2} + m^2\,\Phi = 0. \tag{9.32}$$

The solution of the radial-component equation (9.30) is expressible as

$$\underline{R}(r) = \underline{R}_1\,j_n(\beta r) + \underline{R}_2\,y_n(\beta r), \tag{9.33}$$

with $\underline{R}_1$ and $\underline{R}_2$ being constants.
$j_n(\nu)$ and $y_n(\nu)$ are termed spherical *Bessel* functions, they read

$$j_n(\nu) = (-\nu)^n\left(\frac{1}{\nu}\frac{d}{d\nu}\right)^n\frac{\sin\nu}{\nu}, \tag{9.34}$$

$$y_n(\nu) = -(-\nu)^n\left(\frac{1}{\nu}\frac{d}{d\nu}\right)^n\frac{\cos\nu}{\nu}, \tag{9.35}$$

with $n = 0, 1, 2, \ldots$ being their *order*.
The radial solution is also expressible as follow,

$$\underline{R}(r) = \underline{R}_3\,\underline{h}_n(\beta r) + \underline{R}_4\,\underline{h}_n^*(\beta r), \tag{9.36}$$

with $\underline{R}_3$, $\underline{R}_4$ being constants. $\underline{h}_n(\beta r)$ is termed spherical *Hankel* function, written as

$$\underline{h}_n(\beta r) = j_n(\beta r) + \mathrm{j}\,y_n(\beta r). \tag{9.37}$$

The solutions of the elevation- and azimuth-component equations are

$$\Theta(\theta) = \hat{\Theta}\,P_n^m(\cos\theta), \tag{9.38}$$

$$\underline{\Phi}(\varphi) = \hat{\Phi}\,\mathrm{e}^{\mathrm{j}m\varphi}, \tag{9.39}$$

where $\hat{\Theta}$ and $\hat{\Phi}$ are amplitudes, and $P_n^m(\eta)$ is the associated *Legendre* function

---

[4] Done by adding a "magic-zero" quantity $(m^2 - m^2)/(r^2\sin^2\theta) + [n(n+1) - n(n+1)]/r^2$ to the right-hand side of (9.29).

$$P_n^m(\eta) = (-1)^m (1 - \eta^2)^{m/2} \frac{d^m}{d\eta^m} P_n(\eta), \tag{9.40}$$

with $P_n(\eta)$ being the *Legendre* polynomial

$$P_n(\eta) = \frac{1}{2^n n!} \frac{d^n}{d\eta^n} (\eta^2 - 1)^n. \tag{9.41}$$

For practical convenience, the solutions of the angular (elevation and azimuth) components in (9.38) and (9.39) are combined with specified amplitude values of

$$\underline{Y}_n^m(\theta, \varphi) = \sqrt{\frac{2n+1}{4\pi} \frac{(n-m)!}{(n+m)!}} P_n^m(\cos\theta) e^{jm\varphi}. \tag{9.42}$$

The term $\underline{Y}_n^m(\theta, \varphi)$ is denominated *spherical harmonics*. The integer $n = 0, 1, 2, \ldots$, is called the *order*, and $m = 0, \pm 1, \pm 2, \ldots, \pm n$ is named the *degree (mode)* of the spherical harmonics, respectively.

The spherical harmonics form a set of orthonormal basis functions.[5] They are applicable as expansion of any arbitrary function, $g(\theta, \varphi)$, on the surface of a sphere, namely,

$$g(\theta, \varphi) = \sum_{n=0}^{\infty} \sum_{m=-n}^{n} A_n^m \underline{Y}_n^m(\theta, \varphi), \tag{9.43}$$

with appropriate coefficients, A_n^m. The real-valued spherical harmonics

$$\begin{cases} \sqrt{\frac{2n+1}{2\pi} \frac{(n-m)!}{(n+m)!}} P_n^m(\cos\theta) \cos m\varphi & m > 0, \\[3mm] \sqrt{\frac{2n+1}{2\pi}} P_n(\cos\theta) & m = 0, \\[3mm] \sqrt{\frac{2n+1}{2\pi} \frac{(n-m)!}{(n+m)!}} P_n^m(\cos\theta) \sin|m|\varphi & m < 0, \end{cases} \tag{9.44}$$

up to the order of $n = 3$, are illustrated in Fig. 9.5 with $P_n(\cdot)$ being a *Legendre* polynomial—see (9.41).

Combining the radial component in (9.33) with the above spherical harmonics into a single expression brings

$$\underline{p}(r, \theta, \varphi) = \sum_{n=0}^{\infty} \sum_{m=-n}^{n} [A_n^m y_n(\beta r) + B_n^m j_n(\beta r)] \underline{Y}_n^m(\theta, \varphi) \tag{9.45}$$

---

[5] See the solution to Problem 9.7.

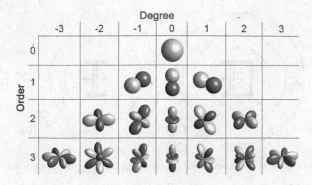

Fig. 9.5 Real-valued pherical harmonics in (9.44) up to the third order ($n = 0, 1, 2, 3$) as in (9.44), and for degrees, m, between $-3 \leq m \leq 3$. The lobes indicate positive and negative values by different shading. For each given order, n, there are $2n + 1$ degrees as listed in each row of the table

satisfies *Helmholtz'* equation (9.29). This condition is convenient for solving traveling-wave problems with A_n^m and B_n^m being (constant) coefficients.

Similarly, combination of the radial component in (9.36) with (9.42)

$$\underline{p}(r, \theta, \varphi) = \sum_{n=0}^{\infty} \sum_{m=-n}^{n} [C_n^m \, h_n(\beta r) + D_n^m \, h_n^*(\beta r)] \, \underline{Y}_n^m(\theta, \varphi), \qquad (9.46)$$

also satisfies the *Helmholtz* equation, which is convenient for solving standing wave problems with C_n^m and D_n^m being coefficients. The term containing C_n^m represents an outgoing wave component from the spherical coordinate origin, whereas the term containing D_n^m represents an incoming component, traveling toward the origin. The values of the coefficients, A_n^m, B_n^m, C_n^m, and D_n^m, are specific for the actual problem concerned and its boundary conditions.

9.4 Higher-Order Spherical Sound Sources

In the preceding section, we already introduced two spherical harmonics, namely, in form of the spherical waves of 0th and 1st order. Spherical waves of higher order are radiated by spheres with surfaces that oscillate with velocities determined by higher-order spherical functions. Figure 9.2c depicts one possible 2nd-order spherical vibration, that is the one which is indexed ($n = 2, m = 2$)—compare as also in Fig. 9.5.

For spherical sound emitters of nth order in the near-field, the resistive part of the field impedance, Re$\{\underline{Z}_f\}$, increases with frequency as follows,

$$\text{Re}\{\underline{Z}_f\} \propto \omega^{2(n+1)}. \qquad (9.47)$$

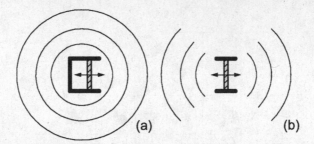

Fig. 9.6 Schematic plots of the sound fields of *breathing*, **(a)**, and *oscillating* sound sources, **(b)**

The increase of Re$\{\underline{Z}_f\}$, with frequency opens the possibility of practical simplifications. If the linear dimensions of the emitter are small compared to the wavelengths, that is, $2\pi r_0 \ll \lambda$, the radiation of higher-order spherical waves is negligible at low frequencies. Thus, a monopole source provides a good low-frequency approximation for all *breathing* sound sources such as a loudspeaker mounted in a box (...cabinet). A dipole sound source serves well to approximate the behavior of *oscillating* sources at low frequencies, for instance, loudspeakers without a baffle—see Fig. 9.6a, b.

9.5   Line Arrays of Monopoles

Arrangements of several sources along a line in space are called *line arrays* (...*linear arrays*). They play a relevant role in practical applications. Because the sound fields of the individual sources interfere with each other, sharply bundled radiation are achieved with these arrays. Common applications of this principle are, for example, line arrays of loudspeakers.

The following discussion considers the directional characteristics of linear arrays composed of monopoles. Hereby we restrict our view to the sound field far away from the array.

Line Array of Identical and Equidistant Monopoles

In an arrangement like the one depicted in Fig. 9.7, we refer to a position in space at a distance of $r_0 \gg 2h$. The sum of the contributions of all monopoles of the array at this reference point is

$$\underline{p}_\rightarrow(r = r_0, \theta) = \frac{\mathrm{j}\,\omega\,\varrho\,\underline{q}_0}{4\,\pi} \sum_{i=1}^{n} \frac{\mathrm{e}^{-\mathrm{j}\beta\,[r_0-(i-1)2d\,\cos\theta]}}{r_0 - (i-1)2d\,\cos\theta} . \tag{9.48}$$

We now neglect the differences in the magnitude of the individual contributions because of $r_0 \gg 2h$. This allows to consider the phase differences only.
This approach leads to the following approximation,

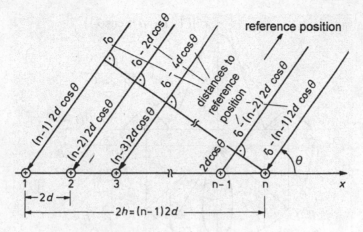

Fig. 9.7 Linear array with monopole sources

$$\underline{p}_{\rightarrow}(r, \theta) \approx \frac{j \omega \varrho \, \underline{q}_0}{4 \pi} \frac{e^{-j\beta r_0}}{r_0} \sum_{i=1}^{n} e^{+j\beta (i-1) 2d \cos \theta} . \tag{9.49}$$

By substituting $\beta d \cos \theta$ with b in the expression for the sum, we obtain an expression with a known series summation,

$$\sum e^{+j(i-1)2b} = 1 + e^{+j2b} + e^{+j4b} + \cdots e^{+j2(n-1)b} = \frac{1 - e^{+j2nb}}{1 - e^{+j2b}} . \tag{9.50}$$

Writing with an expansion using $\sin x = (e^{+jx} - e^{-jx})/2 \, j$, yields,

$$\sum e^{+j(i-1)2b} = \frac{e^{+jnb}}{e^{+jb}} \left(\frac{e^{-jnb} - e^{jnb}}{e^{-jb} - e^{jb}} \right) = e^{+j(n-1)b} \frac{\sin(nb)}{\sin(b)} . \tag{9.51}$$

The term $\sin(nb)/\sin b$ determines the directional characteristic. We discuss it more conveniently in the following paragraph, where a continuously loaded line of monopoles is dealt with.

Continuously Loaded Line Array

First we perform a limit operation by letting the distance between the individual monopoles, $2d$, and, consequently, b, go to zero. With the length of line array, $2h$, kept constant, we then get $n \rightarrow \infty$. To also keep the total source strength, $n \, \underline{q}_0 = \underline{q}'(x) \, 2h$, constant, we normalize by n, with $\underline{q}'(x)$ being a constant velocity load. The result of this operation,

$$\lim_{2d \rightarrow 0} \frac{\sin(nb)}{n \, \sin(b)} = \frac{\sin(nb)}{nb} = \mathrm{si}(nb) , \tag{9.52}$$

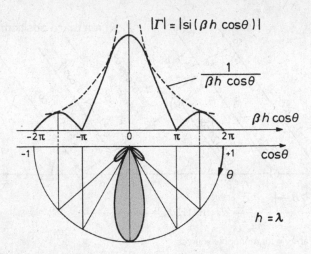

Fig. 9.8 Directional characteristics, Γ, of a line array of the length $2h = 2\lambda$

is a so-called *sinc*- or *si-function* (... sinus cardinalis). With $2h \approx n\,2d$ and, therefore, $nb = \beta h \cos\theta$; it follows that

$$\Gamma = \text{si}\,(\beta h \cos\theta)\,. \tag{9.53}$$

This is the directional characteristic of the in-phase, continuously loaded line array with a constant source-strength load. This formula also loosely covers arrays with a limited number of monopoles.

As an example, Fig. 9.8 illustrates a line array with a length of two wavelengths, $2h = 2\lambda$, and $\beta h = 2\pi$. The upper panel illustrates the directional characteristic in *Cartesian* coordinates. The lower panel shows the typical club-shaped form of the beam in spherical coordinates—vertical section only.

9.6 Analogies to *Fourier* Transform in Signal Theory

In the preceding section, we assumed a continuous source-strength load, $q'(x)$, having the dimension [volume velocity/length]. We continued to presume that each point on the line acts as a (differential) monopole—illustrated in Fig. 9.9.

After the following integration, we calculate the sound pressure at the observation point, $r_0 \gg 2h$. The term in front of the integral is a constant term for a given r_0. The complete expression is

$$\underline{p}_{\rightarrow}(r, \theta) = \underbrace{\frac{j\omega\varrho}{4\pi} \cdot \frac{e^{-j\beta r_0}}{r_0}}_{\text{const.}} \int_{-h}^{+h} \underline{q}'(x)\,e^{+j(\beta\cos\theta)x}\,dx\,. \tag{9.54}$$

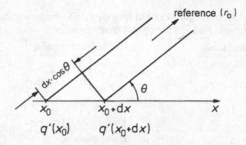

Fig. 9.9 Sound fields of a line array with a continuous volume-velocity load

This expression is isomorphic to the well-known *Fourier* integral from signal theory, also termed *Fourier* transform, which we write as follows,

$$\underline{S}(\omega) = \int s(t)\, \mathrm{e}^{-\mathrm{j}\omega t}\mathrm{d}t\,, \quad \text{or, symbolically, as} \tag{9.55}$$

$$s(t) \quad \circ\!\!\!-\!\!\!\bullet \quad \underline{S}(\omega)\,, \tag{9.56}$$

Here t corresponds to x and ω corresponds to $-\beta \cos\theta$. $\underline{S}(\omega)$ is termed *spectrum* of $s(t)$ in frequency domain, ω.

Disregarding the constant factor, we find the following analogies between the time functions and the source-strength-velocity loads of line arrays.

Time function, $s(t)$ Source-strength load, $q'(x)$

$\underset{\bullet}{\overset{\circ}{|}}$ $\Longleftrightarrow$ $\underset{\bullet}{\overset{\circ}{|}}$

Spectrum, $|\underline{S}(\omega)|$ Directional characteristic, $\Gamma(\theta)$

Table 9.1 lists correspondences of examples that we have treated so far.[6]

At the end of this section. We discuss two additional directional characteristics that are relevant from an application point of view, and which are also obtainable from equivalent relationships in signal theory.

- A source-strength load with a *Gaussian* envelope leads to a *Gaussian* directional characteristic. This is a beam without side lobes—depicted in Fig. 9.10a

- A source-strength load with a phase shift increasing linearly with position,

$$q_2'(x) = q_1'(x)\, \mathrm{e}^{-\mathrm{j}\beta \cos(\theta_\Delta)\, x} \tag{9.57}$$

$$\underset{\bullet}{\overset{\circ}{|}}$$

$$\Gamma_2 = \Gamma_1\left[\,\beta \cos\theta - \beta \cos(\theta_\Delta)\,\right] \tag{9.58}$$

---

[6] For the definition of Γ see (9.27). In the table, the directional characteristics, $\Gamma(\theta)$, have been normalized so that their maxima equal one. $\delta(z)$ is called *Dirac* impulse. It is a special mathematical *distribution* that picks out the value of a function at the position of its argument as follows, $\int_{-\infty}^{+\infty} y(z)\, \delta(z-z_0)\, \mathrm{d}z = y(z_0)$. The area under the *Dirac* impulse is $\int_{-\infty}^{\infty} \delta(z)\, \mathrm{d}z = 1$.

Table 9.1 Some examples of the equivalence of time signal and frequency spectrum verses spatial distribution of source-strength load and directional characteristics

Linear array with constant load	Rectangular impulse
$\underline{q}' = \begin{cases} \text{const} & \text{for} \quad -h < x < h \\ 0 & \text{others} \end{cases}$	$s(t) = \begin{cases} \text{const} & \text{for} \quad -\tau < t < \tau \\ 0 & \text{others} \end{cases}$
$\Gamma = \text{si}\,(-h\,\beta\cos\theta)$	$S(\omega) = 2\tau\,\text{si}\,(\tau\,\omega)$
Monopole	*Dirac* impulse
$\underline{q}'(x) = \delta(x)$	$s(t) = \delta(t)$
$\Gamma = 1$	$S(\omega) = 1$
Dipole	Double *Dirac* impulse
$\underline{q}'(x) = \underline{\mu}_d\,\frac{\mathrm{d}}{\mathrm{d}x}\delta(x) = \underline{\mu}_d\,\delta'(x)$	$s(t) = \frac{\mathrm{d}}{\mathrm{d}t}\delta(t) = \delta'(t)$
$\Gamma = \cos\theta$	$S(\omega) = \omega$

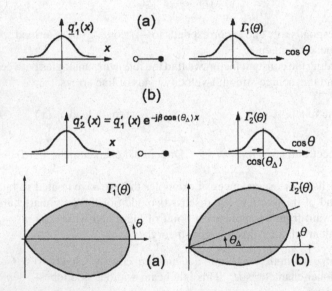

Fig. 9.10 Gaussian distribution of source strength and directional characteristic **(a)** Shaping the directional characteristic **(b)** Shifting the directional characteristic (schematic)

leads to a tilted directional characteristic—schematically shown in Fig. 9.10b. The *shifting theorem* of *Fourier* transform is applied here.[7]

9.7   Directional Equivalence of Sources and Receivers

When a reversible transducer or transducer array is operated as sound emitter, its directional characteristic is equivalent to its directional sensitivity characteristic when operated as a receiver. We find this relationship by using the following two elements,

M... A transducer with arbitrary directional characteristics

X... An auxiliary point source with monopole characteristics

The point source is positioned far away from the transducer concerned. The proof is done in two steps as follows. b

(1) The transducer is fed with an electric current, $\underline{i}_0$. At the position of the auxiliary source we then have

$$|\underline{p}_X| = |\underline{T}_{ip}(\omega, \phi, \theta, r)|\,|\underline{i}_0|. \tag{9.59}$$

(2) The auxiliary point source emits a volume velocity amounting to its source strength, q_0. At the position of the transducer, which is not present at this point, we get

$$|\underline{p}_M| = \frac{\omega\varrho}{4\pi r}|\underline{q}_0|. \tag{9.60}$$

If the transducer is now introduced into the sound field, a voltage, $\underline{u}_1$, appears at its electric output port according to

$$|\underline{u}_1| = |\underline{T}_{pu}(\omega, r, \theta, \phi)|\,|\underline{p}_M|. \tag{9.61}$$

Since the sound field is linear and passive, reversibility according to (4.16) applies as follows,

$$\left|\frac{\underline{i}_0}{\underline{p}_X}\right|_{q=0} = \left|\frac{\underline{q}_0}{\underline{u}_1}\right|_{i=0}, \tag{9.62}$$

and, hence,

$$\left|\frac{T_{ip}}{T_{pu}}\right| = \left|\frac{\omega\varrho}{4\pi r}\right|. \tag{9.63}$$

---

[7] An application of this algorithm, based on the directional equivalence of emitters and receivers—see next section—is the electronic steering of SONAR antennas.

The striking attribute of this equation is that the transfer coefficient of the transmitter in emitting function increases with the frequency concerning the transfer coefficient for receiver operation (sensitivity). This justifies a well known general rule in the field, which is,

Transducers receive low frequencies better than they emit them!

A good example of this rule is a small reversible microphone.

Finally, to show that the directional characteristic, Γ, for emitter operation of a reversible emitter/receiver is identical to its characteristic for receiver operation, we set

$$\frac{|\underline{T}_{ip}(\phi,\theta)|}{|\underline{T}_{pu}(\phi,\theta)|} = \frac{|\underline{T}_{ip}(0,0)|\,\Gamma_{ip}}{|\underline{T}_{pu}(0,0)|\,\Gamma_{pu}} = \left|\frac{\underline{T}_{ip}(0,0)}{\underline{T}_{pu}(0,0)}\right|, \tag{9.64}$$

which results in

$$\Gamma_{ip} = \Gamma_{pu}. \tag{9.65}$$

Regarding the examples dealt with in this book, there are the following correspondences,

– Pressure receiver	$\Longleftrightarrow$	0th-order spherical source
– Pressure-gradient receiver	$\Longleftrightarrow$	1st-order spherical source
– Line microphone	$\Longleftrightarrow$	Line array with a constant load and 90° shifted directional characteristic

9.8   Exercises

Spherical Waves and Line Arrays

Problem 9.1 In the context of the 0th order spherical wave equation,

(a) Show that

$$\frac{1}{r^2}\frac{\partial}{\partial r}\left(r^2\frac{\partial}{\partial r}\right)p = \frac{\partial^2 p}{\partial r^2} + \frac{2}{r}\frac{\partial p}{\partial r} = \frac{1}{r}\frac{\partial^2(rp)}{\partial r^2} = \frac{1}{c^2}\frac{\partial^2 p}{\partial t^2}. \tag{9.66}$$

(b) Given the outward-progressing wave of the 0th-order spherical sound pressure

$$\underline{p}_{\rightarrow}(r) = \frac{\underline{g}_{\rightarrow}}{r}\,e^{-j\beta r}, \tag{9.67}$$

use the *Euler*'s equation to show that

$$\underline{v}_{\rightarrow}(r) = \underline{g}_{\rightarrow}\left(\frac{1}{\varrho c r} + \frac{1}{j\omega\varrho r^2}\right)e^{-j\beta r}. \tag{9.68}$$

(c) Discuss the far-field behavior of the field resistance, $R_f = \mathrm{Re}\left\{\underline{Z}_f\right\}$, of a monopole source, and its field impedance, $\underline{Z}_f$, in the near-field as a function of the angular frequency.

Problem 9.2 A sound source of 0th order in the free field—a breathing sphere—has a radius of a, and the volume velocity on its surface is $\underline{q}(a)$.

(a) Determine the sound-pressure and velocity distribution in the space outside the sphere.

(b) Determine the source strength, $\underline{q}_0$, of a point source which generates the same sound field.

Problem 9.3 Determine and compare the radiated power of spherical sound sources of

(a) 0th order and,

(b) 1st order.

(c) Find the dependency of the sound intensity on the distance.

Problem 9.4 An elementary sound source of 0th order is positioned at the origin of an infinitely-long conical horn with an opening solid angle of Ω. The source is mounted in such a way that all of its volume velocity is released into the horn.

Compare the sound pressure, the velocity, the power, and the intensity in the horn with those of an equivalent source in a free field.

Problem 9.5 Given a dipole with a dipole strength $\underline{M}$,

(a) Sketch the directional characteristics of this dipole.

(b) An additional 0th-order point source, $\underline{q}_0$, is brought to the same position as that of the dipole. Find the magnitude and phase of $\underline{q}_0$ so that in the far-field the entire set-up exhibits the following directional characteristics,

$$\Gamma = \frac{1 + \cos\theta}{2}. \qquad (9.69)$$

(c) Sketch this directional characteristic. What kind of directional characteristic does it represent?

Problem 9.6 Derive the spherical wave equation from the three-dimensional wave equation in *Cartesian* form.

Problem 9.7 The spherical wave equation (9.29) is equal to

$$\frac{1}{r^2}\left[\frac{1}{R}\frac{\partial}{\partial r}\left(r^2\frac{\partial R}{\partial r}\right) + \frac{1}{\Theta\,\sin\theta}\frac{\partial}{\partial\theta}\left(\sin\theta\frac{\partial\Theta}{\partial\theta}\right)\right.$$

$$\left.+ \frac{1}{\Phi\,\sin^2\theta}\frac{\partial^2\Phi}{\partial\varphi^2}\right] + \beta^2 = 0. \qquad (9.70)$$

Add to the right-hand side of the spherical wave equation in (9.70) a zero-quantity

$$\frac{m^2 - m^2}{r^2\sin^2\theta} + \frac{n(n+1) - n(n+1)}{r^2}, \qquad (9.71)$$

with m and n being positive integers.

Show that Eqs. (9.30) and (9.32) is separable into a radial component, r,

$$\frac{1}{r^2}\frac{\partial}{\partial r}\left(r^2\frac{\partial R}{\partial r}\right) + \beta^2\,R - \frac{n(n+1)}{r^2}\,R = 0, \qquad (9.72)$$

and an elevation component, θ,

$$\frac{1}{\sin\theta}\frac{\partial}{\partial\theta}\left(\sin\theta\frac{\partial\Theta}{\partial\theta}\right) + \left[n(n+1) - \frac{m^2}{\sin^2\theta}\right]\Theta = 0, \qquad (9.73)$$

and an azimuth component, φ,

$$\frac{\partial^2\Phi}{\partial\varphi^2} + m^2\,\Phi = 0. \qquad (9.74)$$

Problem 9.8 Given the spherical harmonics, $\underline{Y}_n^m(\theta,\varphi)$, in (9.42) and the *Legendre* functions in (9.40) and (9.41), determine the complex spherical harmonics of the orders $n = 0$, 1, 2, and the degrees of $m = 0$, 1.

Problem 9.9 Given a line array with a source-strength load [volume-velocity per length] evenly distributed over the length, $2h$, of the array, that is,

$$\underline{q}'(x) = \underline{q}'_0\, e^{-x^2/2h^2}, \quad -h \le x \le +h, \qquad (9.75)$$

(a) Determine and sketch the directional characteristics under the far-field assumption that the observation point is much farther away than the finite length of the array.

(b) How does the directional characteristics change when the volume-velocity load of (a) is modified according to the following phase function,

$$\phi = (\beta\,\cos\theta_0)\,x. \qquad (9.76)$$

Chapter 10
Piston Membranes, Diffraction and Scattering

In the preceding chapter, we calculated sound fields of point-source arrays by making use of linear superposition of individual spherical sound fields. For continuously-loaded line arrays, we further substituted the source strength of the individual sources with a length-specific source-strength load, $\underline{q}'_0(x)$. This source-strength load has the dimension [volume-velocity/length].

It suggests itself to extend this method to area radiators like oscillating membranes. Thereby we obtain an area-specific source-strength load, $\underline{q}''_0(x, y) = \underline{v}(x, y)$, with the dimension [volume velocity/area], which is identical with the dimension of particle velocity.

In the case of a line array composed of point sources, reflections and diffractions of the sound field due to the array itself do not exist. Yet, this condition does no longer hold for area radiators because areas may act as both reflectors and diffractors. Computations concerning the sound fields of such radiators can thus become complicated.[1]

However, there is a particular case where reflection and diffraction do not occur. In the current book, we restrict ourselves to just this case, that is, a flat membrane in an infinitely extended, rigid plane baffle. *Huygen's* principle is applicable to this special case in an elementary way by simply superimposing 0th-order point sources.[2] Many practical problems are treatable in this way.

---

[1] The *Kirchhoff–Helmholtz* integral equation is applicable to sound-field calculations in these cases. This equation determines the sound field inside an enclosed space from the sound-pressure and the pressure-gradient distributions on an enclosing surface.

[2] The *Kirchhoff–Helmholtz* integral equation then reduces to the so-called *Rayleigh* or *Huygens–Helmholtz* integral—see Sect. 10.1.

© Springer-Verlag GmbH Germany, part of Springer Nature 2021
N. Xiang and J. Blauert, *Acoustics for Engineers*,
https://doi.org/10.1007/978-3-662-63342-7_10

10.1 The *Rayleigh* Integral

A small vibrating piston results in a 1st-order spherical sound field—see Fig. 9.6b.
However, when hydrodynamic shorting between the front and back of the baffle is
prohibited by placing this piston into an infinitely extended, plane and rigid baffle,
we get a hemispherical sound field of 0th order radiating into half of the space—
illustrated in Fig. 10.1. The hemispherical sound field is

$$\underline{p}_{\rightarrow}(r) = \frac{j \omega \varrho}{2 \pi} \, \underline{q}_0 \, \frac{e^{-j\beta r}}{r} \, , \tag{10.1}$$

where 2π is the spatial angle of a hemisphere. Assuming that the baffle is flat, this
sound field has no normal components in the baffle's plane, and, therefore, the baffle
cannot cause any reflection or diffraction.

Now consider an oscillating membrane with an area A_0 in this infinitely extended,
flat and plane baffle where all area elements oscillate perpendicularly to the area.
We consider this arrangement the superposition of an infinite number of adjacent
monopoles, $d\underline{q}_0 = \underline{v} \, dA$,—depicted in Fig. 10.2. The total sound pressure at an observation point is found by superposition of these monopoles, vis,

$$\underline{p}_{\rightarrow}(r) = \frac{j \omega \varrho}{2 \pi} \int_{A_0} \underline{v}(x, y) \, \frac{e^{-j\beta r}}{r} \, dA \, . \tag{10.2}$$

In acoustics, this integral is usually called *Rayleigh* integral. It is valid at all distances
from the membrane.

It is worth noting that $\underline{v}(x, y)$ needs not be the same everywhere on the membrane.
Its value may depend on the position of the membrane. A finite area A_0 is required for
the *Rayleigh* integral to converge without additional assumptions, such as propagation
losses in the medium. Further, no obstacles are admitted in the hemisphere concerned.

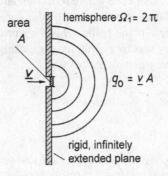

Fig. 10.1 Hemispherical sound field, originating from a point source in a flat and rigid baffle

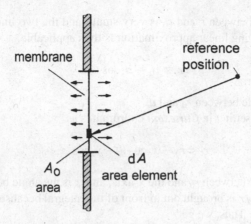

Fig. 10.2 Sound field in front of a membrane in a flat and rigid baffle

10.2 *Fraunhofer*'s Approximation

Fraunhofer's approximation applies when the distance from the reference point to the membrane is very large compared to the linear dimensions of the radiating membrane. Figure 10.3 depicts the situation to be discussed.

The quantity, r_0, is the distance from the reference point to any position on the membrane within the area A_0. That position is preferably the membrane's center of

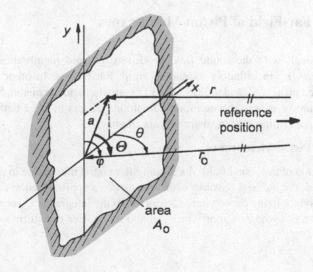

Fig. 10.3 *Fraunhofer*'s approximation

gravity. The angle between r and r_0 is very small, and the two lines are practically parallel. The following linear approximation is thus applicable,

$$r \approx r_0 - a \cos \Theta \,, \tag{10.3}$$

where Θ is the angle between r_0 and a.

We express this setting in *Cartesian* coordinates as

$$r \approx r_0 - x \cos \theta - y \cos \varphi \,, \tag{10.4}$$

where θ is the angle between r_0 and the x-axis, and φ is the angle between r_0 and the y-axis. The factor $1/r$ is brought out in front of the integral because $1/r \approx 1/r_0$—as we saw in Sect. 9.5. Thus, we get

$$\underline{p}_\rightarrow (r, \theta, \varphi) \approx \frac{\mathrm{j} \omega \varrho \, \mathrm{e}^{-\mathrm{j} \beta r_0}}{2 \pi r_0} \times$$

$$\int_{-\infty}^{+\infty} \int_{-\infty}^{+\infty} \underline{v}(x, y) \; \mathrm{e}^{\mathrm{j}(\beta \cos \theta)x} \; \mathrm{e}^{\mathrm{j}(\beta \cos \varphi)y} \; \mathrm{d}x \, \mathrm{d}y \,. \tag{10.5}$$

This expression is again isomorphic to a *Fourier* transform. It is actually a two-dimensional, spatial *Fourier* transform, where $\beta \cos \theta$ is the phase coefficient in x-direction, and $\beta \cos \varphi$ is the phase coefficient in y-direction. Recall that $\beta = 2\pi/\lambda$.

10.3   The Far-Field of Piston Membranes

This section deals with the sound field produced by rigid membranes, so-called *piston membranes*, in infinitely extended, rigid, flat baffles. In other words, we speak of baffled pistons with identical $\underline{v}(x, y)$ everywhere on the piston. Such piston membranes qualify as appropriate models for loudspeakers in large baffles as long as the membranes of the loudspeakers vibrate in phase.

Rectangular Piston Membranes

The computation of the sound field of a rectangular piston membrane in an infinitely extended, rigid, flat baffle is feasible with *Fraunhofer*'s approximation according to Fig. 10.4. By symbolizing the constant factors left of the integral as $\underline{C}_1$ or $\underline{C}_2$, respectively, we write *Fraunhofer*'s approximation and its *Fourier* transform as follows,

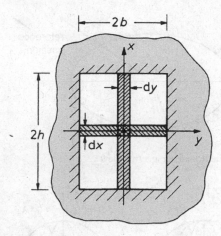

Fig. 10.4 Rectangular membrane

$$\underline{p}_{\rightarrow}(r, \theta, \varphi) = \underline{C}_1 \int_{-b}^{+b} \left[\int_{-h}^{+h} \underline{v} \, e^{j(\beta \cos \theta)x} \, e^{j(\beta \cos \varphi)y} \, dx \right] dy \qquad (10.6)$$

$$= \underline{C}_2 \underbrace{\int_{-h}^{+h} e^{j(\beta \cos \theta)x} \, dx}_{} \underbrace{\int_{-b}^{+b} e^{j(\beta \cos \varphi)y} \, dy}_{}, \qquad (10.7)$$

$$\underline{p}_{\rightarrow}(r, \theta, \varphi) \circ\!\!-\!\!\bullet \underline{C}_2 \, 2h \, \underbrace{\text{si}\,(h \, \beta \cos \theta)}_{\Gamma(\theta)} \, 2b \, \underbrace{\text{si}\,(b \, \beta \cos \varphi)}_{\Gamma(\varphi)} . \qquad (10.8)$$

Because each of the two integrals is a one-dimensional *Fourier* integral, the integration is subject to well-known rules. The result of the integration is (10.8), where the first term on the right-hand side includes the directional characteristic concerning θ, and the second one for φ. The total two-dimensional directional characteristic is formed by multiplying the two one-dimensional characteristics. Each of them represents a continuously loaded line array, one on the x-axis, and another on the y-axis. Thus, we get

$$\Gamma(\theta, \varphi) = \Gamma(\theta) \, \Gamma(\varphi) . \qquad (10.9)$$

The third dimension, r, is not considered because the derivation above only deals with the sound field far away from the membrane.

Circular Piston Membranes

The calculation of circular pistons is slightly more complicated and requires a transformation into polar coordinates—illustrated in Fig. 10.5. The result of the calculation is given below,

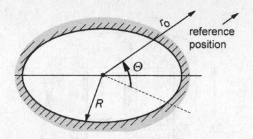

Fig. 10.5 Circular membrane from a perspective view

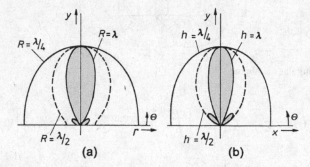

Fig. 10.6 Examples for directional characteristics. **(a)** Circular membrane. `**(b)** Quadratic membrane

$$\Gamma(\Theta) = \frac{2\,\mathbf{J}_1(R\,\beta\,\cos\Theta)}{R\,\beta\,\cos\Theta}\,, \tag{10.10}$$

where Θ is the angle between the membrane and a line leading from the observation point to the middle of the membrane. R is the radius of the membrane, and $\mathbf{J}_1$ is the first-order *Bessel* function of the first kind.

The two functions $\sin x/x = \mathrm{si}\,(x)$ and $\mathbf{J}_1(x)/x$ look very similar. Thus, circular membranes have a directional characteristic similar to that of rectangular ones. The examples shown in Fig. 10.6a, b, illustrate the similarities when choosing $b = h = R$.

10.4   The Near-Field of Piston Membranes

Because the sound field close to a membrane cannot be computed by *Fraunhofer*'s approximation, the *Rayleigh* integral itself must be solved. Discrete numerical methods are commonly used to accomplish this.

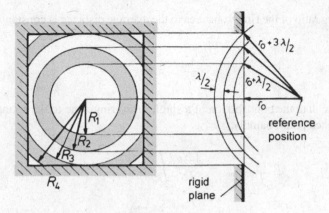

Fig. 10.7 The construction of *Huygens–Fresnel* zones

Zone Construction After *Huygens* and *Fresnel*

In this context, we introduce a traditional method of solving this integral by graphic interpretation of the conditions close to r the membrane. This technique is called *Huygens–Fresnel* zone construction. It is applicable to piston membranes, or more specifically, vibrating plane areas where $\underline{v}$ is constant across the area.

As shown in Fig. 10.7, the radiating area is subdivided into ring-shaped zones. The average difference in radial distance of two adjacent zones is $\lambda/2$. As a result, the average contributions rendered by two adjacent zones have a phase difference of 180°. The ring zones are constructed so that all complete zones make identical contributions, and adjacent complete zones cancel each other out. The magnitude of the resulting sound pressure at the observation point is estimated from the remaining areas' contribution after cancellation.

We now prove that the contributions of the complete zones are identical. On the one hand, the area of a ring zone is

$$A_n = \pi \left(R_n^2 - R_{n-1}^2 \right) . \tag{10.11}$$

By applying *Pythagoras*' law repeatedly, we find

$$A_n = \pi \left[\left(r_0 + n\frac{\lambda}{2} \right)^2 - \left(r_0 + \frac{(n-1)\lambda}{2} \right)^2 \right]$$

$$= \pi\lambda \left[r_0 + (2n-1)\frac{\lambda}{4} \right] . \tag{10.12}$$

On the other hand, the average distance of a ring zone from the reference point is

$$\bar{r}_n = \frac{1}{2} \left[r_0 + \frac{n\lambda}{2} + r_0 + (n-1)\frac{\lambda}{2} \right] = \left[r_0 + (2n-1)\frac{\lambda}{4} \right] . \tag{10.13}$$

The resulting ratio of the ring-zone area to the average distance is constant and equal to

$$\frac{A_n}{\bar{r}_n} = \pi \lambda. \tag{10.14}$$

Now we calculate the contribution of a single (nth) ring zone to the sound pressure at the reference point, namely,

$$\underline{p}_{\rightarrow,\,n\text{thzone}} = \frac{j \omega \varrho}{2 \pi} \int_{A_n} \underline{v}\, \frac{e^{-j\beta r}}{r}\, dA$$

$$= j \omega \varrho \underline{v} \int_{r_0+(n-1)\lambda/2}^{r_0+n\,\lambda/2} e^{-j\beta r}\, dr, \tag{10.15}$$

where $A = \pi(r^2 - r_0^2)$, and $dA = 2\pi r\, dr$. The general solution becomes

$$\underline{p}_{\rightarrow,\,n\text{th zone}} = \varrho c\, \underline{v}\, e^{-j\beta r_0}\, e^{-j n \cdot \pi}(e^{j\pi} - 1) = \pm 2\,\varrho c \underline{v}\, e^{-j\beta r_0}, \tag{10.16}$$

where $\beta \cdot \lambda/2 = \pi$, and the positive sign results from n being odd integers, otherwise negative.

The interpretation of this result follows directly from this expression. The first term on the right side of the equation describes an undisturbed plane wave. The second term stands for a wave that results from diffraction at the outer rim of the ring zone. Hence, we find two times the magnitude of the plane wave's sound pressure at the reference point. In the example of a plane wave being interrupted by an infinitely extended, or at least very large baffle with a circular hole in it—see Fig. 10.10, the sound pressure at the observation point behind the baffle is up to *two times larger*!

The following rules are helpful for graphical evaluation of *Huygens–Fresnel* zone constructions. The contribution of incomplete zones is considered to be approximately proportional to the ratio of the remaining area to the complete area and must assume the appropriate sign. When summing up, one starts by letting the contribution of the first half zone stand while its second half cancels out with the first half of the adjacent zone, and so on. This procedure allows the resulting field to be interpreted as the sum of a plane wave and interfering diffracted waves from the rims of the radiating areas.

On-Axis Sound Pressure and Radiation Impedance of Circular Piston Membranes

As the next step, we discuss the sound pressure of a diverging wave, $|\underline{p}_{\rightarrow}(r_0)|$, on the middle axis of a circular piston membrane. This is an example of the near field of a circular piston membrane in an infinitely extended, rigid, plane baffle with radius R as sketched in Fig. 10.8. The distance from the reference (observation) point to the

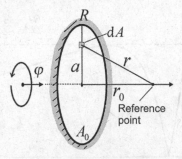

Fig. 10.8 Reference point on the middle-axis of the circular membrane

middle of the circular membrane, r_0, is given. Thereupon follows the distance from the observation point to an elementary area on the membrane with a radial distance to the middle, a.

We now express the the sound pressure at the reference point using *Raleigh*'s integral in (10.2), as follows.

$$\underline{p}_\rightarrow(r) = \frac{j\omega\varrho\underline{v}_0}{2\pi}\int_{A_0}\frac{e^{-j\beta r}}{r}\,dA = \underline{C}\int_0^{2\pi}d\varphi\int_0^R\frac{e^{-j\beta r}}{r}a\,da, \qquad (10.17)$$

where $\underline{C} = j\omega\varrho\underline{v}_0/2\pi$.

Considering $r\,dr = a\,da$, $r^2 = a^2 + r_0^2$, $a\rightarrow R$, and $r\rightarrow\sqrt{R^2+r_0^2}$, with $a\rightarrow 0$ and $r\rightarrow r_0$, we obtain—with $\hat{\underline{p}}_0 = \underline{v}_0\varrho c$,

$$\underline{p}_\rightarrow(r) = 2\pi\underline{C}\int_{r_0}^{\sqrt{R^2+r_0^2}}e^{-j\beta r}\,dr = \hat{\underline{p}}_0\left(e^{-j\beta r_0} - e^{-j\beta\sqrt{R^2+r_0^2}}\right), \qquad (10.18)$$

The following interpretation follows directly from (10.18). The first term represents an undisturbed plane wave, while the second term represents a wave component coming from the outer rim of the circular membrane. The sound field constitutes the superposition of these two wave components. With help from the relation

$$e^{-ja} - e^{-jb} = -2j\sin\frac{a-b}{2}e^{-j\frac{a+b}{2}}, \qquad (10.19)$$

the sound pressure along the middle axis results as

$$\underline{p}_\rightarrow(r_0) = 2j\,\hat{\underline{p}}_0\sin\left[\frac{\beta}{2}\left(\sqrt{R^2+r_0^2} - r_0\right)\right]e^{-j\frac{\beta}{2}(\sqrt{R^2+r_0^2}+r_0)}. \qquad (10.20)$$

From the phase term inside $\sin(\cdot)$ we determine the maxima and minima of the sound pressure under the condition

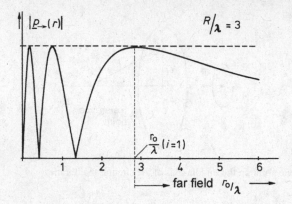

Fig. 10.9 Sound pressure on the axis of a circular piston membrane as a function of distance and wavelength

$$\sqrt{R^2 + r_0^2} - r_0 = (2n - 1)\frac{\lambda}{2}, \tag{10.21}$$

and pressure nodes (zeros) under the condition

$$\sqrt{R^2 + r_0^2} - r_0 = n\,\lambda, \tag{10.22}$$

with n being positive integers.

For a given r_0 and wavelength, λ, the number of *Huygens–Fresnel* zones is equal to

$$\nu_z = \frac{\sqrt{R^2 + r_0^2} - r_0}{\lambda/2}. \tag{10.23}$$

Maxima and minima occur whenever ν_z is exactly an integer number, which is the case for

$$r_m = \frac{\left(\frac{R}{\lambda}\right)^2 - \left(\frac{\nu_z}{2}\right)^2}{\nu_z/\lambda} \quad \text{for } \nu_z \text{ being integer}, \tag{10.24}$$

where odd integers of ν_z result in maxima and even ones in minima.

Exactly one zone exists for $\nu_z = 1$. For distances greater than $r_0|_{\nu_z=1}$, the sound pressure decreases monotonically, which means that we then are in the far-field. The equation cannot be fulfilled for $R < \lambda$, meaning that there is no zero at all. There is a finite number of zeros for $R > \lambda$—see Fig. 10.9.

Calculation of the acoustic power transmitted by a piston in a baffle requires the real component of the radiation impedance, called the radiation resistance, which is

$$r_{\text{rad}} = \text{Re}\{\underline{Z}_{\text{rad}}\}, \quad \text{with} \quad \underline{Z}_{\text{rad}} = \frac{\underline{F}}{\underline{v}} = \frac{\int_{A_0} \underline{p}\,\mathrm{d}A}{\underline{v}}. \tag{10.25}$$

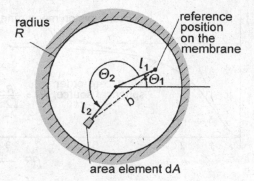

radius R

reference position on the membrane

Θ_2 l_1 Θ_1

l_2 b

area element dA

Fig. 10.10 Coordinates for calculating the radiation resistance of a circular piston

Now we describe the steps of how to calculate this resistance. We perform the calculation using the sound pressure directly on the surface area, A_0, of the radiating piston. Integration across this area is shown in Fig. 10.10.

For a circular piston membrane the sound pressure follows from *Rayleigh*'s integral as follows,

$$\underline{p}_\rightarrow(l_1, \Theta_1) = \frac{\mathrm{j}\,\omega\,\varrho}{2\,\pi} \int_{l_2=0}^{R} \int_{2=0}^{2\pi} \underline{v}\,\frac{e^{-\mathrm{j}\beta b}}{b}\,l_2\,\mathrm{d}\Theta_2\,\mathrm{d}l_2\,, \qquad (10.26)$$

where $b = \sqrt{l_1^2 + l_2^2 - 2\,l_1\,l_2\,\cos\Theta_2}$ is the distance between the reference point and the elementary area, $\mathrm{d}A = l_2\,\mathrm{d}\Theta_2\,\mathrm{d}l_2$.

To derive the force acting on the membrane, that is, $\underline{F} = \int \underline{p}_\rightarrow\,\mathrm{d}A = \int \underline{p}_\rightarrow\,l_1\,\mathrm{d}\Theta_1\,\mathrm{d}l_1$, the following integration must be performed,

$$\underline{F} = \int_{l_1=0}^{R} \int_{\Theta_1=0}^{2\pi} \underline{p}_\rightarrow(l_1, \Theta_1)\,l_1\,\mathrm{d}\Theta_1\,\mathrm{d}l_1\,. \qquad (10.27)$$

Evaluation of this expression is tedious—compare the solution of Problem 10.3. for details. We only present the result at this point, which is

$$r_{\mathrm{rad}} = \mathrm{Re}\,\{\underline{Z}_{\mathrm{rad}}\} = A_0\,\varrho\,c\left[1 - \frac{\mathbf{J}_1(2\beta\,R)}{\beta\,R}\right]\,, \qquad (10.28)$$

where $\mathbf{J}_1$ is the first-order *Bessel* function of the first kind. Figure 10.11 illustrates the results for the circular piston and its area-equivalent 0th-order spherical source—shown in Fig. 8.6b. Their courses are related, except for some overshoots. These overshoots are caused by diffraction happening at the rim of the circular membrane. The results for rectangular pistons look quite similar.

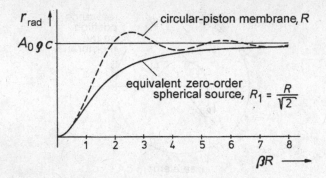

Fig. 10.11 Radiation resistance of a circular-piston membrane as a function of distance and wavelength

10.5 General Remarks on Diffraction and Scattering

The calculations performed above for pistons are expandable to the phenomenon of *diffraction* caused by circular holes in baffles—see Fig. 10.12. We assume that a plane wave hits such a baffle perpendicularly and that there is a constant velocity, $\underline{v}(x, y)$, normal to the plane of the area of the hole. The situation is identical to the piston in a baffle.

We now calculate the sound field behind the baffle with *Rayleigh*'s integral, using the method of superposition of 0th-order point-source waves, keeping in mind their phases and mutual interferences.

Figure 10.12 illustrates three typical cases, namely, $R \ll \lambda$, $R \approx \lambda$, and $R \gg \lambda$. The upper panels represent the sound fields yielded by the *Rayleigh* integral,

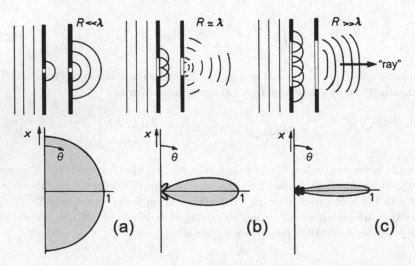

Fig. 10.12 Sound fields and directional characteristics of circular holes with different radii. **Upper panel** Wave-field plots for a circular hole in a baffle when hit by a plane wave. **Lower panel** Corresponding directional characteristics. (**a**) Small compared to the wavelength. (**b**) On the order of the wavelength. (**c**) Large compared to the wavelength

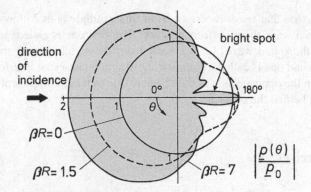

Fig. 10.13 Directional characteristics of the sound field of a rigid sphere exposed to a plane wave

the lower panels show their respective directional characteristics, Γ. For $R \ll \lambda$, diffraction causes the waves to propagate into a full hemisphere behind the baffle without any shadowing effects. For $R \approx \lambda$, both diffraction and shadowing occur. We see pronounced side lobes in addition to the main beam. For $R \gg \lambda$, the waves propagate mainly in the direction of the main beam. Side lobes are present but negligible, resulting in a well-pronounced beam. The sound wave thus propagates straight along the axis of the beam like a *ray*.

Diffraction also occurs when obstacles of finite dimension reside in a sound field. When a sound wave hits such an obstacle, it causes parasitic waves that superimpose themselves upon the original ones. Solving the wave equation consists of the following tasks.

- Formulating the boundary conditions at the surface. For resting, rigid obstacles for example, the normal component of the volume velocity, $v_\perp$, must be zero at the surface.

- Summing up the original and parasite waves results in a sound field that meets the given boundary conditions

- Considering far-field effects, the parasite sound field vanishes at large distances from the obstacle

As an example, Fig. 10.13 illustrates the sound field resulting from the impingement of a plane wave on a rigid sphere. The figure contains data for three different relative sphere sizes, namely, $\beta R = 2\pi (R/\lambda) = 0$, 1.5, and 7. The curves depict the magnitude ratio, $|\underline{p}(\theta)/\underline{p}_0|$, of the sound pressure on the surface of the sphere, p, in relation to the sound pressure in the undisturbed free field, $\underline{p}_0$. The following interesting details become obvious.

On the side facing the incoming wave, the sound accumulates (piles up), whereby the sound pressure increases by a factor of up to two (6 dB) at high frequencies. We must take this effect into account, for instance, when designing microphones. On the opposite side of the obstacle, the sound pressure is nonzero. There is actually a pronounced maximum at $\theta = 180°$, called a *bright spot*. This effect is caused by waves *creeping* around the sphere and adding in phase at $\theta = 180°$.

The conclusion that we draw drawn from this example is as follows. There are two well known wave-theory effects at work. The first one is called *scattering*. It occurs when the sound waves bounce back from the surface facing the incoming wave. The second one is called *diffraction*. Diffraction causes an interference field, building-up on the opposite side of the obstacle due to wave components deflecting into the space behind the obstacle.

10.6 Exercises

Piston Membranes, Diffraction and Scattering

Problem 10.1 Using *Fraunhofer*'s approximation of *Rayleigh*'s integral shows that the directional characteristic of an oscillating plane area is given by

$$\Gamma(\theta, \varphi) = \Gamma(\theta)\, \Gamma(\varphi), \tag{10.29}$$

if $v(x, y)$ is separable as follows,

$$v(x, y) = v_0\, \psi_x(x)\, \psi_y(y), \tag{10.30}$$

where $\psi_x(x)$ and $\psi_y(y)$ are weighting function of unity dimension.

Problem 10.2 Find the directional characteristic of a line array of length $2h$ along the x-direction. The line source consists of elementary sources with cardioid directional characteristic along the y-direction.

Problem 10.3 Calculate the far-field directional characteristics of a circular piston membrane of radius R. Being a radiator, the particle velocity, $\underline{v}_0$, is assumed to be constant over the entire piston.

Problem 10.4 Calculate the radiation resistance of a circular piston membrane with the radius R.

Problem 10.5 A plane wave hits a quadratic opening in a rigid plane wall of infinite size. The edge length of the opening is $a = 6.7\,\lambda$.

Perform a *Huygens–Fresnel* zone construction for a reference point perpendicular to the center of the opening at a distance of $r_0 = 3\,\lambda$.

(a) Determine the zones according to their order, n

(b) Sketch the zones beginning from the reference point.

(c) Show that the sound pressure at the reference point consists of a sum of the contributions of undisturbed zone parts and edge refraction at the rim of the opening.

Chapter 11
Dissipation, Reflection, Refraction, and Absorption

The wave equation as derived in Sect. 7.1 and used so far in this book, is valid for sound propagation in lossless media. The *Helmholtz* form of this equation is

$$\frac{\partial^2 \underline{p}}{\partial x^2} - (\mathrm{j}\beta)^2 \, \underline{p} = 0 . \tag{11.1}$$

Its solution for the forward-progressing plane wave is

$$\underline{p}_{\rightarrow}(x) = \underline{p}_{\rightarrow} \, \mathrm{e}^{-\mathrm{j}\beta x} . \tag{11.2}$$

However, the assumption of a lossless medium is an idealization. There is always some loss of acoustic energy when sound propagates in real media. This is the so-called *dissipation* of sound energy into thermal energy. Dissipation causes *spatial damping* of the sound waves.

The wave equation can account for small dissipation by replacing the imaginary term, the *phase coefficient*,[1] $\mathrm{j}\,\beta$, by a complex one. The complete complex term is the *propagation coefficient*,

$$\underline{\gamma} = \breve{\alpha} + \mathrm{j}\,\beta , \tag{11.3}$$

where $\breve{\alpha}$ is the *damping coefficient* that describes spatial damping.[2] The resulting form of the wave equation is

$$\frac{\partial^2 \underline{p}}{\partial x^2} - \underline{\gamma}^2 \, \underline{p} = 0 , \tag{11.4}$$

---

[1] As mentioned before, we prefer the letter symbol β to k ... wave number.

[2] Recall that Sect. 8.3 had already introduced $\breve{\alpha}$ with the exponential wave. However, there the damping was caused by geometric expansion of the wave and not by dissipation.

© Springer-Verlag GmbH Germany, part of Springer Nature 2021
N. Xiang and J. Blauert, *Acoustics for Engineers*,
https://doi.org/10.1007/978-3-662-63342-7_11

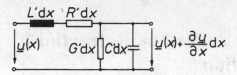

Fig. 11.1 Equivalent circuit for spatially-damped plane-wave propagation

and has the plane-wave solution

$$\underline{p}_{\rightarrow}(x) = \underline{p}_{\rightarrow}\, e^{-\breve{\alpha}x}\, e^{-j\beta x}\,. \tag{11.5}$$

The mathematical description mentioned above is analog to wave propagation in loss-afflicted electric transmission lines. The relevant electric and acoustic equations are homomorphic. Figure 11.1 illustrates the equivalent electric circuit with the resistance load, R', the inductance load, L', the susceptance load, G', and the capacitance load, C'.

We recall the respective wave equation and the complex propagation coefficients from electrical transmission theory as

$$\frac{\partial^2 \underline{u}}{\partial x^2} - \underline{\gamma}^2 \underline{u} = 0 \quad \text{and, consequently,} \tag{11.6}$$

$$\underline{\gamma} = \sqrt{(R' + j\,\omega\,L')(G' + j\,\omega\,C')}\,. \tag{11.7}$$

This leads us to the characteristic impedance of the transmission line, the so-called line impedance, $\underline{Z}_{\mathrm{L}}$, namely,

$$\underline{Z}_{\mathrm{L}} = \sqrt{\frac{R' + j\,\omega\,L'}{G' + j\,\omega\,C'}}\,. \tag{11.8}$$

Note that, in general, the line impedance, $\underline{Z}_{\mathrm{L}}$, becomes complex when losses are involved. This statement holds for acoustic wave propagation as well—as is shown in Sect. 11.2.

In acoustics, the term *dissipation* does not have the same meaning as *absorption*. Absorption means that sound disappears from a specified space through a boundary. The sound energy that leaves the space is called absorbed, no matter whether it dissipates or just transmits into another space. The current chapter deals with absorption in Sect. 11.4. Absorption is a phenomenon of high technical relevance—in architectural acoustics for instance.

Table 11.1 Damping coefficients of air in two meteorological conditions. Values given in Neper—as to Neper, refer to Sect. 1.6

Frequency f	0.5	1	2	4	8	16	kHz
$\breve{\alpha}$ (10 °C, 70%)	0.23	0.42	1.04	3.40	12.30	39.96	10^{-3} Np/m
$\breve{\alpha}$ (20 °C, 50%)	0.32	0.57	1.13	3.20	11.00	38.60	10^{-3} Np/m

11.1 Dissipation During Sound Propagation in Air

The assumptions for lossless sound propagation as put forward in Sect. 7.1 are as follows,

- No inner friction, which means no viscosity
- Negligible thermal conduction
- Behavior as a perfect gas

These assumptions are not strictly valid in real fluids. Indeed, there are three prominent reasons for dissipation, namely,

- *Viscosity* ... Real media, such as air, show at least some viscosity. As a result, the medium's periodical compressions and expansions go along with losses due to friction.

- *Thermal conduction* ... Thermal energy leaks from the compressed and thus warmer zones to the expanded and thus cooler ones. In consequence, the compression is no longer strictly adiabatic and, hence, pressure differences are attenuated. This behavior causes dissipation.

- *Deviation from perfect gas—molecular dissipation.* ... Most gases are composed of multi-atomic molecules. For example, air contains O_2, N_2 and H_2O. These molecules do not only have translatory degrees of motional freedom but also such as vibratory and rotatory ones. Energy from translatory to, for example, vibratory and/or rotatory motion occurs with some delay, an effect which is called *relaxation*. This effect causes deviation from adiabatic compression since energy is taken from translatory motion and returned to it at a different point in time.

It is known that all three kinds of dissipation in free sound propagation in the air are proportional to the second power of the frequency, f^2. The molecular dissipation is paramount in the audio frequency range of 16 Hz–16 kHz, mainly due to vibratory modes of the molecules.

In Table 11.1 we present some estimates for the damping coefficient, $\breve{\alpha}$, in air. According data in the literature show profound variances.

Dissipation During Sound Propagation in Tubes

In contrast to free propagation, viscosity and thermal transfer can gain in relevance when the sound propagates along a duct. In a tube with rigid walls made of heat-conducting material, consider the following two effects.

- *Viscosity* ... The walls of the tubes are rigid compared to the air. Hence the particle velocity, $\underline{v}$, is zero directly on the surface of the wall. Yet, already at a tiny distance from the wall, the velocity behaves as a progressive wave. Therefore, we have a strong velocity gradient in the radial direction within a thin boundary layer. Although the viscosity of air is low, the frictional losses occurring in this way become considerable.

- *Thermal conduction* ... With the walls conducting heat well, dissipation due to heat transfer is likely. These effects are certainly larger than in a free wave as there the compressed and expanded sections are separated by a quarter-wavelength air distances, that is, $\lambda/4$. But now, due to well-conducting walls, these sections are in more direct contact.

Both effects are proportional to, (a), the ratio of the circumference, U, and the cross-sectional area, A, and, (b), the root of the frequency, $\sqrt{\omega}$. *Cremer* has derived the following approximation for so-called *wide tubes*, that is, tubes where the boundary layer is small compared to the diameter,

$$\breve{\alpha} \approx 6.7 \cdot 10^{-6} \, (U/A) \, \sqrt{\omega}/\text{Hz} \quad \dots \text{ for wide tubes}. \tag{11.9}$$

It is known that viscosity accounts for $\approx 2/3$ of the damping coefficient resulting from this formula.[3]

11.2   Sound Propagation in Porous Media

Due to the previous discussion it is obvious that sound propagation in media perforated by many narrow tubes and connected cavities is profoundly damped. Media of this consistency are called porous.[4]

With a set-up as shown in Fig. 11.2, the characteristic flow resistance $\varXi$, of porous media can be measured, though, to be precise, only for the "static" (steady) case of a continuous gas flow. $\varXi$ is defined as follows,

$$\varXi \, v_a = -\frac{p_\Delta}{x_\Delta} \approx -\frac{\partial p}{\partial x}\bigg|_x, \tag{11.10}$$

where v_a is the velocity outside the porous media.

In a strongly simplified model introduced by *Rayleigh*, the porous medium is replicated as a rigid skeleton that is perforated by perpendicular narrow ducts— illustrated in Fig. 11.3. The ratio of the duct area inside the probe, A_i, to the total area outside the probe, A_a, is called *porosity*, σ, namely,

---

[3] In the analog circuit of Fig. 11.1, viscosity corresponds to R' and thermal conduction to G'.

[4] Note that a medium with mutually unconnected cavities is *not* porous. Artificial foams can be produced both ways, either with or without connections between the internal voids (gas bubbles).

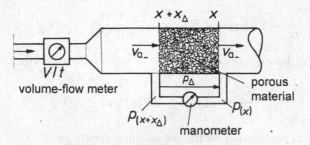

Fig. 11.2 Set-up for measuring the static flow resistance, $\varXi$

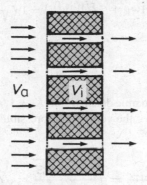

Fig. 11.3 *Rayleigh*'s model of porous materials

$$\sigma = \frac{A_i}{A_a} < 1.$$ (11.11)

The volume velocities inside, q_i, and outside the probe, q_a, are the same due to mass conservation, thus we have $q_i = q_a$. This means that the respective velocities, v_i and v_a, relate as follows.

$$\sigma v_i = v_a \quad \text{and} \quad \sigma \varXi v_i = -\frac{\partial p}{\partial x}.$$ (11.12)

Euler's equation is now complemented by this loss term while the continuity equation stays the same. We thus get the following pair of linear differential wave equations for inside porous media,

$$-\frac{\partial p}{\partial x} = \varrho \frac{\partial v_i}{\partial t} + \sigma \varXi v_i \quad \text{and} \quad -\frac{\partial v_i}{\partial x} = \kappa \frac{\partial p}{\partial t}.$$ (11.13)

In complex notation this reads as

$$-\frac{\partial \underline{p}}{\partial x} = j\,\omega\,\varrho\,\underline{v}_i + \sigma\,\varXi\,\underline{v}_i \quad \text{and} \quad -\frac{\partial \underline{v}_i}{\partial x} = j\,\omega\,\kappa\,\underline{p}\,. \tag{11.14}$$

Combination of these two equations results in the 2nd-order wave equation for porous media,

$$\frac{\partial^2 \underline{p}}{\partial x^2} - (j\,\omega\varrho + \sigma\,\varXi)(j\,\omega\,\kappa)\,\underline{p} = 0. \tag{11.15}$$

For the complex propagation coefficient we subsequently get

$$\underline{\gamma} = \sqrt{(j\,\omega\varrho + \sigma\,\varXi)(j\,\omega\,\kappa)} = j\,\omega\,\sqrt{\frac{\varrho_{eq}}{K}}\,, \tag{11.16}$$

where

$$\varrho_{eq} = \varrho + \frac{\sigma\,\varXi}{j\,\omega}\,, \tag{11.17}$$

is the so-called equivalent dynamic density. and $K = 1/\kappa$ is the bulk (volume) modulus, which is the reciprocal of the volume compressibility, κ. For the characteristic (field) impedance, it follows

$$\underline{Z}_c = \frac{\underline{p}_{i\rightarrow}}{\underline{v}_{i\rightarrow}} = \sqrt{\frac{j\,\omega\,\varrho + \sigma\,\varXi}{j\,\omega\,\kappa}} = \sqrt{\varrho_{eq}\,K}\,. \tag{11.18}$$

For *low frequencies* the complex propagation coefficient is approximately proportional to the root of the frequency, that is

$$\underline{\gamma} \approx \sqrt{\sigma\,\varXi\,\omega\,\kappa}\;e^{j45°}, \quad \text{thus,} \quad \check{\alpha} = \sqrt{\sigma\,\varXi\,\omega\,\kappa/2} \sim \sqrt{f}\,. \tag{11.19}$$

For *high frequencies* the propagation coefficient asymptotically reaches a limiting value that is imaginary, as results from eliminating $\check{\alpha}$ from the formula for $\underline{\gamma}$, so that

$$\underline{\gamma} \approx j\,\omega\sqrt{\varrho\,\kappa} = j\,\frac{\omega}{c} = j\,\beta\,. \tag{11.20}$$

To be sure, the *Rayleigh* model mimics porous media only very roughly. The porosity, σ, is quantifiable via static measurement. Furthermore, we consider that the air in cavities inside the medium increases the volume compliance, κ, but is not accelerated. We account for this compliance increase by a structural factor, $\chi \leq 1$, accessible by dynamic measurement. This factor is introduced into the wave equation by substituting ϱ by $\varrho\chi$.

The equivalent electric circuit of Fig. 11.4 represents the wave equation inside porous media. Yet, this modeling covers viscosity only and not thermal conduction.

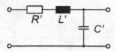

Fig. 11.4 Equivalent circuit for inside a porous medium

11.3 Reflection and Refraction

Reflection and *refraction* are important phenomena in wave propagation. In the following, we treat both of them together by assuming a situation where a wave hits the boundary between two media, Medium 1 and Medium 2, with different speeds of sound, c_1 and c_2. If we take both the boundary and the wave as infinite in space, this situation is presentable in one plane—see Fig. 11.5.

Reflection

The incoming wave splits into two orthogonal components, one of which propagates in $(-y)$–direction and the other one in $(+x)$–direction. The x–component is not affected by the boundary in any way. However, The $(-y)$–component inverts its direction at the boundary from $-y$ to $+y$, due to reflection, while maintaining its speed, c_1.

The components of the phase coefficient in x–direction, β_x, are identical for the incoming and reflected waves, in other words

$$\beta_1 \sin \Theta_1 = \beta_1 \sin \Theta_1' , \qquad (11.21)$$

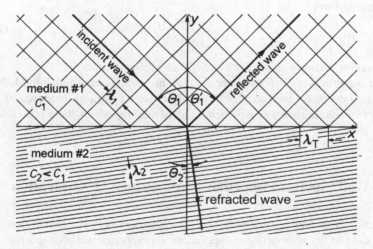

Fig. 11.5 Wave propagation at the boundary between two fluid media

or, with $\beta_1 = 2\pi/\lambda_1$,

$$\frac{\lambda_1}{\sin \Theta_1} = \frac{\lambda_1}{\sin \Theta_1'} = \lambda_T, \tag{11.22}$$

where λ_T is the wavelength of the trace of the wave along the boundary, the so-called *trace wavelength*.

Since, due to the boundary condition, $\underline{p}_1(x) = \underline{p}_2(x)$, the trace wavelength is identical for both the incident and reflected waves, we directly arrive at the law of reflection. This law states that the incoming angle is identical to the outgoing one, namely,

$$\Theta_1 = \Theta_1'. \tag{11.23}$$

Refraction

The non-reflected part of the incoming wave transmits into Medium 2 and propagates there with a different sound speed, c_2, and at a different angle Θ_2. At the boundary, the phase coefficient and, hence, the trace wavelengths are again identical for both the incoming and the transmitted waves. The physical reasons for this are that fluids exhibit no shear, and that the sound pressure is homogeneous at the boundary. Consequently, one can write

$$\beta_1 \sin \Theta_1 = \beta_2 \sin \Theta_2, \tag{11.24}$$

and, thus,

$$\frac{\lambda_1}{\sin \Theta_1} = \frac{\lambda_2}{\sin \Theta_2} = \lambda_T. \tag{11.25}$$

By multiplying with the frequency, which is identical for both waves, we arrive at the following form,

$$\frac{c_1}{c_2} = \frac{\sin \Theta_1}{\sin \Theta_2}. \tag{11.26}$$

This equation is known as the *Snell*'s (*Snellius'*) refraction law.

If the conditions of *Snell*'s law are not met, no sound is transmits into Medium 2. This effect occurs, for instance, for shallowly oblique (grazing) incidence of sound from air to water. In such a case, which is called *total reflection*, there is no refracted wave at all.

11.4 Wall (Surface) Impedance and Degree of Absorption

This section deals with the wave effects at the boundary between two fluid media in more detail and, by appropriate extrapolation, also approximately covers the impact of waves encountering a wall.

Boundary Between Two Fluids

As a starting point, we take a case as depicted in Fig. 11.5. The characteristic impedances of the two media are $\underline{Z}_{w_1} = \varrho_1 c_1$ and $\underline{Z}_{w_2} = \varrho_2 c_2$. We now consider the boundary to Medium 2 as a wall with the wall impedance $\underline{Z}_{wall} = \underline{Z}_{w_2}$. Following Sect. 7.5, the reflectance, $\underline{R}$, for perpendicular incidence from Medium 1 on Medium 2 is found to be

$$\underline{R}_{\perp} = \frac{\underline{Z}_{wall} - \underline{Z}_{w_1}}{\underline{Z}_{wall} + \underline{Z}_{w_1}} = \frac{\varrho_2 c_2 - \varrho_1 c_1}{\varrho_2 c_2 + \varrho_1 c_1}. \tag{11.27}$$

For oblique incidence, only those components of the particle velocity that are perpendicular to the boundary are reflected or transmitted. In acoustics, the wall impedance, $\underline{Z}_{wall}$, is defined as the ratio of the total sound pressure and the normal velocity at the wall surface, which is $y = 0$. Accordingly, in Medium 1 we get

$$\underline{Z}_{wall}(\Theta_1) = \frac{\underline{p}_1}{\underline{v}_1} = \frac{\underline{p}_{1\rightarrow} + \underline{p}_{1\leftarrow}}{(\underline{v}_{1\rightarrow} + \underline{v}_{1\leftarrow})\cos\Theta_1}, \tag{11.28}$$

Referring to Fig. 11.5 yields

$$\begin{aligned}
\underline{Z}_{wall} &= \frac{\hat{p}_0\,e^{-j\beta x \sin\Theta_1} + \underline{R}(\Theta_1)\,\hat{p}_0\,e^{-j\beta x \sin\Theta_1}}{\frac{\hat{p}_0}{Z_{w_1}}\left[e^{-j\beta x \sin\Theta_1} - \underline{R}(\Theta_1)\,e^{-j\beta x \sin\Theta_1}\right]\cos\Theta_1} \\
&= \frac{Z_{w_1}}{\cos\Theta_1}\frac{1 + \underline{R}(\Theta_1)}{1 - \underline{R}(\Theta_1)} = \frac{\varrho_1 c_1}{\cos\Theta_1}\frac{1 + \underline{R}(\Theta_1)}{1 - \underline{R}(\Theta_1)}.
\end{aligned} \tag{11.29}$$

Solving for the reflectance of oblique incidence, $\underline{R}(\Theta_1)$, results in

$$\underline{R}(\Theta_1) = \frac{\underline{Z}_{wall}(\Theta_1) - \frac{\varrho_1 c_1}{\cos\Theta_1}}{\underline{Z}_{wall}(\Theta_1) + \frac{\varrho_1 c_1}{\cos\Theta_1}} = \frac{\underline{Z}_{wall}(\Theta_1)\cos\Theta_1 - \varrho_1 c_1}{\underline{Z}_{wall}(\Theta_1)\cos\Theta_1 + \varrho_1 c_1}. \tag{11.30}$$

This simple case already shows that the wall impedance, $\underline{Z}_{wall}$, in general, is a function of the incoming oblique angle Θ_1. Further, in (11.30) it is weighted with $\cos\Theta_1$.

Locally Reacting Boundaries and Walls

With $c_2 \ll c_1$, the angle of the refracted sound, Θ_2, approaches zero. The refracted wave propagates perpendicularly away from the boundary. The wall impedance, then, no longer depends on the angle of incidence of the incoming sound. This means that $\underline{Z}_{wall} \neq f(\Theta_1)$.

The same holds for wall structures where adjacent wall elements are not coupled to each other because this causes that all movements parallel to the wall are suppressed. Walls that react in this way are said to be *locally reacting*. The *Rayleigh* model of porous materials shows this feature, but also many technologically relevant porous materials and wall constructions do behave approximately the same.

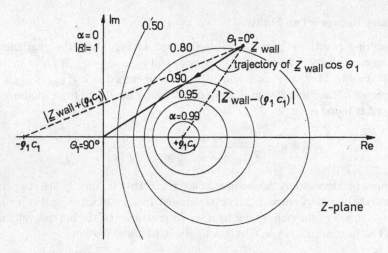

Fig. 11.6 Trajectories of constant degrees of absorption, α, in the complex $\underline{Z}$-plane

Degree of Absorption from Wall Impedances

The acoustic energy that is absorbed at a boundary is described by the *degree of absorption*, α, as introduced in Sect. 7.5 as

$$\alpha = 1 - \left|\underline{R}\right|^2 . \tag{11.31}$$

By plotting the wall impedance, $\underline{Z}_{\text{wall}}$, in the complex $\underline{Z}$-plane—as depicted in Fig. 11.6—we find the trajectories for constant $|\underline{R}|$ and constant α to be so-called *Appollonian* circles.[5]

An *Appollonian* circle is the geometric location of all positions from which the ratio of the distances to two reference points, $\pm \varrho_1 c_1$, is constant. Conveniently, this is what is required by the formula,

$$|R(\Theta_1)| = \left| \frac{\underline{Z}(\Theta_1) - \varrho_1 \, c_1}{\underline{Z}(\Theta_1) + \varrho_1 \, c_1} \right| . \tag{11.32}$$

The trajectories of $\underline{Z}(\Theta_1) = \underline{Z}_{\text{wall}} \cos \Theta_1$, for $0 \leq \Theta \leq 90°$, are straight lines. These straight lines are helpful for determining the degree of absorption, α. Figure 11.7 shows two choices as examples. To understand how they are derived, think of the α-circles as isohypses—that is, lines of equal height.

The degree of absorption, α, approaches zero for wall-parallel incidence. For $|\underline{Z}_{\text{wall}}| > \varrho_1 \, c_1$, there exists an optimum match in the sense that α assumes a maximum at a specific Θ_1. For $|\underline{Z}_{\text{wall}}| < \varrho_1 \, c_1$, the best possible match and, thus, the highest degrees of absorption are achieved for perpendicular sound incidence.

---

[5] By conformal transformation of the complex $\underline{Z}$-plane into the complex $\underline{r}$-plane, we get a so-called *Smith* chart, which is often useful in this context.

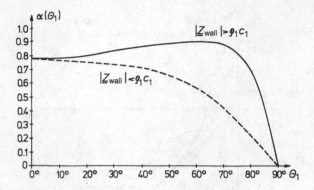

Fig. 11.7 Absorption as a function of angle Θ_1

Practical room acoustics is particularly interested in the degree of absorption averaged over all possible angles of incidence—the so-called *random incidence*, $\overline{\alpha}$. As derived in Sect. 12.5, this random incidence is given by

$$\overline{\alpha} = \int_0^{\pi/2} \alpha(\Theta_1)\sin(2\,\Theta_1)\,d\Theta_1 \,. \tag{11.33}$$

Note that for locally reacting walls this quantity never reaches one.

11.5   Porous Absorbers

Porous absorbers are of high relevance in practice. They are, for example, built from fibrous materials such as fabric, mineral wool or cocos fibre that is compressed into mats or plates as well as from porously extruded artificial foams. To estimate of the absorptive behavior, the *Rayleigh* model—see Sect. 11.2—is useful again. When considering perpendicular sound incidence and substituting $\underline{v}_i$ by $\underline{v}_a$ in (11.18), along with ϱ_{eq} in (11.17), the wall impedance, $\underline{Z}_{wall}$, is

$$\underline{Z}_{wall} = \frac{1}{\sigma}\sqrt{\frac{j\omega\varrho + \Xi\,\sigma}{j\omega\,\kappa}} = \frac{\sqrt{\varrho_{eq}\,K}}{\sigma}\,. \tag{11.34}$$

For high and low frequencies the following approximations hold. For *high frequencies* we get

$$\underline{Z}_{wall} \approx \frac{1}{\sigma}\sqrt{\frac{\varrho}{\kappa}} = \frac{1}{\sigma}\,\varrho\,c\,, \tag{11.35}$$

which is real and does not depend on frequency. However, for *low frequencies*, we find a complex and frequency-dependent relationship, namely,

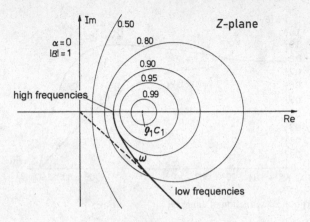

Fig. 11.8 Trajectory of $\underline{Z}_{\text{wall}}$ in the $\underline{Z}$-plane for porous absorbers

$$\underline{Z}_{\text{wall}} \approx \frac{1}{\sigma}\sqrt{\frac{\varXi\,\sigma}{\omega\,\kappa}}\;e^{-j\,45^\circ}\,. \tag{11.36}$$

The trajectory of $\underline{Z}_{\text{wall}}$ in the $\underline{Z}$-plane is shown in Fig. 11.8. To discuss the course of α, it is helpful to think of the α-circles as isohypses again. Accordingly, for $\alpha(\omega)$ one gets a monotonically increasing curve with a maximum of

$$\alpha_{\max} = \frac{4\,\sigma}{(1+\sigma)^2}\,. \tag{11.37}$$

These conditions are valid for an infinitely thick layer of porous material. If the absorber thickness is finite and the material is placed upon a rigid wall, part of the energy is reflected and re-transmitted—after having passed the absorbing material a second time.

The situation can be illustrated by regarding that in front of the wall a standing wave develops. Directly upon the wall, the perpendicular component of the particle velocity is zero, that is, $\underline{v}_\perp = 0$. Consequently, the absorber is ineffective at this point—see Fig. 11.9. Thus, for low frequencies, a finite layer of material provides less absorption than an infinitely thick one. The finite layer is best exploited when positioned at a distance to the reflecting, rigid wall. α arrives at a relative maximum whenever the absorptive layer is in a velocity maximum. This arrangement is frequently used in practice. Figure 11.10 schematically shows the three cases as discussed above.

Real porous absorbers are not well-represented by the *Rayleigh* model, even with the structural factor, χ, considered. For instance, it is often unclear whether the absorbers react locally. Local reaction is enforceable by *cassetting*—sketched in Fig. 11.11—although for compressed mineral wool this is usually unnecessary.

To achieve a high α, enhancing the effective absorbent area is the way to go, for example, with porous wedges. Audience, is also a considerable absorber.

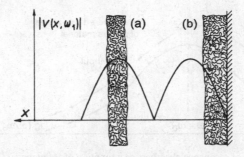

Fig. 11.9 Placements of layers of porous material in front of a hard wall. (**a**) With air-gap. (**b**) Directly on the wall

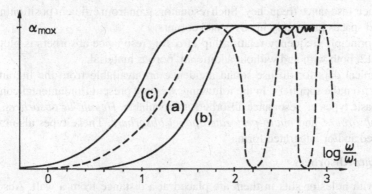

Fig. 11.10 Absorption of porous material as a function of frequency. (**a**) Finite layer with air-gap. (**b**) Finite layer directly on wall. (**c**) Infinite thickness

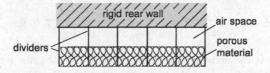

Fig. 11.11 Porous absorber in a cassette structure

11.6 Resonance Absorbers

As shown above, when using absorbent materials the effective layer must be placed a quarter wavelength, $\lambda/4$, in front of the wall. For a 100-Hz frequency with a wavelength of 3.4 m, for instance, this means placement of 85 cm in front of walls. In practice, so much space is usually not available.

Especially for low frequencies, that is, for so-called *bass traps*, a different absorber principle is employed, which is based on *resonance absorption*. To this end, the absorbent wall is covered with acoustic resonators, the input impedance of which is

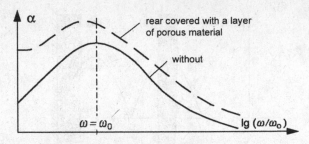

Fig. 11.12 Absorption of resonance absorbers as a function of frequency

low at their resonance frequency. Such resonators gain from efficient positioning—in enclosed spaces preferably in edges and corners.

The principle frequency relationship of α for resonance absorbers is plotted in Fig. 11.12, both with and without additional porous material.

Technical data for diffuse sound incidence are available from the literature or directly from the suppliers. In the following, we only present fundamental concepts. Three basic types of resonance absorbers are available, *Helmholtz absorbers*, *membrane absorbers*, and *micro-perforated panel absorbers*. These types also exist in combined and/or integrated form.

Helmholtz Absorbers

Plates with holes or slits in them are placed at a distance from a wall. Absorbent materials is often put on the rear side of the plates. Figure 11.13a illustrates the arrangement. For the wall impedance we get

$$\underline{Z}_{\text{wall}} = \frac{\underline{p}_\rightarrow}{\underline{v}_\rightarrow} = \underline{Z}_{\text{a}}\, A = \mathrm{j}\,\omega\, m'' + \frac{1}{\mathrm{j}\,\omega\, n''} + r'', \qquad (11.38)$$

where $r'' = \varXi\, b_1$ is the area-specific resistance, $n'' = \kappa\, b_2$ the area-specific compliance, and $m'' = \varrho\,(b_3 + \eta)\,\sigma$ the area-specific mass. Thereby, $\sigma = \pi\, r^2/b_4^2$ is the perforation rate (porosity), and η is a correction factor (mouth correction) considering that at the mouth of a hole or slit more air mass is moved than is actually inside the mouth. An estimate for circular holes is $\eta = 1.6\, r$.

Membrane Absorbers

These absorbers possess co-vibrating membranes—such as plates, foils, or other mass-afflicted materials—in front of an air-gap before a wall. To decrease their resonance frequencies, the membranes are sometimes loaded with additional weight. Absorbent material is used to partly or completely fill the air-space. The arrangement is depicted in Fig. 11.13b. Numerous different built forms exist, including such with more than one membrane layer.

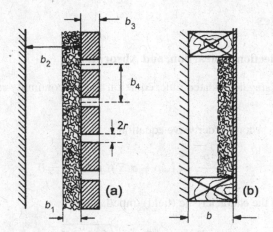

Fig. 11.13 Resonance absorbers. (a) *Helmholtz* absorber. (b) Membrane absorber

Although bending waves of the membranes are certainly possible, for rough calculations, we usually assume wall-perpendicular movements only. The reason for this assumption is that reacting forces due to bending of the material are usually negligible compared to those due to the stiffness of the air cushion. m'' is the area-specific mass of the plate, and $n'' = \kappa\, b$ is its compliance. The area specific resistance, r'', is hard to estimate. It contains losses within the plate.

Micro-perforated Panel Absorbers

A special kind of panel absorbers uses *micro-perforated* panels. These absorbers incorporate perforated thin panels or foils in front of an air-gap before a rigid wall, similar to what Fig. 11.13a shows, but without absorbent materials on the rear side of the panel. The perforations in the thin panel or foil are in the sub-millimeter range (diameter 0.1–1 mm) to provide high acoustic resistance but low area-specific acoustic-mass reactance. This low mass reactance is necessary for broadband absorbers. Besides the micro-perforated panels or foils, no further absorbent material is required.

Micro-perforated panel absorbers are of the resonant type. Single-panel absorbers have bandwidth as wide as 1–2 octaves. Two different resonant frequencies about 20% apart, realizable with double-layered micro-perforated panels, allow for even broader absorption bandwidths. Yet, the most intriguing feature is that micro-perforated panel absorbers are producible from a great variety of panel or foil materials, including thin metal sheets and flexible and/or translucent foils.

11.7   Exercises

Dissipation, Reflection, Refraction, and Absorption

Problem 11.1 Using the updated Euler equation and the continuity equation (11.13) for porous media,

(a) Derive the second-order wave equation

$$\frac{\partial^2 p}{\partial x^2} - (j\,\omega\varrho + \sigma\,\varXi)(j\,\omega\,\kappa)\,\underline{p} = 0. \qquad (11.39)$$

(b) Show that the characteristic (field) impedance is

$$\underline{Z}_c = \frac{\underline{p}_{i\to}}{\underline{v}_{i\to}} = \sqrt{\varrho_{eq}(\omega)\,K}, \qquad (11.40)$$

with $K = 1/\kappa$ and $\varrho_{eq}(\omega)$ being the equivalent dynamic density of the porous media—see (11.17)].

Problem 11.2 A *Kundt*'s tube is considered lossy, that is, with a complex propagation coefficient, $\underline{\gamma} = \alpha + j\,\beta$.

How do losses influence the standing waves and, therefore, the measurement results concerning the envelope's maximum and minimum values and the standing wave ratio, given that the standing-wave-ratio method is applied?

Problem 11.3 A *Kundt*'s tube is lossless with air as the medium, similar to Fig. 7.8, and the specimen under test is a piece of porous material with a thickness, d, which is appropriately fitted in the tube end with a rigid backing. In addition to two microphone positions, M_1, M_2, in front of the material surface, a further microphone, M_3, is placed in the rigid backing right on the back surface of the porous material as shown in Fig. 11.14.

Determine the complex-valued propagation coefficient, $\underline{\gamma}$, and the characteristic impedance, $\underline{Z}_c$, of the porous material.

Problem 11.4 A sound wave is incident with an angle of θ_1 upon a boundary between two different media, $\underline{Z}_1, \beta_1$ and $\underline{Z}_2, \beta_2$.

Determine the reflected and transmitted waves. Express the reflectance (reflection coefficient) and the absorption coefficient.

What conditions lead to maximum absorption?

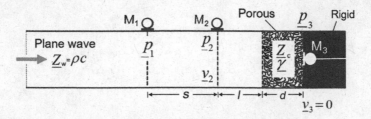

Fig. 11.14 Three microphone method in *Kundt*'s tube for characterizing the porous materials

Problem 11.5 Discuss the angular dependence of the degree of absorption, α, of a wall based on the ratio $|\underline{Z}_{wall}|/Z_{w,\,air}$.

Chapter 12
Geometric Acoustics and Diffuse Sound Fields

So far in this book, we dealt with sound propagation on the basis of the wave equation. However, this approach becomes difficult when treating sound fields inside rooms with complicated shapes like concert halls or churches. An approximate method called *geometrical acoustics* is useful in these cases.

This method considers sound propagation in terms of so-called *sound rays*. We introduced sound rays already in Sect. 10.5, where a ray icon designates the wave bundle that emerges from a circular hole in a rigid wall when $R \gg \lambda$. The idea is that the wave bundle propagates along a straight line like a ray of light.

The concept of rays is mathematically defined as the limiting case of maintaining plane areas of constant phase and letting the wavelength go to zero. In practice, rays serve to approximate propagating waves under the following two conditions. The wavelength is small compared to the linear dimensions of boundary areas and obstacles. Diffraction and interference are negligibly.

The energy density, W'', within a ray is equal to the energy density in a plane propagating wave. To compute its amount, we consider a wave bundle propagating through an area of $1\,\mathrm{m}^2$ for $1\,\mathrm{m}$—see Fig. 12.1 for illustration. The energy density, then, is the active power, $\overline{P}$, times the traveling time, $t_1 = (1/c) \cdot 1\,\mathrm{m}$, divided by the volume of $1\,\mathrm{m}^3$—or, in mathematical terms,

$$W''_{\mathrm{ray}} = \underbrace{\underbrace{|\overrightarrow{I}| \cdot 1\,\mathrm{m}^2}_{\overline{P}} \; \underbrace{\frac{1}{c} \cdot 1\,\mathrm{m}}_{t_1} \; \underbrace{\frac{1}{1\,\mathrm{m}^3}}_{V^{-1}}}_{W} = \frac{|\overrightarrow{I}|}{c} = \frac{\overline{I}}{c}, \tag{12.1}$$

where $\overrightarrow{I}$ is the sound-intensity vector, and $\overline{I}$ is the magnitude of the averaged sound intensity.

© Springer-Verlag GmbH Germany, part of Springer Nature 2021
N. Xiang and J. Blauert, *Acoustics for Engineers*,
https://doi.org/10.1007/978-3-662-63342-7_12

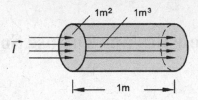

Fig. 12.1 Sound rays propagating through a unit volume

Rays are usually considered incoherent so that their energy densities superimpose when they meet.[1] Summing-up all rays the rays reveals

$$\sum W''_{\text{ray}} = \frac{\sum |\vec{I}|}{c} = \frac{\bar{I}_\Sigma}{c}. \tag{12.2}$$

The assumption that the rays are incoherent is valid for most broadband signals like speech or music, assuming that the rays have traveled different distances from the source. However, this is not the case, for impinging and reflected waves close to reflecting surfaces. We also cannot assume incoherence for narrow-band or pure-tone signals.

12.1 Mirror Sound Sources and Ray Tracing

The behavior of rays at plane reflecting surfaces is particularly relevant for geometrical acoustics. Plane means here that any unevenness of the surface is small compared to the wavelengths of the sound considered. The reflection law, $\Theta_1 = \Theta'_1$, holds, and is even applicable to curved surfaces as long as the curvature is small in relation to the wavelength—see Fig. 12.2.

It is particularly handy to treat sound propagation using rays when studying room acoustics and outdoor sound propagation over longer distances—as often necessary in connection with noise-control problems. When using the concept of sound rays, relevant rules and laws from optics directly apply. The length of a ray is proportional to its traveling time, thus making it possible to not only determine the directions of sound propagation but also the arrival times at given target position.

We depict reflection on plane surfaces by *mirror sources* (*virtual sources*) as shown in Fig. 12.3. The mirror source, $\underline{q}_m$, and the primary source, $\underline{q}_0$, simultaneously send out identical sound fields. The combination of these sound fields on the surface

---

[1] Consult Sect. 1.6 for incoherent superposition.

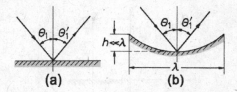

Fig. 12.2 Reflection of sound rays **(a)** At planes. **(b)** At moderately curved plates

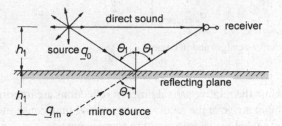

Fig. 12.3 Mirror sound sources emanating from reflection at a plane

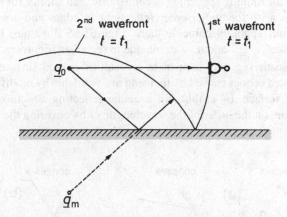

Fig. 12.4 Wave fronts of both the primary and the reflected sounds

produces a reflected wave that fulfills the boundary condition for full reflection, namely, the normal component of the particle velocity, $v_\perp$, being zero.

In Fig. 12.4, we assume that both sources transmit a short sound impulse at the initial time, t_0. The figure shows the wave fronts of both the primary and the reflected sounds and illustrates how the second wavefront arrives at the receiver later than the first one.

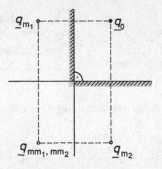

Fig. 12.5 Mirror sources at edges and in corners

Investigations into the relative arrival times of reflections are important, especially since reflections that arrive at the receiver with a delay may cause the perception of disturbing echoes—what we should avoid in room acoustics.

The specific perceptual *echo threshold* is dependent on the character of the sound. It is about 50 ms for running speech, larger for music, and shorter for impulses.

Complications arise for higher-order reflections in edges and corners because, mirror sources may spatially coincide there. Figure 12.5 illustrates how the 2nd-order mirror sources, $\underline{q}_{mm_1}$ and $\underline{q}_{mm_2}$, coincide in rectangular corners.

Focusing or scattering may happen as a result of curved surfaces—shown in Fig. 12.6. Unwanted echoes caused by focusing are avoidable by modifying the form of the reflecting surface, by employing irregular reflecting structures with linear dimensions that are on the order of the wavelength, or by covering the surfaces with sound-absorbent materials.

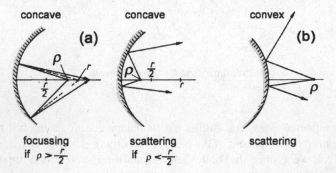

Fig. 12.6 Focusing or scattering effects—ρ denotes the source position (**a**) Concave surface. (**b**) Convex surface

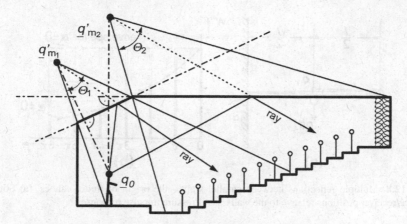

Fig. 12.7 Guiding sound via the ceiling of an auditorium

Figure 12.7 presents an application example for geometrical acoustics. In an auditorium, the sound from a speaker, $\underline{q}_0$, is guided to the audience via ceiling reflections. $\underline{q}_{m_1}$ and $\underline{q}_{m_2}$ are the mirror sound sources representing the tilted and horizontal sections of the ceiling, respectively. Two sample rays are depicted for illustration. The two mirror sources illuminate the spatial sections Θ_1 and Θ_2. The rear wall is made absorbent to avoid audible echoes.

As to the construction of the graph, note that the mirror sources are positioned perpendicularly to the reflecting surfaces at the same distance to the surface as the original source, yet, outside the room concerned. The rays originating from them are restricted to the spatial sector defined by the individual reflecting surfaces.

12.2 Flutter Echoes

Now, we consider a case involving a highly directional sound source, $\underline{p}_1$, between two parallel walls a distance, l, apart. The source emits a short sound-pressure impulse directed perpendicularly toward one of the walls and propagating like a ray—shown in Fig. 12.8a. The two walls be slightly absorbent, characterized by a degree of absorption,[2] α. A microphone close to the position of the source would record a signal as schematically plotted in Fig. 12.8b.

If the interval between the individual impulses at the receiver, $\tau = l/c$, is larger than the echo threshold, the impulses become perceptible as a series of individual echoes, called *flutter echo*. Flutter echoes should be avoided in room acoustics. This is accomplished by a slight tilting of the two walls by $> 5°$, or by making their surfaces absorbent or scattering.

---

[2] α has the dimension one. Thus, it is a degree and not a coefficient.

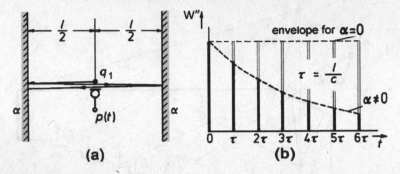

Fig. 12.8 Multiple reflections between parallel walls—the origin of flutter echoes (a) Sound source/receiver positions relative to the walls. (b) Pressure impulse response

The envelope of the impulse series decreases exponentially for $\alpha > 0$, meaning that the ray looses a given percentage of its energy whenever a reflection takes place.[3] The following variables are involved in the flutter echo situation, with W_0'' being the energy density of the impinging ray,

– $W_1'' = (1 - \alpha)\, W_0''$ ... energy density after the 1st reflection

– $W_n'' = (1 - \alpha)^n\, W_0''$ ... energy density after the nth reflection

With the traveled distance between two reflections, l, we arrive at a temporal *reflection density*, n', that is, the number of reflections per time, of

$$n' \approx c/l.\qquad(12.3)$$

Note that n' is correct up to a rest that is negligible when $l \ll 340\,\text{m}$ holds. $340\,\text{m}$ is the distance that sound propagates in air in one second.

For large n, we substitute the expression for W_n'' with a monotonic function as follows,

$$W''(t) = W_0''\,(1 - \alpha)^{n't} = W_0''\,(1 - \alpha)^{(c/l)t}.\qquad(12.4)$$

The discrete energy losses are replaced in this way by continuous spatial damping, which allows the above expression to be written using $y = e^{\ln y}$, namely,

$$W''(t) = W_0''\,\left[e^{\ln(1-\alpha)}\right]^{n't} = W_0''\,e^{n'\ln(1-\alpha)t}.\qquad(12.5)$$

This decreasing exponential function represents an analytical description of the decreasing envelope in Fig. 12.8b.

---

[3] The chunks of sound energy that the source sends out and that subsequently oscillate between the two walls are sometimes dubbed "sound particles".

For small amounts of absorption, $\alpha \ll 1$, the expression can be simplified using *Taylor*'s expansion to

$$-\ln(1 - \alpha) = \alpha + \frac{\alpha^2}{2} + \frac{\alpha^3}{3} + \ldots \approx \alpha \tag{12.6}$$

and by truncating it after the first term, so that we finally get

$$W''(t) \approx W_0'' \, e^{-n'\alpha t} . \tag{12.7}$$

12.3 Impulse Responses of Rectangular Rooms

We now move beyond the case of two parallel walls and consider a rectangular (cuboid) room with six reflecting boundaries, that is, four walls, one floor, and one ceiling. This room illustrates an important rule in room acoustics which states that the reflection density, n', increases with t^2 in many rooms.

Figure 12.9 illustrates this concept. The figure shows the plan view of a rectangular room with a single sound source in it, along with mirrored rooms of nth order with one mirror source in each of them.[4]

Let V be the volume of the cuboid room. Now all of the mirror sources simultaneously transmit a sound impulse at $t = 0$. All impulses that originate from within a hollow sphere with the radius $r_\Delta = c \cdot 1\,s$, arrive at the receiver in the original room within the same interval of 1 s. The number of mirror sources in the hollow sphere is approximately the volume of the hollow sphere divided by the volume of the original cuboids, V, which is

$$n \approx \frac{\frac{4}{3} \pi \left[(r + r_\Delta)^3 - r^3 \right]}{V} . \tag{12.8}$$

For $r \gg r_\Delta$, neglecting the higher-order difference terms, r_Δ^2 and r_Δ^3, this expression approaches

$$n'(t) \approx \frac{4 \pi r^2 r_\Delta}{V \cdot 1\,s} = \frac{4 \pi c^3 t^2}{V} . \tag{12.9}$$

In other words, the density of the impulses arriving at the receiver is increasing with the square of expired time. The reflections also come from more directions over time, resulting in an ever more homogeneous distribution both over time and space.

---

[4] This is the case for rectangular rooms where many mirror sources coincide spatially due to the rectangular corners. In more irregular rooms the situation may become more complicated, particularly, when focusing occurs.

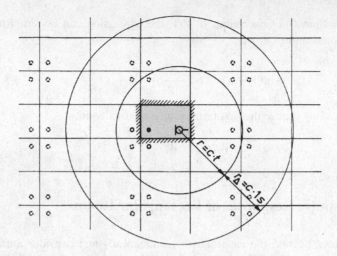

Fig. 12.9 Image sources of one sound source for a rectangular room

Figure 12.10 is a simplified illustration of what is called an *echogram* particulary, an *impulse echogram*,[5] or *energy-impulse response*. In the figure, we see the direct sound and the early, low-order reflections as discrete events. Then the echogram becomes denser and denser, so that individual impulses can no longer be discriminated. This late part of the echogram is called *reverberation tail*.

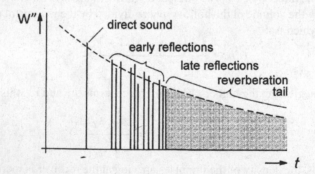

Fig. 12.10 Echogram (energy impulse response)

---

[5] For recording such echograms, the emitting sound source should radiate spherically, such as spherical loudspeaker arrays. In practice, pistol shot are sometimes used to approximately generate spherical sound-impulses.

It is important to consider the intensity characteristics of this situation. Given that the sound source is a spherical one of 0th order, the active intensity of the transmitted sound drops with $1/r^2$. On the one hand, we see this effect for the early reflections of the echogram. On the other hand, the reflection density increases with t^2. The active intensity measured over an interval, t_Δ, is therefore constant.

However, we have to consider that real instruments used to measure echograms always measure with a running time window, t_Δ, because they have low-pass characteristics. Thus, the echogram shows a constant envelope as long as no absorption or dissipation occurs because, in this special, idealized case, the energy in the running time window stays constant over time. Yet, such a case is unrealistic in practice because there are always some losses. In real rooms, the envelope of the reverberation tail decreases exponentially. We justify this statement in the following section.

12.4 Diffuse Sound Fields

From the discussion in Sect. 12.3, it is obvious that it is hardly possible to trace the fate of each sound ray individually, particularly in the reverberation tail. Nevertheless, it is possible to make important statements about the average fate of late reflections. Such an approach is called *statistical room acoustics*. We begin with an idealized model that adequately describes the sound field of the reverberant tail, in the ideal case also called *diffuse sound field*. The model *diffuse sound field* is conceptually characterized by the following assumption, expressed in term of geometrical room acoustics.

A *diffuse sound field* is composed of many rays with the average properties of uniform intensity and uniform spatial distribution

This statement assumes that all rays with uniformly distributed propagation directions have, on average, been reflected the same number of times. It also means that the *mean free-path* length between two reflections is assumed the same for all rays—on average.

The results of statistical room acoustics are independent of the room shape because only the average fate of rays is considered and described by parameters in statistical sense.

Sound Power Impinging Upon the Walls

In a diffuse sound field composed of rays from all possible directions, the magnitude of the intensity is given by

$$\overline{I}_d = \iint_{4\pi} |\vec{I}(\Omega)| \, d\Omega = W_d'' \, c, \tag{12.10}$$

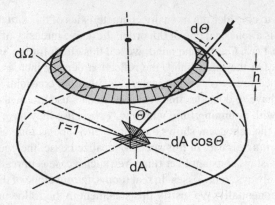

Fig. 12.11 Diffuse field impinging on a wall

where Ω is the spatial angle and W_d'' denotes the energy density in the diffuse sound field. The rays that impinge on a wall from all directions transport sound power onto the wall. In the following, we determine the total power that hits the walls perpendicularly.

We start by computing the active intensity that hits a differential surface element, dA, on the wall. This intensity is obtained by integrating over all differential intensities on a hemisphere, 2π—as depicted in Fig. 12.11.

The wave bundle that arrives from a spatial angle of $d\Omega$ yields an elementary component of the energy-density, namely,

$$W_d'' = \frac{W_d''}{4\pi}\, d\Omega\,, \tag{12.11}$$

where $d\Omega = 2\pi \sin\Theta\, d\Theta$ is derived from $A_{\text{zone}} = 2\pi h$, the area of a spherical zone on a unit-radius sphere. Consequently, we get

$$dI = \frac{W_d'' c}{2} \sin\Theta\, d\Theta\,. \tag{12.12}$$

The perpendicular component of the differential intensity, dI, is obtained as

$$dI_{\text{wall}} = dI \cos\Theta\,. \tag{12.13}$$

To get the total intensity that hits the surface element perpendicularly, we integrate dI_{wall} in (12.13) by substituting (12.12) over the angular range of $0 \le \Theta \le \pi/2$ as follows,

$$I_{\text{wall}} = \frac{W_d'' c}{2} \int_0^{\frac{\pi}{2}} \cos\Theta\, \sin\Theta\, d\Theta$$

$$= \frac{W_d'' c}{2} \frac{1}{2} \int_0^{\frac{\pi}{2}} \sin 2\Theta \, d\Theta = \frac{W_d'' c}{4} . \tag{12.14}$$

which is apparently only 1/4th of the total diffuse intensity.

The total power impinging on a wall with a total surface area, A, thus is

$$\overline{P}_{\text{wall}} = \frac{W_d'' c}{4} A. \tag{12.15}$$

Sound Power Absorbed by the Walls

The power absorbed by the walls is

$$\overline{P}_{\text{wall, abs}} = \frac{W_d'' c}{4} \overline{\alpha} A, \tag{12.16}$$

where $\overline{\alpha}$ is the degree of absorption for random incidence. It can be determined by the equation

$$\overline{\alpha} = \int_0^{\frac{\pi}{2}} \alpha(\Theta) \sin 2\Theta \, d\Theta . \tag{12.17}$$

This expression is known as *Paris'* formula and has already been mentioned in Sect. 11.4. The validity of *Paris'* formula becomes clear by realizing that the absorbed power is equal to

$$dI_{\text{wall, abs}}(\Theta) A = \alpha(\Theta) \, dI_{\text{wall}} A. \tag{12.18}$$

Average Reflection Density and Mean Free-Path Length

The energy that hits the walls during a time span of one second is

$$W_{1s} = \frac{W_d'' c}{4} A \cdot 1 s. \tag{12.19}$$

The total diffuse-field energy present in a room with the volume, V, is

$$W_{\text{room}} = W_d'' V. \tag{12.20}$$

The rays transport this energy to the wall on an average of $\bar{n}'$ times per second, that is

$$\bar{n}' \, W_d'' \, V = \frac{W_d'' \, c}{4} \, A \cdot 1\,s, \quad \text{with} \quad \bar{n}' = \frac{A\,c}{4\,V}. \tag{12.21}$$

This leads to the expression for the mean free-path length as

$$\bar{l} = \frac{c}{\bar{n}'} = \frac{4\,V}{A}. \tag{12.22}$$

12.5   Reverberation-Time Formulas

When inserting the average reflection density, $\bar{n}'$, and the average degree of absorption, $\bar{\alpha}$, into (12.5), which describes the decay of a multiply reflected wave bundle—see Sect. 12.2—we obtain the following expression for the time-dependency of the energy density in the diffuse sound field,

$$W_d''(t) = \sum W_{ray}''(t) = W_d''(t = 0) \, e^{\left[\frac{Ac}{4V} \ln(1-\bar{\alpha})t\right]}. \tag{12.23}$$

In practice, we compute an average degree of absorption as follows,

$$\bar{\alpha} = \frac{\sum_i A_i \bar{\alpha}_i}{\sum_i A_i}, \tag{12.24}$$

where $\bar{\alpha}_i$ is the individual absorption of sub-area A_i.

According to *Sabine*, the time span required for the energy density to decrease to one millionth of its initial value, that is, 10^{-6} or $-60\,dB$, respectively, is called the *reverberation time*, T_{60}. Inserting this reverberation time into (12.23) brings us to

$$10^{-6} \, W_d''(t = 0) = W_d''(t = 0) \, e^{\left[\frac{Ac}{4V} \ln(1-\bar{\alpha}) T_{60}\right]}. \tag{12.25}$$

In air under normal condition, that is, $20\,°C$ temperature, $1000\,hPa$ static pressure, and 50% humidity, the sound speed, c, is about $340\,m/s$. Using this value and solving for T_{60} in (12.25), we obtain

$$T_{60} = 13.8 \frac{4V}{-\ln(1-\bar{\alpha})\,A\,c} \approx 0.163 \cdot \left(\frac{s}{m}\right) \frac{V}{-\ln(1-\bar{\alpha})\,A}, \tag{12.26}$$

an expression that is known as *Eyring*'s reverberation formula.

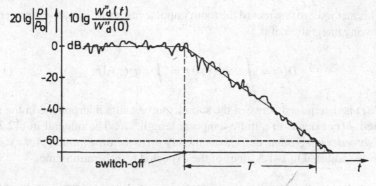

Fig. 12.12 Time trace of a reverberant noise sound after being switched off

For small amounts of absorption, $\overline{\alpha} \ll 1$, this formula simplifies to

$$T_{60} = 13.8 \frac{4V}{\overline{\alpha} A \, c} \approx 0.163 \cdot \left(\frac{s}{m}\right) \frac{V}{\overline{\alpha} A}, \qquad (12.27)$$

which is preferred in practice and known as *Sabine*'s reverberation formula.

There are different ways of measuring the reverberation time, including the *switch-off method*, and the *integrated-impulse-response method*.

Using the switch-off method, a room is excited with a noise source until a steady-state energy-density level is reached. Then the sound source is switched off, and the sound energy-density level, respectively, the sound-pressure level, is recorded as a function of time, producing an energy-density time function like the one graphically presented in Fig. 12.12. The recorded graph is the so-called *energy time curve*.

The time interval between switching-off the source and the instant at which the sound-energy level has decreased by 60 dB is taken as the reverberation time, T_{60}. Besides T_{60} other reverberation quantities are in use referring to less dynamic level ranges, for example, T_{30}, which relates to a decay of 30 dB, yet extrapolated to a decay of 60 dB.[6] In practice, the decay range is often taken from –5 dB downwards.

The switch-off method has the disadvantage that the state of the sound field is not clearly defined at the switch-off instant because the exiting noise is of stochastic nature. The measured decay traces, $D(t)$, thus vary between trials.

This statistical uncertainty is overcome with the *integrated-impulse-response method*, also known as *Schroeder*'s method. Here the excitation is accomplished with a distinct sound impulse.[7]

---

[6] Besides T_{30} and T_{60}, the *early-decay time*, T_{edt}, is a relevant decay parameter, which is defined by the slope of the first 10-dB energy decay after the switch off, but then extrapolated linearly to a decay of 60 dB. T_{edt} is considered to correlate well with the instantaneous perception of reverberance.

[7] There also exist advanced correlation methods using pseudo-random or swipe-sine excitations to determine the impulse responses experimentally.

For this method, we first record the room impulse response, $h(t)$, and then perform the following integration of it,

$$D(t) = \int_t^\infty h^2(\tau)\,\mathrm{d}\tau = \int_t^{t_L} h^2(\tau)\,\mathrm{d}\tau, \tag{12.28}$$

where $h(t)$ is determined between the sound source and a microphone in the room concerned. $h(t)$ extends to a finite temporal length,[8] t_L. The integral in (12.28) is called *Schroeder integration*, carried out merely up to the time limit, t_L, for example, with a value adjusted to 1–1.5 times of the expected reverberation time.

Considering the theory of diffuse sound field predicts that the *diffuse-field* assumption is sufficiently valid within the frequency range that the geometrical acoustics is applicable when the following conditions are met.

- Sound absorption is well distributed about the boundaries of the room.

- The shape of the room is irregular and, in particular, focusing elements are avoided.

- Total sound absorption is small or moderate. A good check is whether the ratio of room volume and equivalent absorbent area—see next paragraph—is not too small, usually > 1 m.

The Equivalent Absorbent Area

The following expression,[9]

$$A = \overline{\alpha}\, A \quad \text{with} \quad \overline{\alpha} \ll 1, \tag{12.29}$$

is understood to be a conceptual area with a unit degree of absorption, that is 100% of absorption or $\overline{\alpha} = 1$, representing the total absorption present in the room. This conceptual area is called *equivalent absorbent area*.

If there are partial areas at the boundary of a room, each with an individual absorption value of $\overline{\alpha}_i$, then the equivalent absorbent area is defined to be

---

[8] To determine frequency-dependent reverberation times, band-pass filters, for example, with octave bandwidth or third-octave bandwidth, are applied to filter the broadband room impulse response, $h(t)$, before *Schroeder*'s integration.

[9] For moderate $\overline{\alpha}$ values ranging approximately $0 < \overline{\alpha} < 0.5$, the absorption term in *Eyring* formula yields a more general expression for the equivalent absorption area, $A = -\ln(1 - \overline{\alpha})\,A$.

$$A = \sum_i A_i \, \overline{\alpha}_i + \underbrace{8 \, \overline{\alpha}_{air} \, V}_{\text{correction}} [\text{m}^{-1}], \qquad (12.30)$$

where A_i are the individual areas with individual degrees of absorption $\overline{\alpha}_i$. The correction component in the formula, not derived here, is only used to account for dissipation in the transmission medium of large rooms. Values of $\overline{\alpha}_{air}$ are given in Table 11.1.

12.6   Application of Diffuse Sound Fields

This section introduces three frequently used applications of the diffuse-sound-field model.

Reverberation-Time Acoustics

The reverberation time, T_{60}, estimated with either *Eyring*'s or *Sabine*'s formula, is one of various relevant parameters of the *quality-of-the-acoustics* of spaces, which includes, among other things, their suitability for specific performances.

Table 12.1 presents preferred values of T_{60} for various performance styles. The values are drawn from the literature and refer to the 500–1000 Hz range. Many acoustical consultants consider a moderate increase toward low frequencies as adequate because it is said to increase the listeners' sense of warmth and envelopment. Yet, there is also dissenting opinion, arguing that an increase of reverberation toward low frequencies affects articulation accuracy. This is an issue, particularly, in spaces for sound recordings such as recording studios.

The following guidance holds for speech. A too-short reverberation time produces high intelligibility but also increases the effort required from the speaker. If T_{60} is too long, auditory smearing takes place and deteriorates intelligibility. 0.8–1.0 s is a reasonable compromise for speech running at normal speed, that is, roughly a rate of 50 syllables/minute.

Reverberation time is undoubtedly a relevant parameter of room-acoustic quality but certainly not the only one. Proper guidance of early reflections and avoidance of echoes are at least as relevant.

Table 12.1 Reverberation times

Speech	Chamber music	Opera houses	Concert halls	Organ music
0.8–1.0 s	1.4–1.6 s	1.5–1.7 s	1.9–2.2 s	2.5 s and more

Measurement of Random-Incidence Absorption

Random-incidence absorption measurements are conducted in *reverberation chambers*. These are special spaces where a diffuse sound field is be realized by using highly reflective, obliquely oriented walls and panels.[10]

The equivalent absorbent area, A_{α_1}, of the empty chamber is be determined beforehand, usually by a reverberation-time measurement for T_1. With a sample of the material to be measured in the chamber, the reverberation-time measurement is then repeated. Using the measured reverberation-time, T_2, then, the equivalent absorbent area of the chamber including the sample is,

$$A_{\alpha_2} = 0.163 \left[\frac{s}{m}\right] \frac{V}{T_{\alpha_2}}. \tag{12.31}$$

Consequently, the equivalent absorption area of the sample results in

$$A_{,sample} = A_{sample}\ \overline{\alpha}_{sample} = A_2 - A_1\,, \tag{12.32}$$

where A_{sample} is the sample size under test. The random-incidence degree of absorption, $\overline{\alpha}_{sample}$, results as follows,

$$\overline{\alpha}_{sample} = 0.163 \left[\frac{s}{m}\right] \frac{V}{A_{sample}} \left(\frac{1}{T_2} - \frac{1}{T_1}\right). \tag{12.33}$$

Measurement of the Total Power of a Sound Source

For measurements of the emitted power of sound sources, diffuse sound fields in reverberation chambers are in use. To this end, the source is brought into the chamber and operated to transmit ongoing sounds.

The sound power transmitted by the source and the sound power absorbed by the equivalent absorbent area of the reverberation chamber reaches a stationary balance as follows.[11]

$$\underbrace{\overline{P}_{source}}_{introduced} = \underbrace{\frac{A_\alpha W_d'' c}{4}}_{absorbed} \quad \text{with} \quad W_d'' = \frac{I_d}{c} = \frac{p_{rms}^2}{\varrho\, c^2}. \tag{12.34}$$

The power of the sound source, consequently, is

---

[10] To measure α for perpendicular sound incidence only, measuring tubes are applied—see Sect. 7.6. Note that α is then termed normal-incidence absorption, $\alpha_\perp$.

[11] p_{rms} is the rms-value of sound pressure, p—see Sect. 16.4 for its definition.

$$\overline{P}_{\text{source}} = p_{\text{rms}}^2 \, \frac{A_\alpha}{4 \, \varrho \, c} . \tag{12.35}$$

The following rules are useful when performing the measurements. A sufficient number of measuring points must be well-distributed across the room but not too close to walls or corners because there is no guarantee for full incoherence of incoming and reflected sounds in such locations. These positions may thus result in too high measured values.

The Critical Radius

A stationary, diffuse sound field has the same average energy density, W_d'', everywhere in space. Close to a sound source, however, the energy density of the direct sound is much higher. The situation is depicted in Fig. 12.13.

The critical radius is the distance from a 0th-order spherical source where the direct and diffuse energy densities are just equal. Equality of the two energy densities is given at

$$W_{\text{direct}}'' = \underbrace{\frac{\overline{P}_{\text{source}}}{4 \, \pi \, r^2} \, \frac{1}{c}}_{I_{\text{direct}}} \overset{!}{=} W_d'' = \underbrace{\frac{4 \, \overline{P}_{\text{source}}}{A_\alpha} \, \frac{1}{c}}_{I_{\text{diffuse}}} , \tag{12.36}$$

which leads to the critical radius as follows,

$$r_c = \sqrt{\frac{A_\alpha}{16 \, \pi}} . \tag{12.37}$$

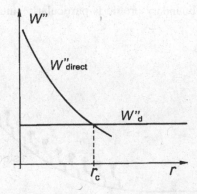

Fig. 12.13 The critical radius—the equilibrium of direct- and diffuse-field energy densities with a 0th-order spherical source

For directional sources the distance of equilibrium of direct and diffuse field is larger in the direction of focused transmission. The r_c for directional sources is called *diffuse-field distance* or *critical distance*. When taking sound recordings, knowing the diffuse-field distance is useful because a microphone placement within this distance primarily receives direct-sound signals. However, placement outside predominantly produces diffuse-sound signals—which are auditorily perceived as *spatial impression*.

12.7 Exercises

Geometrical Acoustics

Problem 12.1 Design a reasonable room boundary for an auditorium as sketched in Fig. 12.14.

Problem 12.2 A rectangular room with linear dimensions of $l_x = 25$ m, $l_y = 16$ m, and $l_z = 6$ m is given. The absorption coefficients of the floor, the ceiling and the walls are $\overline{\alpha}_1 = 0.2$, $\overline{\alpha}_2 = 0.7$, and $\overline{\alpha}_3 = 0.4$, respectively.

(a) Calculate the reverberation time of the unoccupied room.
(b) How does the reverberation time change when the room is occupied by 150 people? (Assume an equivalent absorption area of 0.5 m^2/person).
(c) Find the critical distance of a dipole source for this room.

Problem 12.3 What a boundary profile is particularly suitable for constructing a "whisper gallery"?

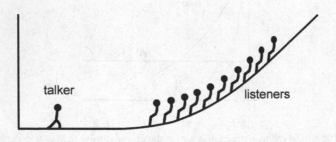

Fig. 12.14 Part of the section of an auditorium—with a talker position and a listeners' area

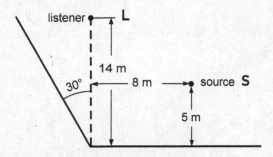

Fig. 12.15 A sound source and a receiver in front of an oblique wall

Problem 12.4 Given a sound source and a receiver in front of an oblique (non-rectangular) corner—illustrated in Fig. 12.15.

(a) Determine the mirror images of 1st and 2nd order for the sound source, **S**.
(b) Estimate the impulse sequences originating from the above mentioned mirror-image sources at the listener's position, **L**, when the source sends out a short impulse. Let the absorption coefficient be $\bar{\alpha} = 0.5$—propagation losses in air be negligible.

Chapter 13
Insulation of Air- and Structure-Borne Sound

Sound insulation is the confinement of sound to a space in such a way that transmission to neighboring spaces is totally or partially prevented. Sound insulation is predominantly based on reflection caused by impedance discontinuities in possible transmission paths. Dissipation and absorption may also play a role, but their influence role is usually minor. Another term for sound insulation is *sound damming* because the sound is, so-to-say, "dammed in".

Sound insulation must not be confused with *sound damping*. Damping of sound means that sound energy has been removed from a sound field through dissipation and/or absorption. The transmission of sound to another space is one possible method of absorption. Thus, absorption is not necessarily dissipation, the latter being transformation of acoustic/mechanic energy into thermal energy.

Measures for airborne and structure-borne sound insulation are of particular technological relevance. Non-porous leaves or walls are typically inserted into airborne transmission paths to achieve insulation, whereas insulation of structure-borne sound is accomplished by inserting elastic elements (springs) or layers (resilient materials, air gaps). Sometimes heavy *interlocking masses* are also used. In every case, the goal is to create impedance discontinuities that result in reflection.

13.1 Sound in Solids—Structure-Borne Sound

An important difference between solids and fluids is that solids experience shear forces. This difference means that solids store energy through both volume changes and changes of form. Thus, in solids, several wave types in addition to longitudinal waves exist. The types of waves that are possible in a specific case are dependent, among other things, on the specific form or the configuration of solid bodies under consideration.

© Springer-Verlag GmbH Germany, part of Springer Nature 2021
N. Xiang and J. Blauert, *Acoustics for Engineers*,
https://doi.org/10.1007/978-3-662-63342-7_13

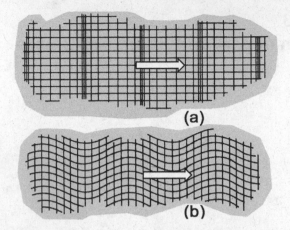

Fig. 13.1 Sound waves in infinitely extended solids (**a**) Longitudinal wave. (**b**) Transversal wave

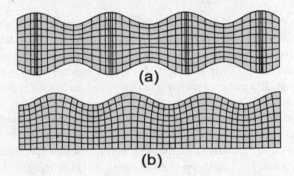

Fig. 13.2 Sound waves in finitely extended solids (**a**) Quasi-longitudinal (dilatational) wave. (**b**) Surface wave

Infinitely extended solids only experience longitudinal density waves and transverse shear waves—shown in Fig. 13.1, but finite solids like rods or plates also carry dilatation, surface, torsion, and bending waves—illustrated in Figs. 13.2, and 13.3.

The different types of waves listed above couple with each other at boundaries, junctions, and points of impact. This coupling means that selective damping of one wave type does not prevent this kind of wave from being excited again somewhere else, eventually in a different form. For example, a rod that is perpendicularly fixed to a plate and carrying an elongation wave excites bending waves at the junction with the plate. Bending waves are also called *flexural* waves.

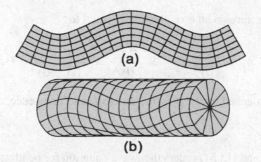

Fig. 13.3 Sound waves in finitely extended solids (plates and bars) **(a)** Bending (flexural) wave. **(b)** Torsion wave

13.2 Radiation of Airborne Sound by Bending Waves

The specific combination of transverse and angular motion that characterizes bending waves results in considerable surface velocities and, consequently, airborne-sound emission. Because of this effect, this type of wave is of specific practical interest.

The Wave Equation for Bending Waves

We now derive the wave equation for a lossless free bending wave in thin plates. Energy storage in these waves is accomplished by the *mass load*, m'' [mass/area], and the *bending stiffness*, B' [N/m^2], which is a stiffness per unit width. Figure 13.4 illustrates the situation.

A bending of the elementary plate with the bending stiffness B', excites a torque (bending moment) per width, $\underline{T}'$. The torque is proportional to the flexion, $\partial^2 \underline{\xi}_x / \partial y^2$.

The following consequently holds,

$$\underline{T}' = -B' \, \frac{\partial^2 \underline{\xi}_x}{\partial y^2} \, , \tag{13.1}$$

where the negative sign results from taking T' positive for power flow in positive x–direction, but this direction of T' is opposite to the one which produces a positive flexural curvature (by a second derivative of $\underline{\xi}_x$ with respect to y).

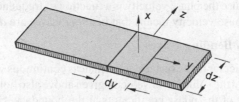

Fig. 13.4 Element of a plate with mass and bending stiffness

The relative pressure on an area element is equal to

$$\underline{p}_{B'} = \frac{\partial^2 \underline{T}'}{\partial y^2} = -B' \frac{\partial^4 \underline{\xi}_x}{\partial y^4} = \frac{-B'}{j\omega} \frac{\partial^4 \underline{v}_x}{\partial y^4}, \tag{13.2}$$

which is counterbalanced by the pressure due to mass persistence, expressed as

$$\underline{p}_{m''} = j\omega m'' \underline{v}_x. \tag{13.3}$$

Combining (13.2) and (13.3) renders the wave equation for bending waves, namely,

$$j\omega m'' \underline{v}_x + \frac{B'}{j\omega} \frac{\partial^4 \underline{v}_x}{\partial y^4} = 0. \tag{13.4}$$

When counting for the damping loss in a plate, consider an additional resistive damping term, $\zeta' \underline{v}_x$,

$$\zeta' \underline{v}_x + j\omega m'' \underline{v}_x + \frac{B'}{j\omega} \frac{\partial^4 \underline{v}_x}{\partial y^4} = 0. \tag{13.5}$$

with ζ' being a loss factor per unit width. Yet, for simplicity, we ignore the damping loss in the following.

The solution for propagating bending waves of this 4th-order differential equation (13.4) is

$$\underline{v}_x(y) = \underline{v}_{x \to} e^{-j\beta_B y}. \tag{13.6}$$

Inserting this solution into the wave equation result in the bending phase coefficient, β_b, which is equal to

$$\beta_b = \sqrt{\omega} \sqrt[4]{\frac{m''}{B'}}. \tag{13.7}$$

This leads directly to the phase velocity of the free bending wave, c_b,

$$c_b = \sqrt{\omega} \sqrt[4]{\frac{B'}{m''}} = \frac{\omega}{\beta_b}. \tag{13.8}$$

This expression shows that different spectral components of the wave propagate with different speeds because the phase velocity is a function of frequency. This frequency dependency of the phase velocity means that bending waves are *dispersive*.

Sound Radiation by Bending Waves

The velocity perpendicular to the surface of the plate is continuous with the velocity of the plate, meaning that the solution for velocity given above also holds for the adjacent layer of air. Equality of the phase coefficients at the boundary, $\beta_b = \beta \sin \theta = \beta_y$, yields

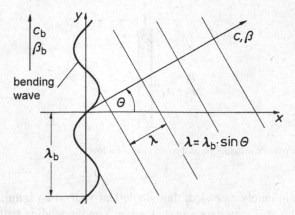

Fig. 13.5 Radiation of airborne sound by bending waves

$$\beta_b = \frac{\omega}{c_b} = \beta \sin \theta = \frac{\omega}{c} \sin \theta, \tag{13.9}$$

from which follows

$$c = c_b \sin \theta. \tag{13.10}$$

This equation is similar to the law of refraction—as introduced in Sect. 11.3.

The radiation due to bending waves is expressible by the phase coefficient along the x–direction, that is,

$$\beta_x = \beta \cos \theta = \sqrt{\beta^2 - (\beta \sin \theta)^2} = \omega \sqrt{\frac{1}{c^2} - \frac{1}{c_b^2}}. \tag{13.11}$$

This expression includes two distinguishable cases.

- *Case I* For $c < c_b$ or, equivalently, $\beta > \beta_b$ or $\lambda < \lambda_b$, the resulting quantity inside the square root in (13.11) is positive. The Equation (13.11) results in a real (non-complex) solution for β_x. Equation (13.10) is fulfilled, resulting in radiation of airborne sound at an angle of θ—shown in Fig. 13.5. The result is that a component of the phase coefficient exists along the x–direction.

- *Case II* For $c > c_b$, β_x in (13.11) becomes imaginary and results in a real damping coefficient, $\alpha_x = j(j\beta_x)$. Equation (13.10) cannot be fulfilled in this case. Thus, we observe an exponential decrease in x–direction, which physically amounts to a hydrodynamic short-circuiting near the surface. This situation equates the *total reflection* in refraction.

13.3 Sound-Transmission Loss of Single-Leaf Walls

We restrict ourselves to non-porous walls in this section. In acoustical terms, *single leaf* refers to a panel in which the cross-sectional particle velocity, v_x, is identical at all points inside the leaf. This is the case in thin, solid leaves. Longitudinal waves inside the leaf are thus negligible.

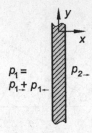

Fig. 13.6 Sound-pressure distribution at the two sides of a leaf

Consider an infinitely extended, thin single-leaf wall in an infinitely extended fluid like air. The wall may have a mass load, m'', and a bending stiffness, B'. An enforced bending wave is excited by the sound-pressure distribution in front of and behind the leaf. The pressure difference between the two sides of the leaf controls this excitation—see Fig. 13.6.

Inserting this pressure difference into the bending wave equation as the exciting term that represents the pressure balance, we obtain the following expression,

$$\left[\underline{p}_{1\rightarrow} e^{-\mathrm{j}\beta y \sin\theta} + \underline{p}_{1\leftarrow} e^{-\mathrm{j}\beta y \sin\theta} \right] - \underline{p}_{2\rightarrow} e^{-\mathrm{j}\beta y \sin\theta}$$

$$= \mathrm{j}\,\omega\, m''\, \underline{v}_{\mathrm{x}} + \frac{B'}{\mathrm{j}\omega}\frac{\partial^4 \underline{v}_{\mathrm{x}}}{\partial y^4}\,. \quad (13.12)$$

Inserting the solutions

$$\underline{v}_{\mathrm{x}\rightarrow}(y) = \underline{v}_{\mathrm{x}\rightarrow} e^{-\mathrm{j}\beta y \sin\theta} \quad \text{with} \quad \underline{v}_{\mathrm{x}\rightarrow}(y) = \underline{v}_{2\rightarrow}(y)\cos\theta\,, \quad (13.13)$$

yields

$$[\,(\underline{p}_{1\rightarrow} + \underline{p}_{1\leftarrow}) - \underline{p}_{2\rightarrow}\,]\, e^{-\mathrm{j}\beta y \sin\theta}$$

$$= \left(\mathrm{j}\,\omega\, m'' + \frac{B'}{\mathrm{j}\omega}\,\beta^4 \sin^4\theta \right) e^{-\mathrm{j}\beta y \sin\theta} \overbrace{\underline{v}_{2\rightarrow}\cos\theta}^{\underline{v}_{\mathrm{x}\rightarrow}(y)}\,. \quad (13.14)$$

Now, since

$$\underline{v}_{2\rightarrow} = \frac{\underline{p}_{2\rightarrow}}{\varrho c} \quad \text{and} \quad \underline{v}_{2\rightarrow} = \underline{v}_{1\rightarrow} + \underline{v}_{1\leftarrow} = \frac{\underline{p}_{1\rightarrow}}{\varrho c} - \frac{\underline{p}_{1\leftarrow}}{\varrho c}\,, \quad (13.15)$$

we eliminate $\underline{p}_{1\leftarrow}$ and $\underline{v}_{2\rightarrow}$ in (13.14) and obtain

$$\frac{\underline{p}_{1\rightarrow}}{\underline{p}_{2\rightarrow}} = 1 + \frac{\cos\theta}{2\varrho c}\left(\mathrm{j}\,\omega\, m'' + \frac{B'}{\mathrm{j}\omega}\,\beta^4 \sin^4\theta \right)\,. \quad (13.16)$$

Rewriting in logarithmic terms yields the so-called *transmission loss, R*,

$$R = 20 \lg \left| \frac{\underline{p}_{1\rightarrow}}{\underline{p}_{2\rightarrow}} \right| \quad \text{dB}$$

$$= 20 \lg \left| 1 + \frac{j \cos \theta}{2 \varrho c} \left(\omega m'' - \frac{B'}{\omega} \beta^4 \sin^4 \theta \right) \right| \quad \text{dB} . \qquad (13.17)$$

Perpendicular Sound Incidence

In the case of perpendicular sound incidence, that is, for $\sin \theta = 0$, the 2nd term in the parentheses of (13.17) vanishes. This also holds for leaves that are very compliant for bending waves, that is, for $B' \to 0$. The transmission loss, R, now approximately becomes

$$R \approx 20 \lg \left(\frac{\omega m''}{2 \varrho c} \right) \quad \text{dB} . \qquad (13.18)$$

In this case, the transmission loss only depends on the mass load, m'', and increases at a rate of $6\,\text{dB}$ per doubling of the mass load, This relation is called the *mass law* for walls.

Oblique Sound Incidence

An important special case relates to oblique sound incidence. It is given when the entire parenthetical term vanishes, and we thus have $R = 0$. This is the case when the y–component of the phase coefficient of the sound wave in air is equal to the phase coefficient of the bending wave in the leaf, such that

$$\beta^4 \sin^4 \theta = \omega^2 \frac{m''}{B'} \quad \text{and, thus,} \quad (\beta \sin \theta)^4 = (\beta_b)^4 , \qquad (13.19)$$

and, finally,

$$\beta \sin \theta = \beta_b . \qquad (13.20)$$

The equality of the phase coefficients is called *trace matching*. This match is kind of a spatial resonance. For instance, a lossless wall like a window grate becomes completely transparent when this condition occurs.

It follows from (13.7) and (13.20) that

$$\frac{\omega}{c} \sin \theta = \sqrt{\omega} \sqrt[4]{\frac{m''}{B'}} , \qquad (13.21)$$

which in turn leads to

$$\omega_c = 2\pi f_c = \sqrt{\frac{m'' c^4}{B' \sin^4 \theta}} \quad \dots \text{for} \quad \sin \theta < 1 . \qquad (13.22)$$

This critical frequency, $f_c = \omega_c/(2\pi)$, is called *coincidence frequency*. The inclusion of f_c generates the following formula for R,

$$R = 20 \lg \left| 1 + \frac{j \cos \theta}{2 \varrho c} \, \omega m'' \left(1 - \frac{f^2}{f_c^2} \right) \right| \quad \text{dB} \,. \tag{13.23}$$

Below a *limiting coincidence frequency*, $f_{c,\,\text{lim}}$, trace matching becomes impossible because

$$\lambda_{\text{air}} > \lambda_{\text{b}} \,. \tag{13.24}$$

This limiting frequency is determined by $\lambda_{\text{air}} = \lambda_{\text{b}}$, as follows,

$$f_{c,\,\text{lim}} = \frac{1}{2\pi} \sqrt{\frac{m'' c^4}{B'}} \quad \ldots \text{ for } \sin \theta = 1 \,, \text{ namely, } \theta = 90° \,. \tag{13.25}$$

The following advice is relevant for practical application. The coincidence frequency is usually set well above or below the frequency range under consideration for achieving a high transmission loss. Leaves made of solid brick or concrete have coincidence frequencies in the range of 50–100 Hz. Plywood and dry-plaster panels show coincidence frequencies of about 1–3 kHz. The limiting coincidence frequency increases when the panel is loaded with additional mass or when slits are cut into it, which make the panel more compliant for bending waves. Absorbent coatings or viscous internal *sandwich layers* are adequate means for dampening bending waves.

The principal frequency relationship of R as a function of f for both directional and random (diffuse-field) sound incidence is shown in Fig. 13.7. Below f_c, the amount of insulation afforded by single-leaf walls is essentially proportional to the mass load and, thus, proportional to the frequency. This proportionality indicates an increase of 6 dB/oct. Above f_c, the stiffness term becomes dominant and proportional to ω^3, amounting to an 18 dB/oct increase. However, such a slope is rarely achieved in practice, amongst other reasons, due to how the wall is clamped, and due to bypasses—refer to Fig. 13.12.

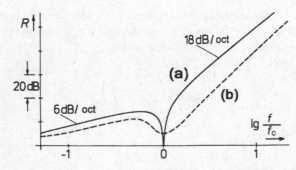

Fig. 13.7 Transmission loss as a function of frequency for a single-leaf wall (**a**) Directed oblique incidence. (**b**) Random incidence

13.4 Sound-Transmission Loss of Double-Leaf Walls

As just discussed, the sound-transmission loss of single-leaf walls is governed by the mass load. This behavior certainly holds below the coincidence frequency but is also usually sufficient above it since the bending stiffness, B', of many wall materials is proportional to the mass load, m''.

In cases where the mass of walls is structurally limited, double-leaf walls—pictured schematically in Fig. 13.8 (b, lower panel)—are employed to increase sufficient sound insulation.

In terms of electro-acoustic analogies, a single-leaf wall experiencing perpendicular sound incidence represents a two-port element—depicted in Fig. 13.8a. A double-leaf wall with an air gap inside represents a T-shaped two-port circuit—see Fig. 13.8 (b, upper panel). In terms of network theory, both wall types act as low-pass filters of the 1st and 3rd order, respectively.

With a low-pass filter of the 3rd order, a transmission loss of three times 6 dB/oct or 18 dB/oct is theoretically achievable. Yet, this amount of transmission loss only holds above the fundamental *drum resonance*,[1] ω_0.

The drum resonance is determined by the equation

$$\omega_0 = \frac{1}{\sqrt{n''\, m''_{\text{total}}}},\tag{13.26}$$

in which the total effective mass is equal to

$$m''_{\text{total}} = \frac{m''_1\, m''_2}{m''_1 + m''_2}.\tag{13.27}$$

Below its fundamental resonance, a double-leaf wall behaves like a single-leaf one because the two leaves couple more or less rigidly via the air gap. When the linear dimension of the air gap matches the wavelength of sound waves in air, cavity resonances arise that reduce the transmission loss. This effect is reducible with loose fills of absorbent material like mineral wool in the air space between the leaves.

Coincidence effects occur in the case of oblique sound incidence, but their negative impact is reduced by making the mass loads of the two leaves different.

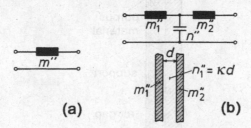

(a)   **(b)**

Fig. 13.8 Equivalent circuits (**a**) Single-leaf wall. (**b**) Double-leaf wall

---

[1] This configuration resembles the two-mass resonator, discussed in Sect. 3.7.

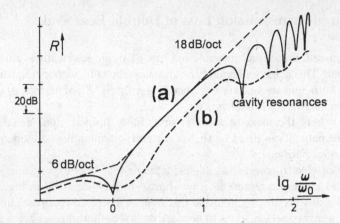

Fig. 13.9 Transmission loss as a function of frequency for a double-leaf wall (**a**) Directed oblique incidence. (**b**) Random incidence

Figure 13.9 depicts the principal course of R as a frequency function for directional and diffuse (random) sound incidences. The cavity resonances are indicated. The advantage of a double-leaf wall over a single-leaf one lies in the region above the drum resonance where R increases with a slope of up to 18 dB/oct—and this at a much lower weight than comparable single-leaf walls!

In real walls in buildings, the sound transmission does not only occur through the wall itself but also via the clamping at its rim. Transmission through this path is reduced substantially by making at least one of the two leaves very compliant for bending waves by setting its coincidence frequency, $f_{c,\,lim}$, above the frequency range concerned. In this case, bending waves transfer into the compliant leaf without being emitted as airborne sound. Such bending-wave-compliant leaves consist, for instance, of gypsum board, metal sheets, or heavy foils. Figure 13.10 sketches a common construction.

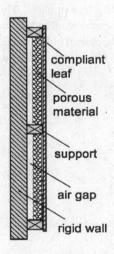

Fig. 13.10 Arming a single-leaf wall with an additional, bending-compliant leaf

13.5 The Weighted Sound-Reduction Index

In architectural acoustics and related fields, the sound-insulation capability of a wall is characterized by an internationally-standardized single-number index called the *Weighted Sound-Reduction Index*, R_w. This index is specific to the wall element considered and independent of the actual installation situation.

The procedure for measuring R_w assumes random sound incidence and a relevant frequency range of 100–3150 Hz (ISO 10140), or 125–4000 Hz (ASTM E90).

Figure 13.11 shows the measurement set-up. The sending room is excited by noise, and the random-incident sound-pressure, L_{ds}, is determined. The receiving room has a known equivalent absorption area, $A_{\alpha,r}$—refer to Sect. 12.5 for this quantity. The sound-pressure level, L_{dr}, is measured in the diffuse field of this room.

The sound-reduction index, R, is defined as

$$R = 10 \lg \left(\frac{I_{ds}}{I_{dr}} \right) = 10 \lg \left(\frac{P_{ds}}{P_{dr}} \right) \quad \text{dB} , \tag{13.28}$$

where the sound power impinging on the wall, P_{ds}, is

$$P_{ds} = \frac{S \, W_{ds}'' \, c}{4} = \frac{S \, p_{ds,\,rms}^2}{4 \, \varrho \, c} , \tag{13.29}$$

W_{ds}'' is energy density in the source room—see (12.15)—and $p_{ds,\,rms}$ is the RMS-value of the averaged sound pressure in the source room.

The sound power transmitted through the wall into the receiving room, P_{dr}, is given by

$$P_{dr} = \frac{A_{\alpha,r} \, W_{dr}'' \, c}{4} = \frac{A_{\alpha,r} \, p_{dr,\,rms}^2}{4 \, \varrho \, c} . \tag{13.30}$$

We combine these terms to the ratio,

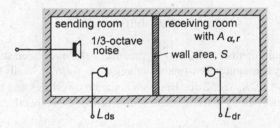

$\llcorner L_{ds}$ $\llcorner L_{dr}$

Fig. 13.11 Measurement set-up for the transmission loss of walls, windows, etc.

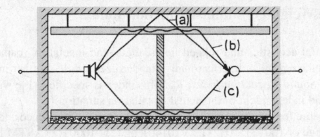

Fig. 13.12 Possible paths of sound transmission between two enclosed spaces

$$\frac{P_{\text{ds}}}{P_{\text{dr}}} = \frac{S}{A_{\alpha,\text{r}}}\,\frac{p^2_{\text{ds, rms}}}{p^2_{\text{dr, rms}}}\,,\tag{13.31}$$

and rewrite it in logarithmic terms as

$$R = L_{\text{ds}} - L_{\text{dr}} + 10\lg\left(\frac{S}{A_{\alpha,\text{r}}}\right)\;\text{dB}\,.\tag{13.32}$$

Be aware that in real installations, sound may be transmitted from one room to another via other paths besides the wall itself. Common by-passes are shown in Fig. 13.12. The following measures reduce the effect of these by-passes. The flanking walls are built sufficiently heavy, the air gap above a suspended ceiling is "compartmentalized" and damped, and the covering floor does not continue beneath the wall but is interrupted there.

According to the standard, R is measured in one-third-octave bands. To obtain a single-number criteria, the frequency curve of the measured R is compared to a reference curve that defines a so-called *Weighted-Sound-Reduction-Index* value of $R_{\text{w}} = 52\,\text{dB}$.[2] The measured curve is then shifted parallel to itself and toward the reference curve in 1 dB steps until the sections below reference remain on average $\leq 2\,\text{dB}$—depicted in Fig. 13.13. The amount of plus or minus shifting in dB is then added to 52 dB. The resulting *Weighted Sound-Reduction Index* of the wall is

$$R_{\text{w}} = (52 \pm \text{shifting})\,[\text{dB}]\,.\tag{13.33}$$

Minimum requirements for R_{w} and R'_{w}, respectively, have been standardized but vary from country to country. Reasonable values are as follows. Walls between apartments should have an $R'_{\text{w}} \geq 52\,\text{dB}$. Between separate dwellings and rooms used for activities that do not belong to the apartment concerned, the respective value is 62 dB.

---

[2] If R_{w} has been measured in a specific test set-up with standardized by-passes, this is indicated by an apostrophe, namely, R'_{w} instead of R_{w}.

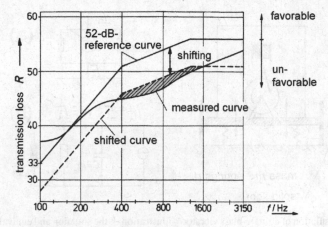

Fig. 13.13 Determination of the *Weighted Sound-Reduction Index*, ISO 140, or *Sound-Transmission Class*, ASTM E90. The reference curve is considered between 100 Hz and 3150 Hz (ISO 140) or 125 Hz and 4000 Hz (ASTM E90)

13.6 Insulation of Vibrations

Insulation of structure-borne sound often turns out to be difficult in practice. On the one hand, materials like steel or concrete have only very low damping coefficients for structure-borne sound waves. On the other hand, it is often impossible to construct adequate insulation measures like soft, resilient inlays in optimal positions. For these reasons, it is best to take care of preventing structure-borne sound from entering structures like buildings, vehicles, and engines in the first place. Directly insulating the vibration source reduces this undesired effect, usually accomplished by providing an elastic support for the source. We now elaborate on how elastic support works acoustically.

Since sources of structure-borne sound are usually small compared to the wavelengths of structure-borne sound waves, we illustrate the fundamental principle of vibration insulation with concentrated elements. Further, we restrict ourselves to an example with only one degree of freedom. If there are more degrees of freedom involved, comparable measures are taken for each of them.

The example with a single-mass vibrator pictured in Fig. 13.14 is known as the *engine-support problem*. The figure shows a sketch of the situation and a mechanic-circuit diagram with an equivalent electric circuit of the 1st-kind analogy.

In our example, we is assume that the exciting force, $\underline{F}_0$, is a constant sinusoidal force.[3] The task at this point is to tune the system in such a way that the force draining

---

[3] Alternatively, a constant velocity, $\underline{v}_0$, could be assumed.

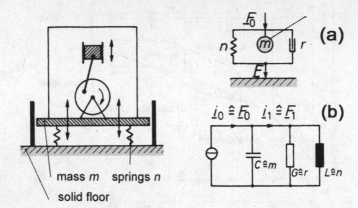

Fig. 13.14 Insulation of a single-mass vibrator—illustration of the situation and equivalent circuits (a) Mechanic equvalent. (b) Electric equivalent

into the ground, $\underline{F}_1$, is minimized. To this end, an *Insulation Index*, R_I, is defined as follows,

$$R_I = 20 \lg \left| \frac{\underline{F}_0}{\underline{F}_1} \right|. \tag{13.34}$$

After a short calculation—based on Fig. 13.14—one gets

$$\frac{\underline{F}_0}{\underline{F}_1} = \frac{r + j\omega m + \frac{1}{j\omega n}}{r + \frac{1}{j\omega n}}. \tag{13.35}$$

and, with

$$\omega_0 = \frac{1}{\sqrt{mn}} \quad \text{and} \quad Q = \frac{\omega_0 m}{r} = \frac{\omega_0}{\omega_\Delta}, \tag{13.36}$$

where $\omega_\Delta = r/m$ is the (angular) frequency bandwidth, we obtain the following expression,

$$R_I = 20 \lg \sqrt{\frac{\left(\frac{\omega}{\omega_0}\right)^2 + Q^2 \left[\left(\frac{\omega}{\omega_0}\right)^2 - 1\right]^2}{\left(\frac{\omega}{\omega_0}\right)^2 + Q^2}} \quad \text{dB}. \tag{13.37}$$

Figure 13.15 presents a plot of this function for various values of the quality factor, Q. Asymptotic values for the slopes of the curves are as follows.

– $Q > 1$ After the vanishing damping at $\omega_0/\omega = 1$ the damping becomes positive beyond $\omega_0/\omega = \sqrt{2}$ (13.37). By dividing by Q^2 and neglecting tiny members of the sums, we get $R_I \simeq 20 \lg \sqrt{(\omega/\omega_0)^4} \sim \omega^2$. This means that the slope approximates 12 dB/oct.

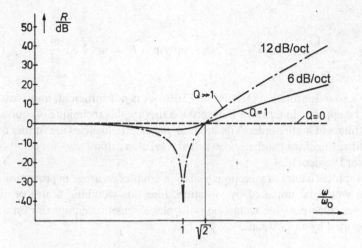

Fig. 13.15 Transmission loss of a single-mass vibrator as a function of frequency

- $Q = 1$ This is a reasonably-damped oscillating case. Neglecting small sum members again, we get $R_\mathrm{I} \simeq 20\lg\sqrt{(\omega/\omega_0)^2} \sim \omega$. The slope, beyond $\omega_0/\omega = \sqrt{2}$ is 6 dB/oct.

Note that R_I only becomes positive for values of $\omega/\omega_0 > \sqrt{2}$. To insulate a source of structure-borne sound, the support must therefore be tuned to a resonance frequency well below the operating frequency, which is usually the frequency of the exciting force. This technique is called *low-end tuning*. Usually one aims for a frequency ratio of 1/5–1/10.

The insulation index decreases with increased damping of the system because the damping acts as a *sound bridge*. However, moderate damping is still important because it limits the "*dancing*" effect experienced by vibrating system at the resonance frequency or close to it. This dancing effect also occurs when a rotating engine starts up and runs slowly through the resonance point of the support. It is thus wise to limit the movement at these frequencies with properly adjusted dashpots or even hard mechanic boundaries.

Many types of elastic elements are available used for vibration insulation, including steel springs, rubber-metal-compound elements, rubber, and synthetic-foam plates. Progressive spring characteristics like those provided by rubber help avoiding dancing of the source.

If the elastic element in use is known to have a linear characteristic, the resonance frequency, ω_0, is determinable from the amount, x_Δ, that the element compresses under a load. This is a handy on-site check of whether the right elastic elements have been installed or if there are accidental sound bridges. The relevant formula is

$$\frac{f_0}{\mathrm{Hz}} \approx \frac{5}{\sqrt{x_\Delta/\mathrm{cm}}}, \qquad (13.38)$$

with

$$\omega_0 = \frac{1}{\sqrt{m\,n}}\,, \quad n = \frac{x_0}{F}\,, \quad \text{and applying }\; F = m \cdot 9.81\,\frac{m}{s^2}\,. \qquad (13.39)$$

When a sound-insulation increase of 12 dB/oct is not sufficient, multi-layer supports are be applied. In terms of transmission theory, this technique is equivalent to low-pass filters of higher order. Vibrations at distinct frequencies are further reduced with tuned resonating absorbers, also called *vibration extinctors*—for details see the solution for Problem 13.4.

The reciprocal nature of vibration insulation enables its usage in protecting equipment that would be impaired by vibration, like fine-weighing scales or electron microscopes. One possible technique is to place sensitive equipment on a heavy table supported by air cushions.

13.7   Insulation of Floors with Regard to Impact Sounds

In architectural acoustics *tapping sound*, that is, structure-borne sound excited by walking on floors (*footfall*), is of particular relevance. Such sound is transmitted to adjacent rooms, especially below the floor, and then radiated as airborne sound. This effect also holds for other approximately point-wise sources of excitation like knocking, falling items, pushed chairs, or household appliances.

Insulation against this kind of sound are accomplished with resilient floor coverings, such as carpets, plastics layers, or wooden plates, under-packed with a compliant layer. We estimate the achieved insulation in the same manner as described earlier. The quantity used as "mass" in such an estimation is that part of the total source mass that is in direct contact with the floor, and the value used as compliance is the one given by the contact area.

The *floating* or *resilient floor* is another efficient method for reducing tapping sounds. As sketched in Fig. 13.16, such a floor consists of a load-distributing plate, possibly made of cast concrete, supported by a compliant layer, usually made of mineral wool or synthetic elastic foam.

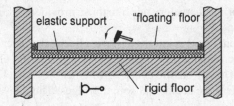

Fig. 13.16 Floating-floor construction for insulation against tapping sounds

The assembly consists of two plates in which bending waves may propagate, the floor itself and the concrete plate on top of it. The latter two couple complexly with the elastic under-packing. Evidently, the bending stiffness of the top plate negligible when certain conditions are met. These conditions are: The top plate is considerably thinner than the floor itself, and the under-packing has a high enough flow resistance to dampen propagation in it.

In most cases, the same formulas are applicable that we have already derived for insulation with concentrated elements—see Sect. 12.5. The resonance frequency of the floating floor is thus determined by

$$f_0 = \frac{1}{2\pi}\sqrt{\frac{s''}{m''}} \quad \text{with} \quad s'' = \frac{1}{n''}, \tag{13.40}$$

where m'' is the mass load and s'' the dynamic stiffness, which is the inverse of the dynamic compliance, n''. The dynamic stiffness of a resilient layer is measured according to international standards. It contains the stiffness of the enclosed air, which accounts for the fact that isothermal compression dominates within the layer.

In architectural acoustics, the resonance frequency, f_0, should be chosen to be well below 100 Hz, so that human speech is sufficiently insulated. Extreme care must be taken to prevent rigid contact of the floating plate with the rigid floor or the walls. Such contacts, called *sound bridges*, impair the insulation significantly. Just one small sound bridge easily reduces the insulation by more than 10 dB.

13.8 Exercises

Bending Waves and Sound Insulation

Problem 13.1 Bending waves are propagating on a single-leaf wall of infinite size.

(a) Derive a differential equation for bending waves.

(b) Calculate the sound-transmission loss of a single-leaf wall for a plane wave of oblique-incidence by applying the above-derived differential equation.

(c) Plot and discuss the sound transmission loss, R, in (13.17) as a function of frequency.

Problem 13.2 Show the transmission loss of a single-leaf wall as in (13.17) with a mass load, m'', a bending stiffness, B', and exhibiting a coincidence frequency of

$$f_c = \frac{1}{2\pi}\sqrt{\frac{m'' c^4}{B' \sin^4 \theta}}. \tag{13.41}$$

(a) What is the lower limit of the coincidence frequency?

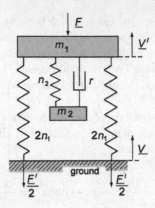

Fig. 13.17 Dynamic absorber

Table 13.1 Sound transmission losses of two different wallboards, measured in standard test chambers

f/Hz	100	125	160	200	250	315	400	500	630
TL_1/dB	24.4	24.1	22.9	20.7	23	25.3	26.9	30.6	32.2
TL_2/dB	22.5	23.5	23.2	21.1	23.6	22.6	21.2	21.7	22.9
f/Hz	800	1000	1250	1600	2000	2500	3150	4000	5000
TL_1/dB	34.7	39.9	43.9	47.4	47.3	46.4	45.1	44.9	47.4
TL_2/dB	24.1	27.7	31.3	34.4	35.6	35	34.2	31.7	35.2

(b) Use the normal incidence case or bending-compliant walls to explain the *mass law*.

Problem 13.3 Develop an equivalent electric circuit for a double-leaf wall and determine the transmission loss. How does the transmission loss behave above the main resonance, the so-called *drum resonance*?

Problem 13.4 To isolate a foundation against vibrations of mass M, a dynamic absorber according to Fig. 13.17 is used.

Calculate the transfer ratio $\underline{F}/\underline{F}'$ when the mass is excited to oscillate.

Problem 13.5 A bending wave, $\underline{v}_z(y) = \hat{\underline{v}}\,e^{-j\beta_b\,y}$, propagates on a sheet of infinite extension along y–direction. The width of the sheet is l.

(a) Determine the sound radiated from the sheet at a large distances from the sheet.

(b) What kind of directional characteristic is shown by the radiation?

Problem 13.6 Two different wallboards were tested in standard sound-transmission-loss chambers. Table 13.1 lists their sound transmission losses.

Determine the *Sound-Transmission Class* in compliance with the ASTM-E90 standard.

Chapter 14
Noise Control—A Survey

So far in this book, we dealt with sound sources that convert electric energy into mechanic/acoustic energy. These sound sources are mainly used to radiate desired sound events because they are conveniently controlled by electric signals.

However, in addition to desired sound events, there are undesired sounds that one would reduce or even eliminate if possible. This kind of sound is called *noise*.[1] A reasonable definition of noise in acoustics must cover various aspects, such as the following one.

Noise is audible sound that disturbs quietness, or the reception of intentional sound, or leads to damage, annoyance or health impairment.

In our engineered and industrialized environment, noise has become an increasing problem because it tends to be the by-product or "garbage" of technological processes. Industrial machines and appliances that are not treated with special noise-control measures radiate 1–10^0/$_{00}$ of their driving power as sound. Although it is only a small fraction of the total power which is converted into sound, considerable sound-power values may still be reached. Table 14.1 lists the sound power of some typical sound sources.

The sound-power level, L_p, is a temporal and spatial average. In a free sound field, we determine it by integrating across all directions of radiation. In a reverberant space, the diffuse-field is measured as described in Sect. 12.4. Although the sound-pressure level at a certain point in space is generally derivable from the sound-power level of the source, this is usually not a simple process.

---

[1] In signal theory, the same term is used for stochastic signals that are not prone to carry information.

© Springer-Verlag GmbH Germany, part of Springer Nature 2021
N. Xiang and J. Blauert, *Acoustics for Engineers*,
https://doi.org/10.1007/978-3-662-63342-7_14

Table 14.1 Sound power of some typical sources

Source type	$\overline{P}$	L_w re 10^{-12} W
Space rocket (Saturn)	$>10^7$ W	>190 dB
Jet airplane	10^4 W	160 dB
Large brass orchestra	10 W	130 dB
Large machine tool	1 W	120 dB
Passenger cars on highway	10^{-2} W	100 dB
Normal speech	10^{-5} W	70 dB
Soft whispering	10^{-9} W	30 dB

14.1   Origins of Noise

Reasons why noise is generated are manifold. Some representative examples of noise sources are listed below—ordered according to their sound excitation types.

Airborne-sound sources

– Excitation by explosion or implosion $\longrightarrow$ impulsive sounds
– Excitation by turbulent flow $\longrightarrow$ non-periodic sounds
– Excitation by intermittent flow, for example, siren, car exhaust $\longrightarrow$ periodic or quasi-periodic sounds

Structure-borne-sound sources

– Excitation by stroke or knock $\longrightarrow$ impulsive sounds
– Excitation by friction $\longrightarrow$ non-periodic sounds
– Excitation by periodic forces, such as magnetic or electric ones $\longrightarrow$ periodic or quasi-periodic sounds

In airborne-sound sources, forced or free vibrations are excited in air volumes. structure-borne sound sources are excited by forced or free vibrations. Such structural vibrations are capable of radiating airborne sound, particularly when large plates like walls, shells or housings are vibrating.

14.2   Radiation of Noise

Radiation of noise follows the general laws of sound radiation as discussed throughout this book, especially in Chaps. 9, 10, and 11. Radiation occurs according to specific directional characteristics that depend on the form of the air volumes or structures and the type of vibration they experience.

The following section discusses sound radiation by line arrays of non-coherent sound sources and the role of meteorological conditions, two items that we did not

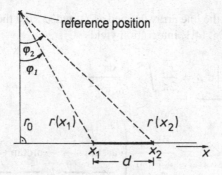

Fig. 14.1 Sound radiation from a line array to a reference position

consider so far. Both items also relate to intentional-sound radiation, but there they are less present.

Line Arrays of Incoherent Sources

The significant difference between arrays of coherent and incoherent sound sources is that no interference occurs in the latter case. For this reason, it is unnecessary to consider the phase of partial sound pressures when computing the sound field. Partial intensity magnitudes add up instead—explained in Sect. 1.6.

We consider, as an example, a line array that is densely occupied by incoherent point sources of the 0th order, positioned on a reflecting surface. Such arrays serve, for instance, as models for heavily used roads.

Figure 14.1 shows a schematic plot of a line array. We assume that it carries a constant load of active sound power, $\overline{P'}$, which is equivalent to a constant power per length.

We now choose a reference position in the far-field at which we want to determine the sound intensity. The total intensity at this point in space, composed of the contributions of all differential line sections, $\mathrm{d}x$, is

$$\frac{p_{\text{rms}}^2}{\varrho\,c} = \left|\vec{I}\right| = \overline{I}\,. \tag{14.1}$$

Each element, $\mathrm{d}x$, contributes a partial intensity of

$$\mathrm{d}\left(\int \left|\vec{I}\right| \mathrm{d}\varphi\right). \tag{14.2}$$

Because the reflective floor restricts us to the upper hemisphere, that is, a spatial angle of 2π, the expression above extends into

$$\mathrm{d}\left(\int \left|\vec{I}\right| \mathrm{d}\varphi\right) = \frac{\overline{P'}\,\mathrm{d}x}{2\,\pi\,r^2(x)} = \frac{\overline{P'}\,\mathrm{d}x}{2\,\pi\,(r_0^2 + x^2)}\,. \tag{14.3}$$

With the end points of the line array defined as x_1 and x_2, and the length being equal to d as illustrated in Fig. 14.1, integration yields

$$\underbrace{\int_{\varphi_1}^{\varphi_2} \left| \vec{I} \right| \mathrm{d}\varphi}_{I_\Sigma} = \frac{\overline{P'}}{2\pi} \int_{x_1}^{x_2} \frac{\mathrm{d}x}{r_0^2 + x^2}$$

$$= \frac{\overline{P'}}{2\pi} \frac{1}{r_0} \left(\underbrace{\arctan \frac{x_2}{r_0}}_{\varphi_2} - \underbrace{\arctan \frac{x_1}{r_0}}_{\varphi_1} \right). \qquad (14.4)$$

Very-long and very-short arrays are two especially noteworthy cases. We discuss them in the following.

Very-long line array

In the case of $r_0 \ll d$, the difference of the apparent angles, $\varphi_2 - \varphi_1$, becomes constant and independent of r_0. An important case is $\varphi_2 - \varphi_1 = \pi$, for which we obtain

$$I_\Sigma = \frac{\overline{P'}}{2} \frac{1}{r_0}, \qquad (14.5)$$

amounting to $-3\,\mathrm{dB}$ per distance doubling. This is half the attenuation rate of $-6\,\mathrm{dB}$ associated with a spherical source. The reason is that the situation equates cylindrical radiation, and the shell area of a cylinder increases proportionally to the distance, r, from the center axis.

Very-short line array

Figure 14.2 illustrates that in the case of a very-short array, that is, $r_0 \gg d$, the following approximation applies,

$$\tan(\varphi_2 - \varphi_1) \approx \varphi_2 - \varphi_1 \approx \frac{d}{r_0} \cos^2 \varphi. \qquad (14.6)$$

Insertion yields

$$I_\Sigma \approx \frac{\overline{P'} d}{2\pi r^2}, \qquad (14.7)$$

indicating a decrease of $6\,\mathrm{dB}$ per distance doubling, with r being the average distance. We thus deal with very-short line arrays as 0th-order spherical sources with a sound power of $\overline{P} = \overline{P'} d$.

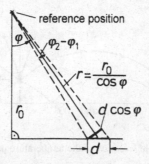

Fig. 14.2 Sound radiation from a very short line array to a reference position

Influence of Meteorological Conditions

Temperature and wind profiles are the most relevant meteorological conditions. We discuss them in the following.

Temperature profiles

The speed of sound, c, is the speed at which a wave front propagates. It depends on the temperature of the medium according to

$$c \sim \sqrt{T} \sim \sqrt{1/\varrho}. \tag{14.8}$$

Refraction always deflects sound into the colder and thus denser medium—discussed in Sect. 11.3. The Figs. 14.3, 14.4, and 14.5 illustrate three typical cases.

In Fig. 14.3, we see a normal temperature profile as may go along with a sunny day, that is, with the ground heated up. The sound is reflected upwards, creating a shadow zone. Sound from sources beyond an *acoustic horizon* does not come across.

Figure 14.4 shows a case of *inversion*, which is characterized by a reversed sign in the gradient of the temperature profile. A situation like this one may occur on a clear summer night when the ground has already radiated its heat while the air is still warm. Then sound propagates over very long distances. It is under such conditions that African *drum telephony* operates.

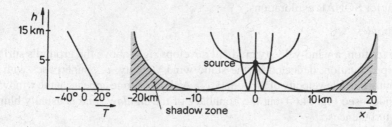

Fig. 14.3 Sound propagation in a normal temperature profile

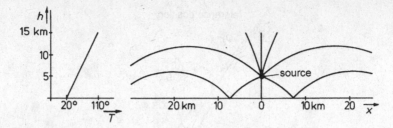

Fig. 14.4 Sound propagation while *inversion* of the temperature profile occurs

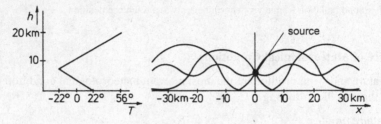

Fig. 14.5 Sound propagation in a *sound channel*

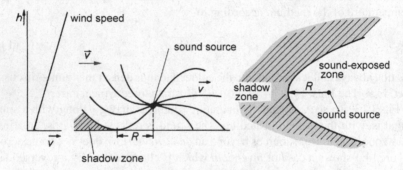

Fig. 14.6 Sound propagation under the influence of wind

Figure 14.5 illustrates the *sound-channel* effect, which may occur early in the morning. Sound channels are frequent in underwater sound environments. They are relevant for SONAR explorations.

Wind profiles

Due to friction, a wind-velocity gradient develops right above the ground's surface. Sound propagation, deflected by the static wind velocity, v_-, superposes with the alternating particle velocity of the sound, $v_\sim$. A shadow zone appears at the windward (luff) side—see Fig. 14.6—but the boundary of the *shadow zone* is usually blurred due to turbulences.[2]

---

[2] It is worth noting that the influence of wind leads to violation of the principle of reciprocity.

14.3 Noise Reduction as a System Problem

Noise, generated and radiated from one or more sound sources propagates over one or multiple paths before arriving at one or more receivers (sinks). It seems natural to discuss this situation as a *system* in transmission-theory terms. Noise control in such systems benefits from knowledge of the structure of the system and the characteristics of their components, including sources, transmission channels and receivers.

Figure 14.7 illustrates the situation of a person exposed to noise from a rotating engine on the floor of an enclosed machine room. Sound emission happens, (1), by airborne sound radiation from vibrating surfaces, (2), by turbulent and intermittent gas flows and, (3), by the vibrating floor. The sound arrives at the person via several different propagation paths—schematized in Fig. 14.8. The planning of noise-control usually includes three categories of tasks, namely, system analysis, goal setting, and decision making. More details are listed below.

System Analysis

The main steps in system analysis are

– Identification of the sources, investigation of the sound-generation mechanisms, and determination of sound powers and directional characteristics

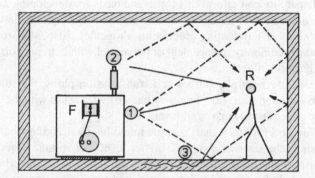

Fig. 14.7 Different kinds of noise radiation and different propagation paths indoors

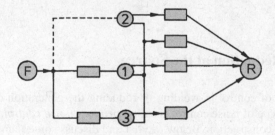

Fig. 14.8 The topology of the system—illustrated for system analysis

– Identification and surveying of the sound-transmission paths. In practice, it is often useful to explore airborne-sound and structure-borne-sound paths separately.
– Investigation of the noise effect on the exposed receiver.

Receiver-Related Limits of Exposure

If the exposed receivers are human beings, the following criteria are important when settling on limiting values.

– Just acceptable interference with communication
– Just acceptable annoyance
– Protection against hearing damage
– Protection against damage of further organs

Applicable laws, standards, and work-protection codes set limits. Yet, it is often difficult to predict the reliability of such limits, particularly in cases of annoyance due to noise that conveys information.

Adequate Noise-Control Measures

When making decisions, it is useful to list all possible measures and weight them according to feasibility and effort. This forms the basis for developing an appropriate battery of noise-control measures. In this context, it is reasonable to consider replacing machines or terminating production altogether. Also, staff-related solutions like frequent temporary replacement of personnel with extreme noise exposure help mitigate harmful effects.

When considering multiple sources and transmission paths, the rules of level addition—introduced in Sect. 1.6—requires that we always start with the component that contributes the most to the overall level.

Annoying low-level sound signals become inaudible by masking them with more pleasant sounds at the same or a slightly higher level. For instance, an open-office space, the noise of the air-conditioning system serves to mask disturbing voices. Also, the diffuse bubble of voices as is typical for restaurants, improves the situation for people at one table by making talk from other tables unintelligible, thus providing a sense of *privacy*.

14.4   Noise Reduction at the Source

The best method of control is avoiding or reducing the generation of noise in the first place. This type of noise control is called *primary noise control*. It starts at the earliest phases of construction. Below, we list and discuss some relevant aspects.

Avoidance/Reduction of the Excitation of Airborne Noise

- *Explosions* and *implosions* are usually unavoidable, but their temporal and spectral noise properties may be alterable, for example, through redesigning combustion chambers.
- *Turbulent flow* creates noise levels that increase with flow speed with a proportionality of up to the 8th power. Reduction of flow speed is therefore of paramount importance. It is accomplished by using ducts with larger cross-sections and avoiding constrictions and edges. Flow-compliant profiles are applied instead when possible.
- *Intermittent flows* should not be coupled to resonant cavities.

Avoidance/Reduction of Excitation of Structure-Borne Noise

- *Jolts* are avoidable with constructive measures that make the kinetic sequences steady up to their 3rd derivatives. If this is impossible, the moving masses and their velocities might be reducible. Further relevant precautions include insertion of elastic layers, allowing for some slack in power trains, reduction of slackness where components hit each other, balancing of rotation elements, and provision for phase compensation by dynamic vibrators—for details of such absorbers, see the solution of Problem 13.4.
- *Friction* is reducible with lubrication and/or high-quality surfaces. When choosing materials, consider their acoustic characteristics. Materials with elasticity and high internal losses are usually preferable.
- *Periodic forces* are minimized by careful balancing, provided that they are mechanically induced. When excited by electric or magnetic forces, proper mechanical construction and electric-signal processing are capable of reducing such forces.

Significant progress in sound control is often achieved by modifying processes. Thus, this alternative is worthy of considering.

Avoidance/Reduction of the Radiation of Structure-Borne Noise as Airborne Noise

Engine elements and housings are preferably designed for low airborne-sound radiation. Minimizing and perforating vibrating surfaces is effective. It also helps divide surfaces so that knots of vibration result. Further, resonances are avoided by adding mass, using plates that are heavy but compliant for bending waves, applying coatings with absorbent layers and inserting viscous layers for dampening.

14.5 Noise Reduction Along the Propagation Paths

Noise control along propagation paths is called *secondary noise control*. We now give an overview with respect to air- and structure-borne noise, organized according to the three main noise-reducing effects on the propagation paths, namely, distribution, reflection and absorption.

Reduction by Distribution

During propagation, sound usually spreads to fill out geometrically larger areas. As a result, the intensity and sound-pressure level decrease with increasing distance from the source. Airborne noise displays this behavior not only for free-field propagation but also for guided propagation as occurs in branching ducts. Diffraction and scattering may also support distribution by breaking up focused beams. Geometric distribution is also a factor for structure-borne noise, where it occurs in plates or at the edges and joints of structures. In canals or rods with constant or decreasing cross-sectional area and in enclosed spaces, distribution does not take place. Beyond the *diffuse-field distance*, the sound field in enclosed spaces is diffuse and, thus, statistically stationary. The diffuse-field distance depends on the directional characteristic of the source—refer to Sect. 12.6. The more the sound is focused, the longer the critical distance.

The maximum possible level decrease beyond the critical distance due to geometric distribution occurs with 0th-order spherical sources in the free field. The decrease then amounts to 6 dB per distance doubling. However, only 3 dB per distance doubling is achieved for extended line arrays and for small sources in shallow rooms because of the geometry of the cylindrical wavefronts that they produce.

Whenever the situation allows for it, consider proper geometric placement of the sound sources relative to the receivers as an effective approach to noise control.

The farther, the better!

Reduction by Reflection

Reflection provide insulation from airborne noise by creating shadow zones behind occlusions like walls and barriers. Refraction at boundaries and inhomogeneities also supports insulation as do meteorological conditions. Structure-borne sound is reflected at boundaries where the impedance changes. The most effective noise-reduction measure during propagation is reflection, implemented by encapsulating the source with non-porous walls and other dividers for airborne sound, or with elastic layers for structure-borne sound.

Complete encapsulation,[3] however, is often not feasible, especially outdoors or in open-space offices. In these cases, noise barriers are less effective. The shadowing ability of such barriers is reduced by diffraction, so that their maximum insertion loss for airborne sound is only about 20 dB.

Diffraction about a barrier is mathematically described by assuming a line array at the rim of the barrier, but calculation of the effect is tedious. For this reason, many noise-control standards provide approximate procedures for estimating the insertion loss. The insertion loss, D_i, increases monotonically with the relative lengths of the detour around the barrier and the wavelength of the sound, λ. An example for an empirical formula corresponding to Fig. 14.9 is

$$D_i = 10 \lg (20.4 \, k_f + 3) \ \ \text{dB} \,, \tag{14.9}$$

where k_f is the *Fresnel number*

---

[3] To increase their efficiency, capsules should have some sound-absorbent material inside.

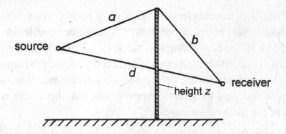

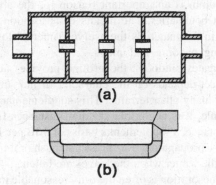

Fig. 14.9   Sound barrier

Fig. 14.10   Two muffler types (**a**) Reflection muffler. (**b**) Comb-filter muffler

$$k_f \doteq \frac{(a+b)-d}{\lambda/2}.$$ (14.10)

Formula (14.10) requires that the sound source be a 0th-order spherical source, that $k_f > -0.1$, and that the by-pass around the edges of the barrier, y, is long enough. Concerning the necessary length the following rule is helpful—with z as the barrier height, we require

$$\text{either} \quad y > 5z \quad \text{or} \quad y > z + \lambda.$$ (14.11)

On the side of the barrier facing the source, a level increase of up to 6 dB may occur due to reflection from the barrier. This effect is avoidable by equipping the barriers with absorbent surfaces.[4]

An example of a case where barriers are not viable is often given by ducts that carry flowing media, like exhaust systems or air-conditioning ducts. Non-porous walls cannot be applied there because the media flow must not be interrupted. Impedance changes—and thus reflection—may still be implemented in several ways, including coupled, cascaded, or branching resonators, detours, by-passes, and steps in the cross-sectional areas—refer to Sects. 8.5 and 8.6.

---

[4] Barriers in enclosed spaces are only sensible if the ceiling above them is absorbent.

The design of mufflers benefits from the theory of transmission-line and reactance filters. Reflection and interference are employed to engineer mufflers to spectral specification. See Fig. 14.10 for two examples, namely, a low-pass and a comb filter. In reflection mufflers, different chambers with different volumes increase the effective bandwidth. Comb-filter mufflers are delicate since interference minima (comb-filter minima) are narrow and hard to adjusts. Appropriate placing of the mufflers in duct systems is relevant for their effectiveness.[5]

Reduction by Absorption

Dissipation in the medium is an important reason for the absorption of airborne sound, in particular at boundaries. The means of absorption for structure-borne sound are dissipation in the materials themselves (internal losses) and absorption due to absorbent coating, padding, and inlays.

When sound propagates outdoors, the ground provides additional absorption. The nature of this effect depends on the character of the surface, which may be lawn, bushes, trees, or many other terrains. This supplementary absorption is often overestimated. As a rule, it is negligible at small distances from the source, $r <$ 100 m. Beyond that distance, a supplementary loss of 8 dB per 100 m is a reasonable estimate for traffic noise propagating over grass and bush or through woodland. This loss is smaller during the winter when the leaves are fallen.

Noise reduction by absorption is often the only reasonable measure to be taken in enclosed spaces with distributed sound sources, like machine-floor halls. For diffuse sound fields we refer to (12.35), according to which

$$p^2_{\mathrm{d, rms}} = \frac{4 \overline{P} \varrho c}{A}, \tag{14.12}$$

holds for the sound pressure. This means that a doubling of the equivalent absorbent area results in a 3 dB decrease of the sound-pressure level in the diffuse sound field.

Additional absorption is more effective the more reverberant the considered space initially was!

In order to dampen structure-borne sound, panels are either coated with an absorbent *anti-boom* material or provided with an internal absorbent layer, producing what we call a *sandwich panel*.

To abate sound propagation in ducts, *absorption mufflers* are applied. Figure 14.11 shows two examples. The absorbent material is either placed on the walls or in the duct as absorbent dividers.[6] In the following, we approximate the sound propagation in a duct with absorbent walls by assuming that the cross-sectional areas are small compared to the wavelengths. The boundaries are taken as locally-reacting—refer to Sect. 11.4.

---

[5] Note that due to reflection the sound level may increase at muffler's input ports.

[6] Filling the duct with absorbent (porous) material is a condition called a *throttling muffler*, accepting that media flow is affected.

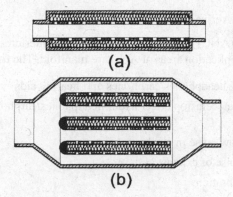

Fig. 14.11 Absorption in mufflers (**a**) Duct with absorbent lining. (**b**) Duct with absorbent dividers

If U is the perimeter and A the cross sectional area, the sound power transmitted into the walls is

$$\mathrm{d}\overline{P}_{\text{wall}}(x) = \frac{p_{\text{rms}}^2(x)\, U\, \mathrm{d}x}{\text{Re}\{\underline{Z}_{\text{wall}}\}} = I_{\text{wall}}(x)\, U\, \mathrm{d}x\,. \tag{14.13}$$

The transmitted power is equal to the incremental loss of axially propagating power, $-\mathrm{d}(I\,A)$, so that

$$\frac{p_{\text{rms}}^2(x)\, U\, \mathrm{d}x}{\text{Re}\{\underline{Z}_{\text{wall}}\}} + \frac{\mathrm{d}[\,p_{\text{rms}}^2(x)]\, A}{\varrho\, c} = 0\,. \tag{14.14}$$

Rewriting this expression yields

$$\frac{\mathrm{d}[\,p_{\text{rms}}^2(x)]}{\mathrm{d}x} + \frac{\varrho\, c}{\text{Re}\{\underline{Z}_{\text{wall}}\}}\, \frac{U}{A}\, p_{\text{rms}}^2(x) = 0\,, \tag{14.15}$$

which, for the forward progressing wave, has the solution

$$p_{\text{rms}}^2(x) = p_{\text{rms}\,\rightarrow}^2\, \mathrm{e}^{-2\check{\alpha}x}\,. \tag{14.16}$$

We now approximate the complex solution for harmonic functions as

$$\underline{p}_{\rightarrow}(x) = \underline{p}_{\rightarrow}\, \mathrm{e}^{-\check{\alpha}x}\mathrm{e}^{-\mathrm{j}\beta x}\,, \tag{14.17}$$

with

$$\check{\alpha} = \frac{1}{2}\, \frac{\varrho\, c}{\text{Re}\{\underline{Z}_{\text{wall}}\}}\, \frac{U}{A}\,. \tag{14.18}$$

This expression denotes a spatially-damped sound wave.

Noise Reduction by Active Noise Control (ANC)

Active noise control (ANC) generates "anti-noise" that interferes destructively with the original noise. Application areas of ANC are manifold. The following lists a few.

- Noise-canceling headphones/earphones and hearing aids
- Reduction of the occlusion effect under headphones/earphones and hearing aids
- Reduction of flow noise in air-ducts
- Reduction of noise of exhaust system
- Reduction of vibrations
- Reduction of the humming sound of electrical transformers
- Provision of quite zones in noisy spaces,[7] for example, inside aircraft or passenger cars

Figure 14.12 shows two typical application examples with *feed-forward control*— see Fig. 14.13a. The microphone (1) picks up the noise and, after phase inversion and an eventual delay, the processed noise is added again at the point in space, (2), at which the suppression aims. Forward suppression is particularly effective against low-frequencies. Up to about 1.5 kHz, reductions of −10 to −20 dB are typical. However, the method is sensible to unexpected variation of the setting, such as the individual fit of headphones or hearing aids. Slight deviations in amplitude and phase already result in a severe performance degradation.

To overcome these issues, adaptive methods are available—see Fig. 14.13b. They require an additional microphone, (3), at the point of intervention. The output of

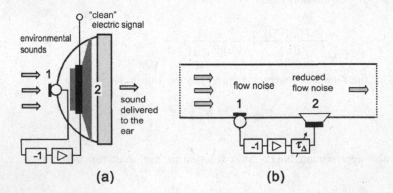

Fig. 14.12 Examples of active noise control (**a**) Sound-canceling headphone. (**b**) Ventilation duct with active noise reduction

---

[7] Active noise control in 3–D spaces requires multi-sensor and multi-loudspeaker settings.

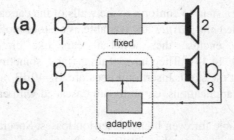

Fig. 14.13 Active-noise-control structures (a) Feed-forward. (b) Adaptive

microphone (3) is feed into an adaptive filter that, after comparison with the noise at point (1), renders an output that is optimized to reduce the noise at point (2).[8]

Noise Reduction at the Receivers' End

Due to reciprocity, the techniques discussed in the current section are also applicable for reducing noise at the receiving end. Human beings are critical receivers in this context.[9]

For them, personal protection against noise often provides the quickest and cheapest noise-control solution. Unfortunately, there is an issue of acceptance since wearing hearing protectors may cause discomfort. In the following, we list common types of personal hearing protectors.

Personal Hearing Protectors

Personal hearing protectors come in a variety of forms—see Fig. 14.14.

Earplugs are a customary product for insertion losses of 20–30 dB. They are made of synthetic foam (vinyl or polyurethane), silicone, or plastic. The user compresses the foam ones by twisting them between the fingers before insertion into the ear canal. There they expand again to fit. The silicone ones consist of a tapered set of

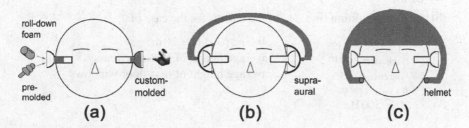

Fig. 14.14 Personal hearing protectors (a) Ear plugs. (b) Ear muffs. (c) Helmets.

---

[8] Adaptive filters process statistical parameter of the noise. For example, least-mean-square, recursive-least-square and *Kalman* filters are in use.

[9] Long-time exposure to noise levels above 85 dB in the mid-frequency range causes permanent impairment of hearing. These levels are not at all uncommon in daily life, especially in places like discotheques or when using headphones. Inner organs are prone to levels above 130 dB causing symptoms like nausea and vertigo.

flanges spaced along a stem to conform to the walls of the ear canal when inserted. The plastic ones, called *oto-plastics* are available as custom-molded—Fig. 14.14a.

Earmuffs usually enclose the   external ears like circum-aural (closed) headphones—see Fig. 14.14b. They have sealing rings, sometimes filled with liquid to tighten against the skull. Insertion losses are 10–40 dB, increasing with frequency. For specific applications, earmuffs are cascaded with earplugs to enhance the insertion loss.

Helmets are available for even higher insertion losses. Special implementations achieve insertion losses of more than 50 dB at higher frequencies—see Fig. 14.14c.

Sound- and vibration-protective suits are disposable to protect inner organs. For instance, astronauts carry them during the start phase of their spacecraft.

To increase their insertion loss by an additional 10–25 dB, active-noise-control is applicable to hearing protectors—see the ANC section above.

14.6   Exercises

Noise Reduction

Problem 14.1   Find a formula for the sound intensity level of

(a) A very long incoherent line source

(b) A very short incoherent line source

assuming that the line sources carry a constant power load.

Problem 14.2   Consider a noise-barrier arrangement as depicted in Fig. 14.15.

(a) For a, $b \gg h_{barrier} - \min(h_{sender}, h_{receiver})$, find an adequate approximation formula for the insertion loss, D_i, for example, from applicable standards.

(b) Find the insertion loss, D_i, of the barrier for the case of

$a = 14$ m   ...4–lane road $b = 20$ m
$h_{sender} = 0.6$ m   ...average height of passenger cars
$h_{receiver} = 2$ m   ...average height of first-floor windows
$h_{barrier} = 4$ m
$f = 500$ Hz

(c) Find a formula to express the variation of D_i in dB when increasing the frequency by a factor of two or even more. Is this formula valid for averaged levels as well?

Problem 14.3   Consider the noise-protection arrangement of Fig. 14.15 for the following case. The source is an infinitely long line-array, and the noise barrier extends parallel to this line source.

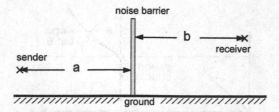

Fig. 14.15 Noise barrier

(a) Calculate the insertion loss D_i of this arrangement.
(b) Calculate the insertion loss D_i for the data listed under Problem 14.2(b) and compare both results

Problem 14.4 The wall of a square tube—see Fig. 14.16—is covered by porous material, namely, either cotton or mineral wool. The following material specification apply.

For *cotton*   …flow resistance, $\Xi = 1\,\mathrm{g\,cm^{-3}\,s^{-1}}$, porosity, $\sigma = 0.95$

For *mineral wool* …flow resistance, $\Xi = 100\,\mathrm{g\,cm^{-3}\,s^{-1}}$, porosity, $\sigma = 0.75$
Calculate the sound damping in the tube for the following conditions.

– The cross-section of the remaining space in the tube is small in comparison with the wavelengths such that the sound pressure across the sectional area is approximately constant.
– Any sound reflected from the rigid tube wall that passes twice through the porous material is considered negligible.
– The frequency of the sound to be damped is 1 kHz.

How thick must the porous-material lining be to fulfill the 2nd condition as to which the sound should at least be attenuated by 20 dB when passing the porous material twice?

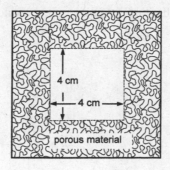

Fig. 14.16 Square tube with a lining of porous material

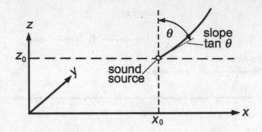

Fig. 14.17 Refraction of sound rays

Problem 14.5 Refraction is described by *Snellius'* law (*Snell's* law),

$$\frac{c}{\sin \theta} = c_{\text{trace}} \, . \tag{14.19}$$

(a) On the basis of this law, derive a general integral-equation for the trajectory of sound rays, $x = \int f(z)\,dz$, for the case that the sound speed is a sole function of height, that is, $c = c(z) \neq c(x, y)$. As sketched in Fig. 14.17, the source is in the position $\{x_0, z_0\}$. The sound ray has an initial slope of

$$\frac{dx}{dz}(z_0) = \tan \theta. \tag{14.20}$$

(b) Solve the integral-equation for the practice-relevant case of $c(z) = c_0 - \xi z$ with $\xi \geq 0$ or $\xi \leq 0$, and $x_0 = 0$.

(c) Calculate the acoustic horizon for $\xi > 0$, that is, the distance from the source from which on the source on the ground is no longer audible. Use the following data.

$\xi = 1 \cdot 10^{-2}/\text{s}$
$c_0 = 343.6\,\text{m/s}$
$z_0 = 2\,\text{m}$

Problem 14.6 A feed-forward noise-control loop as shown in Fig. 14.13a with an inherent lag of $57.5\,\mu\text{s}$ and a loop gain of $G = 1$ is given.

Determine the $-3\,\text{dB}$ limiting frequency up to which noise suppression is effective.

Chapter 15
Solutions to the Exercise Problems

15.1 Chapter 1

Problem 1.1.

Proposed Approach

Euler's formula reads as follows.

$$e^{j\phi} = \cos\phi + j\sin\phi. \tag{15.1}$$

A sinusoidal time-function with an amplitude $\hat{z}$ is expressible in terms of a complex-valued signal in exponential form as

$$\underline{z}(t) = \hat{z}e^{j(\omega t + \phi)}. \tag{15.2}$$

With *Euler*'s formula, we split the exponential part of $\underline{z}$ into a real part plus an imaginary part, that is,

$$\underline{z}(t) = \hat{z}[\cos(\omega t + \phi) + j\sin(\omega t + \phi)]. \tag{15.3}$$

Taking the real part of $\underline{z}(t)$, we obtain

$$Re\{\underline{z}(t)\} = \hat{z}\cos(\omega t + \phi), \tag{15.4}$$

which, as stated in (15.2), is a form in which any arbitrary sinusoidal signal is expressible.

Problem 1.2.

Proposed Approach

Start with applying *Ohm*'s law in complex form, which reads

© Springer-Verlag GmbH Germany, part of Springer Nature 2021
N. Xiang and J. Blauert, *Acoustics for Engineers*,
https://doi.org/10.1007/978-3-662-63342-7_15

$$\underline{u} = \underline{z} \cdot \underline{i}, \tag{15.5}$$

and, in terms of $\underline{i}$, becomes

$$\underline{i} = \frac{\underline{u}}{\underline{z}}. \tag{15.6}$$

First exert the complex notation to $u(t)$ to obtain $\underline{u}$,

$$\underline{u} = \hat{u}\, e^{j(\omega t + \phi_u)}. \tag{15.7}$$

Then denote the complex impedance of a resistor, z_R, of an inductor, $\underline{z}_L$, and of a capacitor, $\underline{z}_C$,

$$z_R = R, \quad \underline{z}_L = j\omega L, \quad \underline{z}_C = \frac{1}{j\omega C}. \tag{15.8}$$

Summation of these components because they are in series leads to

$$\underline{z} = R + j\left(\omega L - \frac{1}{\omega C}\right) = \hat{z}\, e^{j\phi_z}, \tag{15.9}$$

where the magnitude and the phase of the impedance are

$$\hat{z} = \sqrt{R^2 + \left(\omega L - \frac{1}{\omega C}\right)^2}, \tag{15.10}$$

$$\phi_z = \arctan \frac{\omega L - \frac{1}{\omega C}}{R}. \tag{15.11}$$

Substitution of (15.9) into (15.6) leads to

$$\underline{i} = \frac{\underline{u}}{R + j\left(\omega L - \frac{1}{\omega C}\right)}. \tag{15.12}$$

Decomposition of both $\underline{u}$ and $\underline{z}$ into their respective magnitudes, $\hat{u}, \hat{z}$, and their phase angles, ϕ_u, ϕ_z, renders

$$\underline{i} = \frac{\hat{u}\, e^{j(\omega t + \phi_u)}}{\sqrt{R^2 + (\omega L - \frac{1}{\omega C})^2}\; e^{j\phi_z}}. \tag{15.13}$$

Dividing the angular component of $\underline{u}$ by $\underline{z}$ results in

$$\underline{i} = \frac{\hat{u}}{\sqrt{R^2 + (\omega L - \frac{1}{\omega C})^2}}\, e^{j(\omega t + \phi_u - \phi_z)}. \tag{15.14}$$

Finally, the real part of $\underline{i}$ provides the electric current, $i(t)$, which is,

$$i(t) = Re\{\underline{i}\} = \frac{\hat{u}}{\sqrt{R^2 + (\omega L - \frac{1}{\omega C})^2}} \cos(\omega t + \phi_\mathrm{u} - \phi_\mathrm{z}). \qquad (15.15)$$

Problem 1.3.

Proposed Approach

The mechanic impedance is defined as

$$\underline{Z}_\mathrm{mech} = \frac{\underline{F}}{\underline{v}}, \qquad (15.16)$$

but practical testing often directly measures the acceleration, $\underline{a}$, at the point of interest via a suitable acceleration meter. The complex-valued velocity is found through the relations in (16.6), namely,

$$\underline{a} = \mathrm{j}\,\omega\,\underline{v}, \quad \underline{v} = \frac{\underline{a}}{\mathrm{j}\,\omega}. \qquad (15.17)$$

Now we express the force and acceleration in its complex form, that is,

$$\underline{F} = \hat{F}\,\mathrm{e}^{\mathrm{j}(\omega t + \phi_\mathrm{F})} = \hat{F}\,\mathrm{e}^{\mathrm{j}\omega t} \cdot \mathrm{e}^{\mathrm{j}\,\phi_\mathrm{F}}, \qquad (15.18)$$

with $\hat{F}$ and ϕ_F being the peak value and the phase of the force, respectively, and

$$\underline{a} = \hat{a}\,\mathrm{e}^{\mathrm{j}(\omega t + \phi_\mathrm{a})} = \hat{a}\,\mathrm{e}^{\mathrm{j}\omega t} \cdot \mathrm{e}^{\mathrm{j}\,\phi_\mathrm{a}}, \qquad (15.19)$$

with $\hat{a}$ and ϕ_a being the peak value and the phase of the acceleration.
By use of

$$\underline{v} = \frac{\hat{a}\,\mathrm{e}^{\mathrm{j}\omega t} \cdot \mathrm{e}^{\phi_\mathrm{a}}}{\mathrm{j}\,\omega}, \qquad (15.20)$$

the mechanic impedance becomes

$$\underline{Z}_\mathrm{mech} = \mathrm{j}\,\omega\,\frac{\hat{F}}{\hat{a}}\,\mathrm{e}^{\mathrm{j}(\phi_\mathrm{F} - \phi_\mathrm{a})} = \frac{\omega\,\hat{F}}{\hat{a}}\,\mathrm{e}^{\mathrm{j}(\phi_\mathrm{F} - \phi_\mathrm{a} + \frac{\pi}{2})}. \qquad (15.21)$$

Problem 1.4.

Proposed Approach

We apply as reference pressure

$$p_{0,\mathrm{rms}} = 20\ \mu\mathrm{N/m}^2 = 2 \cdot 10^{-5}\ \mathrm{N/m}^2. \qquad (15.22)$$

The sound-pressure level (SPL), p_pr, is given by

$$p_{pr} = 20 \, \log_{10} \frac{p_{rms}}{p_{0,rms}} \,. \tag{15.23}$$

With these, we obtain

(a) For $0.1 \, N/m^2$,

$$L_p = 20 \log_{10} \frac{0.1}{(2 \cdot 10^{-5})} = 73.98 \, dB, \tag{15.24}$$

For $10^2 N/m^2$,

$$L_p = 20 \log_{10} \frac{10^2}{(2 \cdot 10^{-5})} = 133.98 \, dB. \tag{15.25}$$

(b) For $0.3 \, N/m^2$,

$$L_p = 20 \log_{10} \frac{0.3}{(2 \cdot 10^{-5})} = 83.52 \, dB, \tag{15.26}$$

Consequently, for $3 \cdot 10^2 N/m^2$, the result is

$$L_p = 20 \log_{10} \frac{3 \cdot 10^2}{(2 \cdot 10^{-5})} = 143.52 \, dB. \tag{15.27}$$

Problem 1.5.

Proposed Approach

The logarithmic-frequency ratio for semitones is defined as

$$\Psi_{semitone} = 12 \, ld \left(\frac{f_1}{f_0} \right), \tag{15.28}$$

where $ld(\cdot) = \log_2(\cdot)$ is the *logithmus dualis*.

To find the frequency ratio, f_1/f_0, for one semitone from

$$\Psi_{semitone} = 1 = 12 \, ld \left(\frac{f_1}{f_0} \right), \tag{15.29}$$

we write

$$\frac{1}{12} = ld \left(\frac{f_1}{f_0} \right), \tag{15.30}$$

leading to

$$\frac{f_1}{f_0} = 2^{\frac{1}{12}}. \tag{15.31}$$

Thus, the frequency ratio of a 1-semitone interval comes out as $2^{\frac{1}{12}}/1$.

For a 3-semitone interval the frequency ratio is

$$\left(\frac{2^{\frac{1}{12}}}{1}\right)^3 = \frac{2^{\frac{1}{4}}}{1} = \frac{1.189}{1}. \tag{15.32}$$

The logarithmic-frequency interval for a 3-semitone interval then results in

$$\Psi_{\text{semitone}} = 12 \, \text{ld}(1.189) = 12 \frac{\log(1.189)}{\log 2} = 3. \tag{15.33}$$

For a 12-semitone interval the frequency ratio is

$$\left(\frac{2^{\frac{1}{12}}}{1}\right)^{12} = \frac{2}{1}. \tag{15.34}$$

and the according logarithmic-frequency interval is

$$\Psi_{\text{oct}} = 12 \, \Psi_{\text{semitone}}. \tag{15.35}$$

Problem 1.6.

Proposed Approach

The logarithmic-frequency interval for a 3rd-octave band, using the notation $\text{ld} = \log_2$, is

$$\Psi_{\text{3rd oct}} = 3 \, \text{ld}(f_2/f_1) = 1, \tag{15.36}$$

and, therefore,

$$\Psi_{\text{oct}} = \text{ld}(f_2/f_1) = 1. \tag{15.37}$$

From these ratios the limiting frequencies of the corresponding bands are determined using a ratio of $\sqrt{2}$, regarding the center frequency, f_c. For the logarithmic-frequency interval, it follows,

$$(f_2/f_1)_{\text{3rd oct}} = 2^{\frac{1}{3}}, \tag{15.38}$$

and

$$(f_2/f_1)_{\text{oct}} = 2. \tag{15.39}$$

Using these ratios, the upper limiting frequency, f_u, for a third-octave band results as

$$f_u = f_c \cdot 2^{\frac{1}{3}} \cdot 2^{\frac{1}{2}} = f_c \cdot 2^{\frac{1}{6}}, \tag{15.40}$$

and the lower limit frequency, f_1, as

$$f_1 = f_c/(2^{\frac{1}{3}} \cdot 2^{\frac{1}{2}}) = f_c \cdot 2^{-\frac{1}{6}}. \tag{15.41}$$

Building the frequency ratio from the 1st-octave-band limiting frequencies yields

$$\frac{f_u}{f_1} = \frac{f_c \cdot 2^{\frac{1}{6}}}{f_c \cdot 2^{-\frac{1}{6}}} = 2^{\frac{1}{3}}, \tag{15.42}$$

This confirms that the limiting frequencies of a third-octave band, determined in this way, result in the expected frequency ratio. Similarly, for an octave band, the upper limiting frequency comes out as

$$f_u = f_c \cdot 2^{\frac{1}{2}}, \tag{15.43}$$

and the lower limiting frequency as

$$f_1 = f_c \cdot 2^{-\frac{1}{2}}. \tag{15.44}$$

Problem 1.7.

Proposed Approach

Addition of incoherent sources is accomplished with (1.12) as follows,

$$L_\Sigma = 10 \log\left(\frac{|I_1| + |I_2| + \ldots + |I_n|}{I_0}\right). \tag{15.45}$$

For equal sound-source levels and equal distances to the receiver, the total sound-intensity level is

$$L_\Sigma = 10 \log\left(\frac{n\,|I_1|}{I_0}\right) = 10 \log\left(\frac{|I_1|}{I_0}\right) + 10 \log n\,, \tag{15.46}$$

resulting in

$$L_\delta = 10 \log n\,. \tag{15.47}$$

Table 15.1 Level differences for a number of n incoherent sources of identical sound pressure level

n	1	2	3	4	5	6	7	8	9	10
L_Δ (dB)	0	3	5	6	7	7.8	8.5	9	9.5	10

Table 15.1 lists the level differences for each number of sources by substituting the number of sources, n, into the above equation.

Problem 1.8.

Proposed Approach

Assuming that all talkers have a spherical directional characteristic, and given that the SPL at 1 m distance is 65 dB, it follows that at 2 m the SPL is 60 dB, and at 4 m, that is, at the inner ring, the SPL is 55 dB. At the radius of $2a$, at the the outer ring, the SPL contribution is 50 dB.

Multiple incoherent sources are added according to (1.12) as follows,

$$L_\Sigma = 10 \log \left(\frac{|\mathbf{I}_1| + |\mathbf{I}_2| + ... + |\mathbf{I}_n|}{I_0} \right) \text{ dB}, \qquad (15.48)$$

where I_0 is given as 10^{-12} W/m^2.

One possible way of finding the receiver position at 4 m starts with first calculating I_1 with $I_0 = 10^{-12}$ W/m^2, namely,

$$50\,\text{dB} = 10 \log \frac{I_2}{I_0} \text{ dB}, \qquad (15.49)$$

$$10^{5.5} = \frac{I_1}{I_0}, \qquad (15.50)$$

$$I_1 = 10^{5.5} \cdot 10^{-12} = 10^{-6.5} \text{ W/m}^2. \qquad (15.51)$$

We apply the procedure for the intensity I_2 at 8 m (or $2a$), that is,

$$50\,\text{dB} = 10 \log \frac{I_2}{I_0} \text{ dB}, \qquad (15.52)$$

$$10^5 = \frac{I_2}{I_0}, \qquad (15.53)$$

Table 15.2 Sound pressure level for the talkers $1 \le n \le 6$

n	1	2	3	4	5	6
L_Δ (dB)	0	3	5	6	7	7.8
SPL (dB)	55	58	60	61	62	62.8

Table 15.3 Sound pressure level for the talkers $7 \le n \le 18$

Talker	7	8	9	10	11	12	13	14	15	16	17	18
n	1	2	3	4	5	6	7	8	9	10	11	12
L_Δ (dB)	0	3	5	6	7	7.8	8.5	9.3	9.5	10	10.4	10.8
SPL (dB)	50	53	55	56	57	57.8	58.5	59.3	59.5	60	60.4	60.8

$$I_2 = 10^5 \cdot 10^{-12} = 10^{-7} \text{ W/m}^2 . \tag{15.54}$$

For the first six sources with $1 \le n \le 6$ along the inner circle, (15.48) delivers the total sound-pressure level. However, it is a more elegant way to take advantage of Table 15.1, starting at 55 dB for a single talker. Accordingly, we obtain the total sound-pressure level for the talkers along the outer circle by extending Table 15.2, using $10 \log_{10} n$. The results are shown in Table 15.3.

In general, given two (incoherent) sound-pressure levels, L_1, L_2 with $L_2 > L_1$, and $L_\Delta L = L_2 - L_1$, the following holds,

$$\begin{aligned} I &= I_0 \left(10^{L_1/10} + 10^{L_2/10}\right) \\ &= I_0 \, 10^{L_2/10} \left(10^{-L_\Delta/10} + 1\right), \end{aligned} \tag{15.55}$$

such that

$$\begin{aligned} L_\Sigma &= 10 \log_{10}\left(10^{L_2/10}\right) + 10 \log_{10}\left(10^{-L_\Delta/10} + 1\right) \\ &= L_2 + 10 \log_{10}\left(10^{-L_\Delta/10} + 1\right). \end{aligned} \tag{15.56}$$

For example, for the total of seven sources, since the 7th source has a contribution of 50 dB, and the total first six talkers result in 62.8 dB, we find

$$L_\Sigma(7) = [62.8 + 10 \log_{10}(10^{-1.3} + 1)] = 63.01 \text{ dB}. \tag{15.57}$$

Finally, for the total of 18 sources, we get

$$L_\Sigma(18) = 62.8 + 10 \log_{10}(10^{-0.2} + 1) = 64.9 \text{ dB}. \tag{15.58}$$

15.2 Chapter 2

Problem 2.1.

Proposed Approach

The mechanic impedance of a constant mass, m, is expressed by

$$\underline{Z}_m = j\,\omega\, m = \omega\, m\, e^{j\pi/2}, \tag{15.59}$$

and the mechanic impedance of a constant compliance, n, is expressed by

$$\underline{Z}_n = \frac{1}{j\,\omega\, n} = \frac{1}{\omega\, n} e^{j3\pi/2} = \frac{1}{\omega\, n} e^{-j\pi/2}. \tag{15.60}$$

Task (**a**) The mechanic impedance of the mass is solely imaginary as a function of frequency. Its magnitude, $\omega \cdot m$, increases linearly with frequency—illustrated in Fig. 15.1a. Its phase angle equals $\pi/2$—see Fig. 15.1c.

Task (**b**) The mechanic impedance of the compliance is solely imaginary as a function of frequency. Its magnitude, $1/(\omega \cdot n)$, decreases with frequency—illustrated in Fig. 15.1b. The phase angle is $-\pi/2$—see Fig. 15.1d.

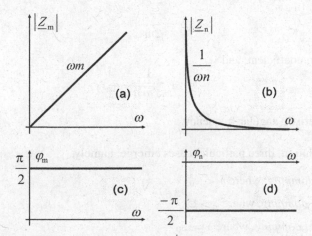

Fig. 15.1 Mechanic impedances as a function of frequency (**a**) Magnitude function of the mass impedance (**b**) Magnitude function of the compliance impedance (**c**) Phase function of the mass impedance (**d**) Phase function of the compliance impedance

Problem 2.2.

Proposed Approach

Either the displacement or the velocity is applicable for describing the mechanic parallel oscillator. For ease of discussion, we express the differential equation for harmonic-force excitation in terms of the displacement, ξ, that is,

$$\hat{F}_0 \cos \omega t = m \frac{d^2\xi}{dt^2} + r \frac{d\xi}{dt} + \frac{1}{n}\xi . \tag{15.61}$$

With no external excitation at the time $t \geq 0$, the differential equation takes on its homogenous form, written as

$$0 = m \frac{d^2\xi}{dt^2} + r \frac{d\xi}{dt} + \frac{1}{n}\xi . \tag{15.62}$$

With the trial solution $\xi = e^{\underline{s}t}$, the homogenous differential equation becomes

$$m \underline{s}^2 + r \underline{s} + \frac{1}{n} = 0 . \tag{15.63}$$

Solving this quadratic equation leads to

$$\underline{s}_{1,2} = -\frac{r}{2m} \pm \sqrt{\frac{r^2}{4m^2} - \frac{1}{mn}} = -\delta \pm \sqrt{\delta^2 - \omega_0^2} , \tag{15.64}$$

where

$$\delta = \frac{r}{2m} \tag{15.65}$$

is the damping coefficient, and

$$\omega_0 = \frac{1}{\sqrt{mn}} \tag{15.66}$$

is the characteristic angular frequency.

Out of this solution, three particular cases emerge, namely,

   (1) *Weak damping*, where $\delta < \omega_0$

   (2) *Strong damping*, where $\delta > \omega_0$

   (3) *Critical damping*, where $\delta = \omega_0$

In the case of weak damping, $\underline{s}$ becomes complex-valued

$$\underline{s}_{1,2} = -\delta \pm j\sqrt{\omega_0^2 - \delta^2} . \tag{15.67}$$

This equation has the following two complex roots.

$$\underline{\xi} = \underline{\xi}_1 e^{-\delta t}\, e^{-j\omega t} + \underline{\xi}_2 e^{-\delta t}\, e^{+j\omega t}\,, \tag{15.68}$$

where $\omega = \sqrt{\omega_0^2 - \delta^2}$.

For the case of $\underline{\xi}_1 = \underline{\xi}_2 = \underline{\xi}_{1,2}/2$, we get

$$\underline{\xi} = \underline{\xi}_{1,2} e^{-\delta t}\, \frac{e^{-j\omega t} + e^{+j\omega t}}{2} = \underline{\xi}_{1,2}\, e^{-\delta t}\, \cos{(\omega t)}\,. \tag{15.69}$$

This solution represents a damped oscillation at the angular frequency

$$\omega = \sqrt{\omega_0^2 - \delta^2}\,, \tag{15.70}$$

with an exponential decaying envelope, $\underline{\xi}_{1,2} e^{-\delta t}$.

In the case of strong damping, the solution takes on two real-valued roots, that is,

$$\underline{s}_{1,2} = -\delta \pm \sqrt{\delta^2 - \omega_0^2}\,, \tag{15.71}$$

which lead to the following two exponential decays (damped decays),

$$\underline{\xi} = \underline{\xi}_1 e^{-(\delta - \sqrt{\delta^2 - \omega^2})t} + \underline{\xi}_2 e^{-(\delta + \sqrt{\delta^2 - \omega^2})t}\,. \tag{15.72}$$

In the case of critical damping, the second term of $\underline{s}$ vanishes, resulting in the single real root,

$$\underline{s}_{1,2} = -\delta\,, \tag{15.73}$$

which produces the following expression for the displacement,

$$\underline{\xi} = (\underline{\xi}_1 + \underline{\xi}_2)\, e^{-\delta t}\,. \tag{15.74}$$

For excitation with $F_0(t) = \hat{F}_0 \cos{\omega t}$, we use the differential equation

$$\hat{F}_0 \cos{\omega t} = m \frac{d^2\xi}{dt^2} + r \frac{d\xi}{dt} + \frac{1}{n}\xi\,. \tag{15.75}$$

Substitution of F by its complex term, $\underline{F}$, results in

$$\underline{F} = -\omega^2\, m\, \underline{\xi} + j\omega\, r\, \underline{\xi} + \frac{1}{n}\,\underline{\xi}\,. \tag{15.76}$$

The solution for $\underline{\xi}$ is

$$\underline{\xi} = \frac{F}{-\omega^2 m + j\omega r + \frac{1}{n}} . \tag{15.77}$$

Expressing phase and magnitude and re-arranging generates

$$\underline{\xi} = \frac{|F|}{\sqrt{(\frac{1}{n} - \omega^2 m)^2 + \omega^2 r^2}} e^{j(\omega t - \phi)} , \tag{15.78}$$

where

$$\phi = \arctan \frac{\omega r}{\frac{1}{n} - \omega^2 m} . \tag{15.79}$$

By taking the real part of $\underline{\xi}$, the solution finally results as

$$\xi = \mathrm{Re}\{\underline{\xi}\} = \frac{|F|}{\sqrt{(\frac{1}{n} - \omega^2 m)^2 + \omega^2 r^2}} \cos(\omega t - \phi) . \tag{15.80}$$

Problem 2.3.

Proposed Approach

The task concerns forced mechanic parallel oscillation. We write the excitation as

$$\underline{F}(t) = F_0 e^{j\omega t}. \tag{15.81}$$

The relation of velocity and force excitation is expressed using (2.24),

$$\underline{F} = j\omega m \underline{v} + r \underline{v} + \frac{1}{j\omega n} \underline{v}, \tag{15.82}$$

which, when factoring out $\underline{v}$ and dividing by $\underline{F}$, renders the complex velocity as

$$\underline{v} = \frac{F}{j\omega m + r + \frac{1}{j\omega n}} . \tag{15.83}$$

For $\underline{F}$ being constant, using (2.25) indicates $\underline{v} = \underline{F}/\underline{Z}_m = \underline{F} \cdot \underline{Y}_m$. This means that $\underline{v}$ is also proportional to the mechanic admittance (mobility), $\underline{Y}_m$. The trajectory of $\underline{Y}_m$ as a function of frequency helps us to discuss the behavior of the velocity—see Fig. 15.2.

The quality (sharpness-of-resonance) factor, Q, is defined as

$$Q = \frac{\omega_0}{2\delta} = \frac{\omega_0 m}{r} . \tag{15.84}$$

Fig. 15.2 Trajectory of the mechanical admittance as a function of frequency

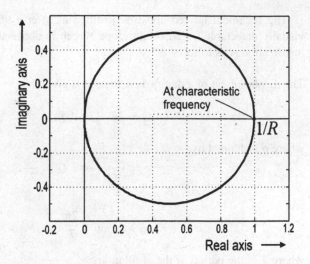

According to (2.25), at the –3 dB points along the resonance curve, the imaginary part of the mechanic impedance of a parallel oscillator is equal to its real part, namely,

$$\left(\omega m - \frac{1}{\omega n}\right)^2 = r^2, \quad \text{or} \quad \omega m - \frac{1}{\omega n} = \pm r, \tag{15.85}$$

with $2\delta = r/m$ and $\omega_0^2 = 1/(m\,n)$, so that

$$\omega^2 \mp \frac{r}{m}\omega - \frac{1}{m\,n} = \omega^2 \mp 2\delta\,\omega - \omega_0^2 = 0. \tag{15.86}$$

Using a 2nd-order polynomial of the form $a\,x^2 + b\,x + c$, the solutions for (angular) frequencies that meet the condition of –3 dB (or ±45 degree) are

$$\omega_{-45} = -\delta + \sqrt{\delta^2 \pm \omega_0^2}, \tag{15.87}$$

$$\omega_{+45} = +\delta + \sqrt{\delta^2 \pm \omega_0^2}. \tag{15.88}$$

The bandwidth at –3 dB is defined as

$$\omega_\Delta = \omega_{+45} - \omega_{-45}. \tag{15.89}$$

Substitution of the two equations above in the bandwidth formula leads to $\omega_\Delta = 2\,\delta$, which implies

$$Q = \frac{\omega_0}{\omega_\Delta}. \tag{15.90}$$

The statement quoted above is justified as after a 96% decrease, what is still visually detectable on an oscilloscope screen is above this threshold. The formal justification is as follows.

The displacement of a damped oscillation is given by

$$\underline{\xi} = \hat{\xi}e^{-\delta t}\cos(\omega t + \phi), \tag{15.91}$$

where δ is defined by

$$\delta = \frac{\omega_0}{2Q} \quad \text{or} \quad Q = \frac{\omega_0}{2\delta}, \tag{15.92}$$

and

$$\omega_0 = \frac{2\pi}{T}, \tag{15.93}$$

where T is the period of the oscillation.

To evaluate the case of a *96% reduction of the starting amplitude*, it is sufficient to focus on the envelope, $|\underline{\xi}|(t)/\hat{\xi}$, without looking at the fine structure of the oscillations, $\cos(\omega t + \phi)$.

Substituting δ in (2.21), (15.92) and (15.93) back into $|\underline{\xi}(t)|/\hat{\xi}$, results in

$$\frac{|\underline{\xi}|}{\hat{\xi}} = e^{-\delta t} = e^{-\frac{\omega_0}{2Q}t} = e^{-\frac{\pi}{QT}t}, \tag{15.94}$$

and substituting the time variable, $t = QT$, which represents the time for the oscillation to undergo a number of "Q" periods leads to

$$\frac{|\underline{\xi}|}{\hat{\xi}} = e^{-\frac{\pi QT}{QT}} = e^{-\pi} \approx 0.04. \tag{15.95}$$

Problem 2.4.

Proposed Approach

The resonance frequency of the given parallel mechanic oscillator is

$$f_0 = \frac{1}{2\pi\sqrt{mn}} = \frac{1}{2\pi\sqrt{0.63\,(\text{Ns}^2/\text{m})\cdot 0.027\,(\text{m/N})}} = 1.22\,\text{Hz}. \tag{15.96}$$

The damping coefficient is

$$\delta = \frac{r}{2m} = \frac{6.02\,(\text{Ns/m})}{2 \cdot 0.63\,(\text{Ns}^2/\text{m})} = 4.78\,\text{s}^{-1}. \tag{15.97}$$

The quality (sharpness-of-resonance) factor is

$$Q = \frac{\omega_0}{2\delta} = \frac{\pi\,f_0}{\delta} = \frac{\pi \cdot 1.22\,(\text{Hz})}{4.78\,\text{s}^{-1}} = 0.8, \tag{15.98}$$

and the decay time is

$$T_\text{d} = \frac{6.9}{\delta} = \frac{6.9}{4.78\,\text{s}^{-1}} = 1.44\,\text{s}. \tag{15.99}$$

The angular frequency is

$$\omega_0 = 2\,\pi \cdot f_0 = 2\,\pi \cdot 1.22\,\text{s}^{-1} = 7.66 > \delta = 4.78\,\text{s}^{-1}. \tag{15.100}$$

Consequently, the system concerned oscillates.

Problem 2.5.

Proposed Approach

We write the input energy as

$$W = \int F(t)\,\text{d}z = \int F(t)\,\frac{\text{d}z}{\text{d}t}\,\text{d}t = \int F(t)\,v(t)\,\text{d}t. \tag{15.101}$$

The transferred instantaneous power, $P(t)$, is then given by

$$P(t) = \frac{\text{d}}{\text{d}t}\int F(t)\,v(t)\,\text{d}t = F(t)\,v(t), \tag{15.102}$$

where

$$F(t) = \hat{F}\,\cos\,(\omega\,t + \phi_F), \tag{15.103}$$

and

$$v(t) = \hat{v}\,\cos\,(\omega\,t + \phi_v). \tag{15.104}$$

Substituting this into the instantaneous-power expression leads to

$$P(t) = \hat{F}\,\cos\,(\omega\,t + \phi_F)\,\hat{v}\,\cos\,(\omega\,t + \phi_v), \tag{15.105}$$

and, applying the multiplication rule

$$\cos \alpha \cos \beta = \frac{1}{2} \left[\cos (\alpha + \beta) + \cos (\alpha - \beta) \right] \qquad (15.106)$$

results in

$$P(t) = \frac{\hat{F} \hat{v}}{2} \left[\cos (2 \omega t + \phi_F + \phi_v) \right] + \frac{\hat{F} \hat{v}}{2} \left[\cos (\phi_F - \phi_v) \right], \qquad (15.107)$$

where the first term alternates with the angular frequency, while the second term remains constant due to the initial phase angles. The average power is given by the second term only.

Complex notation allows for the concise expression of amplitude and phase information in the following form,

$$\underline{F} = \hat{F} e^{j \phi_F}, \qquad (15.108)$$

and

$$\underline{v} = \hat{v} e^{j \phi_v}. \qquad (15.109)$$

To obtain the corresponding expression for P(t), the complex conjugate, $\underline{v}^*$, is required, that is,

$$\underline{P} = \frac{1}{2} \left[\underline{F} \, \underline{v}^* \right]. \qquad (15.110)$$

This is the justification for the complex definition of power.

Expanding $\underline{P} = \bar{P} + j Q$ separates $\underline{P}$ into the so-called *active power* and the so-called *reactive power*. By using the complex conjugate, $\underline{v}^*$, the reactive power of the mass counts positive. This is a common convention.

Problem 2.6.

Proposed Approach

Task (**a**) The input impedance, $\underline{Z}$, is

$$\underline{Z} = j \omega L + R + \frac{1}{j \omega C}, \qquad (15.111)$$

and the input admittance, $\underline{Y}$, is

$$\underline{Y} = \left(j \omega L + R + \frac{1}{j \omega C} \right)^{-1}. \qquad (15.112)$$

The resonance frequency, ω_0, is reached when $j[\omega L - 1/(\omega C)] = 0$. This is the point at which the input impedance becomes real.

$$\omega_0 = \frac{1}{\sqrt{LC}}, \tag{15.113}$$

The input impedance at this point is called the *characteristic resistance*,

$$\underline{Z}_0 = R. \tag{15.114}$$

Task (b) The bandwidth is defined as $\omega_\Delta = \omega_{+45} - \omega_{-45}$. Thereby, ω_{+45} and ω_{-45}, are given by the difference of the frequency points where the real part and the imaginary part of the impedance become equal, that is,

$$R = \left| \omega_{\pm 45} L - \frac{1}{\omega_{\pm 45} C} \right|, \tag{15.115}$$

leading to

$$\omega_{\pm 45}^2 LC \mp \omega_{\pm 45} RC - 1 = 0. \tag{15.116}$$

Solving this quadratic equation for $\omega_{\pm 45}$, and taking the positive solution, that is,

$$\omega_{\pm 45} = \frac{\pm R}{2L} + \sqrt{\frac{R^2}{4L^2} + \frac{1}{LC}}. \tag{15.117}$$

leads to

$$\omega_\Delta = \omega_{+45} - \omega_{-45} = \frac{R}{L}. \tag{15.118}$$

The quality (sharpness-of-resonance) factor, Q, is defined as $Q = \frac{\omega_0}{\omega_\Delta}$. Substitution of (15.113) and (15.118) yields

$$Q = \frac{\omega_0}{\omega_\Delta} = \frac{1}{R}\sqrt{\frac{L}{C}}. \tag{15.119}$$

Task (c) For graphing the input impedance, $\underline{Z}$ in a double-logarithmic fashion as a function of frequency see Fig. 15.3.

Problem 2.7.

Proposed Approach

For a sinusoidal force with constant amplitude over frequency, according to Eq. 2.15, the following holds,

$$\frac{\underline{\xi}}{\underline{F}} = \frac{1}{-\omega^2 m + j\omega r + \frac{1}{n}}, \tag{15.120}$$

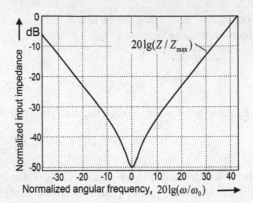

Fig. 15.3 Input impedance, $\underline{Z}$, of the electric series circuit shown in Fig. 1.3

and the characteristic frequency is given by

$$\omega_0 = \frac{1}{\sqrt{m\,n}} . \tag{15.121}$$

Assuming $\underline{F} = \hat{F}\cos(\omega_0 t + \phi)$ and substituting ω_0 into $\underline{\xi}/\underline{F}$ produces

$$\frac{\underline{\xi}}{\underline{F}} = \frac{1}{-\frac{m}{mn} + \frac{1}{n} + j\frac{r}{\sqrt{mn}}} . \tag{15.122}$$

The first two terms cancel out, leaving

$$\frac{\underline{\xi}}{\underline{F}} = \frac{\sqrt{mn}}{j\,r} = -j\frac{\sqrt{mn}}{r} . \tag{15.123}$$

This indicates that the phase of $\underline{\xi}$, concerning the excitation, $\underline{F}$, becomes $-\pi/2$ at the characteristic frequency, ω_0.

15.3   Chapter 3

Problem 3.1.

Proposed Approach

The analogy #2 is based on the equivalences of $\underline{F} \,\hat{=}\, \underline{i}$ and $\underline{v} \,\hat{=}\, \underline{u}$. The Figs. 3.1 and 3.3 are helpful for the construction of the analog circuit. We start by labeling some key nodes by (1)–(6) in Fig. 15.4.

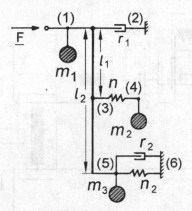

Fig. 15.4 Mechanic system as given in Fig. 3.10, with some key nodes labeled as **(1)–(6)**

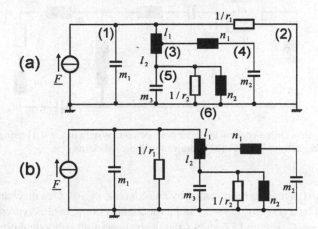

Fig. 15.5 Equivalent circuit constructed using analogy #2 **(a)** Direct translation from Fig. 15.4 **(b)** Redrawing the circuit for improved clarity

In analogy #2, a mass is represented by a *capacitor*, $m \rightarrow C$, a mechanic damper is represented by an *admittance*, $r \rightarrow 1/r$, and a mechanic compliance is represented by an *inductance*, $n \rightarrow L$. Furthermore, a mechanic lever is represented by an *ideal single-coil transformer* as shown in Fig. 3.3.

Figure 15.5a illustrates a direct translation of the mechanic circuit into an electric one. Figure 15.5b depicts the so constructed analog circuit based on Fig. 15.4 for improved clarity, where the mechanic force is represented as a *current source*.

Problem 3.2.

Proposed Approach

The analogy #1 is based on $\underline{F} \hat{=} \underline{u}$ and $\underline{v} \hat{=} \underline{i}$. Particularly, the idling state of a mechanic node corresponds to $\underline{v} = 0$, being equivalent to $\underline{i} = 0$.

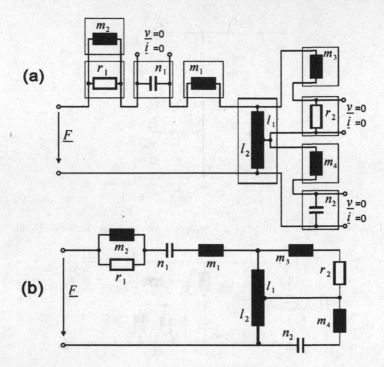

Fig. 15.6 Equivalent circuit constructed from the mechanic system in Fig. 3.11 using analogy #1 (a) Direct translation from Fig. 3.11 (b) Redrawn for improved clarity

In analogy #1, a mass is represented by an *inductor*, $m \rightarrow L$. a mechanic damper is represented by a *resistor*, $r \rightarrow r$, and a mechanic compliance is represented by a *capacitor*, $n \rightarrow C$. Figure 15.6 shows the derivation of an equivalent circuit in two steps.

Problem 3.3.

Proposed Approach

For the mechanic system in Fig. 3.12, analogy #2 is used to derive an equivalent electrical circuit as shown in Fig. 15.7 to explore the functions in question. In analogy #2, the mechanic velocity, $\underline{v}$, represents the electric voltage, $\underline{u}$.

Task (**a**) Figure 15.7 shows the velocity quantities, $\underline{v}$ and $\underline{v}_0$. There ratio is expressible as the ratio of the *voltage* across element n and the total voltage,

$$\frac{\underline{v}_0}{\underline{v}} = \frac{j\omega n}{j\omega n + \frac{1}{j\omega m_0}} , \tag{15.124}$$

with

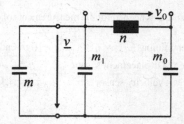

Fig. 15.7 Equivalent circuit constructed from the mechanic system in Fig. 3.12

$$\omega_0^2 = \frac{1}{m_0\, n}\,,$$
(15.125)

We now write (15.124) as

$$\underline{v}_0 = \frac{\omega^2}{\omega^2 - \omega_0^2}\, \underline{v} = \frac{1}{1 - \frac{\omega_0^2}{\omega^2}}\, \underline{v}.$$
(15.126)

Furthermore, the harmonic displacement, $\underline{\xi}_0$, is related with the velocity as follows,

$$\underline{\xi}_0 = \frac{1}{j\,\omega}\, \underline{v}_0.$$
(15.127)

Substituting (15.126) into (15.127) leads to

$$\underline{\xi}_0 = \frac{1}{j\,\omega}\, \frac{1}{1 - \frac{\omega_0^2}{\omega^2}}\, \underline{v}.$$
(15.128)

Task (**b**) The relationships in (15.126) and (15.128) facilitate discussions of two separate cases, namely,

(i) For $\omega \gg \omega_0$, (15.128) becomes

$$\underline{\xi}_0 \approx \frac{1}{j\,\omega}\, \underline{v} = \underline{\xi},$$
(15.129)

and (15.126) brings us to

$$\underline{v}_0 \approx \underline{v}.$$
(15.130)

(ii) For $\omega \ll \omega_0$, (15.128) becomes

$$\underline{\xi}_0 \approx \frac{1}{j\,\omega}\, \frac{1}{-\frac{\omega_0^2}{\omega^2}}\, \underline{v} = j\, \frac{\omega}{\omega_0^2}\, \underline{v}.$$
(15.131)

Table 15.4 Different tuning settings as needed for the measurements of displacement, acceleration, velocity, and jerk

Measurement	Low-end tuning ($\omega \gg \omega_0$)	High-end tuning ($\omega \ll \omega_0$)
Displacement	$\underline{\xi}_0 = \underline{\xi}$ ...Displacement sensor	$\underline{a}/\omega_0^2$ ...Acceleration sensor
Velocity	$\underline{v}_0 = \underline{v}$ ...Velocity sensor	$\underline{\dot{a}}/\omega_0^2$ ...Jerk sensor

Substituting $\underline{a} = j\omega\,\underline{v}$ into (15.131) leads to

$$\underline{\xi}_0 \approx \frac{\underline{a}}{\omega_0^2}, \tag{15.132}$$

and (15.126) for this case becomes

$$\underline{v}_0 = -\frac{\omega^2}{\omega_0^2}\,\underline{v}. \tag{15.133}$$

Further, substituting

$$\underline{a} = j\omega\,\underline{v} \tag{15.134}$$

and

$$j\omega\,\underline{a} = \frac{d\,a(t)}{dt} \tag{15.135}$$

into (15.133), renders

$$\underline{v}_0 = \frac{j\omega\,\underline{a}}{\omega_0^2} = \frac{\underline{\dot{a}}}{\omega_0^2}. \tag{15.136}$$

Note that the device shown in Fig. 3.12 is applicable as a sensor for either displacement, acceleration, velocity, or jerk. To this end, the device is tuned specifically for each measure. Table 15.4 summarizes the relevant settings.

Problem 3.4.

Proposed Approach

Figure 3.13 illustrates an acoustic system consisting of two sets of similar elements. For discussion convenience, these elements are labeled in Fig. 15.8. One set consists of a narrow tube, (1), connected to one pair of parallel cavities, (I) and (I'). Another one consists of a narrow tube, (2), connected to a further pair of parallel cavities, (II) and (II'). The outlet at $\underline{p}_2$ is an open end.

According to the electroacoustic analogy with

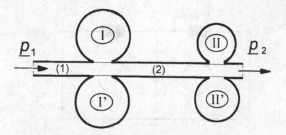

Fig. 15.8 Different elements in the acoustic circuit of Fig. 3.13, labeled as **(1)**, **(2)**, **(I)**, **(I')**, **(II)** and **(II')**

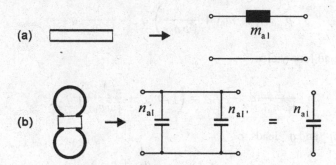

Fig. 15.9 Electroacoustic analogy of the two types of elements contained in the acoustic system in Fig. 15.8 **(a)** Acoustic mass **(b)** Acoustic springs in parallel

$$\underline{p} \,\hat{=}\, \underline{u} \quad \text{and} \quad \underline{q} \,\hat{=}\, \underline{i}, \tag{15.137}$$

a narrow tube is equivalent to a *mass*, and cavities are the equivalents of springs, that is, of *compliances*.

In Fig. 15.9b combination of two parallel springs leads to a lumped compliance,

$$\frac{1}{n_{\mathrm{al}}} = \frac{1}{n_{al}} + \frac{1}{n_{al'}}, \tag{15.138}$$

resulting in

$$n_{\mathrm{al}} = \frac{n_{al}\, n_{al'}}{n_{al} + n_{al'}}. \tag{15.139}$$

The electroacoustic analogy of the acoustic system shown in Fig. 15.8 is equivalent to the electric circuit illustrated in Fig. 15.10.

The first loop in Fig. 15.10 leads to

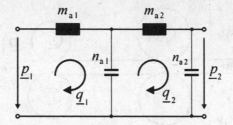

Fig. 15.10 Equivalent circuit of the acoustic system shown in Fig. 15.8

$$\underline{p}_1 = \left(j\omega\, m_{a1} + \frac{1}{j\omega\, n_{a1}}\right)\underline{q}_1 - \frac{1}{j\omega\, n_{a1}}\underline{q}_2, \qquad (15.140)$$

and the second loop yields

$$\frac{1}{j\omega\, n_{a1}}(\underline{q}_1 - \underline{q}_2) = \left(j\omega\, m_{a2} + \frac{1}{j\omega\, n_{a2}}\right)\underline{q}_2. \qquad (15.141)$$

Separating $\underline{q}_1$ and $\underline{q}_2$ leads to

$$\underline{q}_1 = j\omega\, n_{a1}\left(j\omega\, m_{a2} + \frac{n_{a1} + n_{a2}}{j\omega\, n_{a1}\, n_{a2}}\right)\underline{q}_2. \qquad (15.142)$$

Substituting (15.142) into (15.140) results in

$$\underline{p}_1 = \left[(1 - \omega^2\, m_{a1}\, n_{a1})\left(j\omega\, m_{a2} + \frac{n_{a1} + n_{a2}}{j\omega\, n_{a1}\, n_{a2}}\right) - \frac{1}{j\omega\, n_{a1}}\right]\underline{q}_2. \qquad (15.143)$$

Figure 15.10 also indicates that

$$\underline{p}_2 = \frac{1}{j\omega\, n_{a2}}\underline{q}_2. \qquad (15.144)$$

Substituting (15.144) into (15.143) leads to

$$\underline{p}_1 = \left[(1 - \omega^2\, m_{a1}\, n_{a1})\left(\frac{n_{a1} + n_{a2}}{n_{a1}} - \omega^2\, m_{a2}\, n_{a2}\right) - \frac{n_{a2}}{n_{a1}}\right]\underline{p}_2, \qquad (15.145)$$

and

$$\frac{\underline{p}_2}{\underline{p}_1} = \left[(1 - \omega^2\, m_{a1}\, n_{a1})\left(\frac{n_{a1} + n_{a2}}{n_{a1}} - \omega^2\, m_{a2}\, n_{a2}\right) - \frac{n_{a2}}{n_{a1}}\right]^{-1} \qquad (15.146)$$

$$= \frac{1}{\omega^4\, m_{a1}\, n_{a1}\, m_{a2}\, n_{a2} - \omega^2[m_{a1}(n_{a1} + n_{a2}) + m_{a2}\, n_{a2}] + 1}. \qquad (15.147)$$

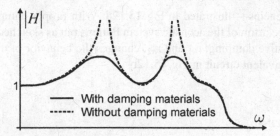

Fig. 15.11 Comparison of transfer functions of the acoustic system shown in Fig. 15.8 before and after introducing damping materials

Note that the following holds,

$$\omega_{01}^2 = \frac{1}{m_{a1}\,n_{a1}} \quad \text{and} \quad \omega_{02}^2 = \frac{1}{m_{a2}\,n_{a2}}. \tag{15.148}$$

Substituting (15.148) into (15.146) results in

$$|H(\omega)| = \frac{|\underline{p}_2|}{|\underline{p}_1|} = \left|\left(1 - \frac{\omega^2}{\omega_{01}^2}\right)\left(\frac{n_{a1} + n_{a2}}{n_{a1}} - \frac{\omega^2}{\omega_{02}^2}\right) - \frac{n_{a2}}{n_{a1}}\right|^{-1}. \tag{15.149}$$

The transfer function given in (15.147) and (15.149) implies that

$$\omega \to 0 \quad |H(\omega)| \to 1$$
$$\omega \to \infty \quad |H(\omega)| \to 0 \tag{15.150}$$

Besides, there are two resonance frequencies (two poles). Since there is no significant damping in the acoustic system—compare Fig. 15.8—we have resonances at two poles with high quality (sharpness-of-resonance) factors. Figure 15.11 conceptually depicts the magnitude spectrum of the acoustic system. Damping materials is inserted in the narrow tube near the entrances to the cavities to reduce the magnitudes at

Fig. 15.12 Treatment of the acoustic system shown in Fig. 15.8 for a smoother magnitude spectrum (**a**) Insertion of damping materials within the narrow tube near the cavity entrances (**b**) Equivalent circuit after insertion of the damping materials

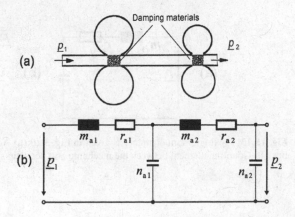

resonance frequencies—illustrated in Fig. 15.12a. With proper damping materials, the magnitude spectrum of the acoustic system flattens out as sketched in Fig. 15.11. Insertion of resistive damping, r_{a1} and r_{a2}, changes the behavior of the system. See the modified equivalent circuit in Fig. 15.12b.

15.4   Chapter 4

Problem 4.1.

Proposed Approach

The analogy #2 requires

$$\underline{F} \mathrel{\hat{=}} \underline{i} \quad \text{and} \quad \underline{v} \mathrel{\hat{=}} \underline{u} \tag{15.151}$$

for the mechanic subsystem. The loudspeaker box itself is considered as an acoustic subsystem. The electroacoustic analogy is given in (3.6) as

$$\underline{p} \mathrel{\hat{=}} \underline{u} \quad \text{and} \quad \underline{q} \mathrel{\hat{=}} \underline{i}. \tag{15.152}$$

Figure 15.13a illustrates the acoustic mass, spring, and damper as given by the loudspeaker box. Furthermore, the coupling between the mechanic and the acoustic subsystem is described by a gyrator, using the analogy #2—see Fig. 15.13b.

Figure 15.14 schematically illustrates the mechanic system of the loudspeaker. The constant coil force, $\underline{F}_4$, is applied to the coil mass, m_4, connected to the spring element, n_6, with a damper, r_6. Simultaneously, it is connected in parallel to the speaker cone, m_5, which appears as a mass linked with a spring element, n_7, again loaded with a damper, r_7.

Figure 15.14a depicts the connected elements in an analog circuit. Note that the speaker frame—labeled 8—is considered motionless, that is, $\underline{v} \mathrel{\hat{=}} \underline{u} \mathrel{\hat{=}} 0$. Figure 15.14b illustrates a simplified circuit with combined elements where

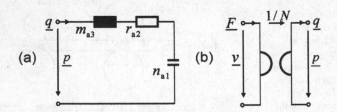

Fig. 15.13 Analog circuit elements as given in Fig. 3.10 **(a)** Acoustic system for the loudspeaker box **(b)** Coupling element between the mechanic and acoustic system

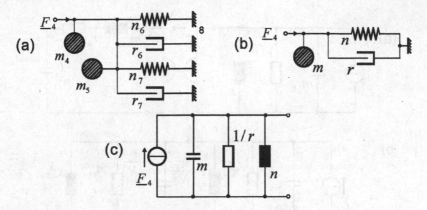

Fig. 15.14 Analog circuit elements as given in Fig. 3.10 **(a)** Mechanic system for the loudspeaker components (4–8) **(b)** Analog circuit with combined elements for the mechanic system **(c)** Analog circuit derived from **(b)**

$$m = m_4 + m_5, \quad \frac{1}{n} = \frac{1}{n_6} + \frac{1}{n_7}, \quad \text{and} \quad r = r_6 + r_7. \tag{15.153}$$

In analogy #2, a mass is represented by a *capacitor*, $m \rightarrow C$, a spring is represented by an *inductor* $n \rightarrow L$, and a damper is represented by a *resistor* with an admittance, $1/r$. The force is represented by a *current source*. Fig. 15.14c illustrates the analog circuit for the mechanic system of the loudspeaker as shown in Figs. 3.10 and 15.14a, and 15.14b.

Figure 4.7 indicates that the electromechanic analogy #2 renders the ideal gyrator as the analogy for the electric-field transducer. Both the electroacoustic, Fig. 15.13a, and the electromechanic subsystems, Fig. 15.14c, take advantage of this fact. Using the ideal gyrator as in Fig. 15.13b, the mechanic force and the mechanic velocity are related to the acoustic pressure and the corresponding volume velocity through (15.151) and (15.152) as follows.

$$\underline{F} = N\underline{i} = N\underline{q} \quad \text{and} \quad \underline{v} = \frac{1}{N}\underline{u} = \frac{1}{N}\underline{p}. \tag{15.154}$$

The relationship of (15.154) shows that the coupling via the ideal gyrator converts the acoustic admittance to the mechanic impedance

$$\underline{Z}_{\text{mech}} = \frac{\underline{F}}{\underline{v}} = N^2 \frac{\underline{q}}{\underline{p}} = N^2 \underline{Y}_{\text{a}}. \tag{15.155}$$

Figure 15.15a presents both equivalent subsystems of Figs. 15.14c and 15.13 in coupled form. Note that the ideal gyrator converts the serial circuit to the parallel one. The conversion performed by the gyrator coupling implies that

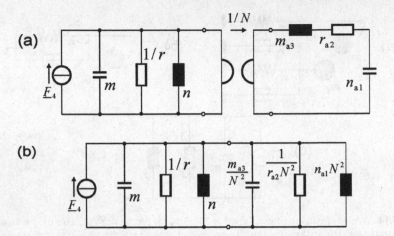

Fig. 15.15 Equivalent circuits of the entire system as shown in Fig. 3.10 **(a)** Analog electric circuit with a gyrator coupling between the mechanic and acoustic subsystems **(b)** Equivalent circuit with simplification when removing the gyrator

- An acoustic spring converts to an inductive element by $n_{a1} N^2$
- An acoustic damper becomes a resistive admittance $1/(r_{a2} N^2)$
- An acoustic mass converts to a capacitive element by m_{a3}/N^2

Figure 15.15b illustrates the two subsystems after coupling.

Problem 4.2.

Proposed Approach

The directional characteristic of a sound source is defined as

$$\Gamma(\theta) = \frac{p(\theta)}{p_{\max}}. \tag{15.156}$$

The directional characteristics of a monopole and a dipole spherical sound source are given by

$$\Gamma_m(\theta) = 1, \quad \text{and} \quad \Gamma_d(\theta) = \cos\theta. \tag{15.157}$$

The combined two directional characteristics of those two types of sound sources is achieved by linear combination, that is,

$$\Gamma_s(\theta) = \alpha\,\Gamma_m(\theta) + (1-\alpha)\,\Gamma_d = \alpha + (1-\alpha)\cos\theta, \tag{15.158}$$

with $0 \le \alpha \le 1$.

Figure 15.16 illustrates the directional characteristics in polar form for $\alpha = 0, 0.25, 0.5, 0.75,$ and 1. The resulting characteristics are *dipole, $\alpha = 0$, hypercar-*

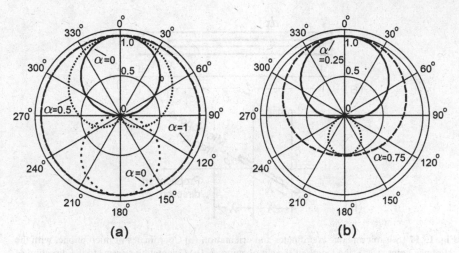

Fig. 15.16 Directional characteristics of combined spherical sound sources **(a)** Dipole characteristic, $\alpha = 0$, cardioid characteristic, $\alpha = 0.5$, and monopole characteristic, $\alpha = 1$ **(b)** *Hypercardioid* characteristic, $\alpha = 0.25$, and *hypocardioid* characteristic $\alpha = 0.75$

dioid, $\alpha = 0.25$, *cardioid*, $\alpha = 0.5$, *hypocardioid*, $\alpha = 0.75$, and *monopole*, $\alpha = 1$, respectively.[1]

Problem 4.3.

Proposed Approach

Figure 15.17a provides a sketch of a slit microphone with the slit opening of length L. Note that slit microphones are oriented at an angle of φ relative to the plane-wave propagation—as shown in Fig. 15.17b.

We now express the contribution of the slit element, dx, to the sound arriving at the reference point, x_{ref}, as

$$d\underline{p}_{\text{ref}} = C\,\underline{p}(x)\,e^{-j2\pi\frac{x_{\text{ref}}-x}{\lambda}}\,dx , \tag{15.159}$$

where C is a sensitivity coefficient, specific to a particular microphone.

The apparent length, λ', of the slit microphone when oriented at an angle, φ, relative to the propagation direction of plane waves, is

$$\lambda' = \frac{\lambda}{\cos\varphi} . \tag{15.160}$$

---

[1] The *hypercardiod* realizes minimum diffuse-field sensitivity and, thus, the highest front/diffuse-field ratio. Further particular characteristics are the *supercardiod*, $\alpha = 0.366$, which shows the highest front/rear-hemisphere ratio, and the *wide cardioid*, $\alpha = 0.7$.

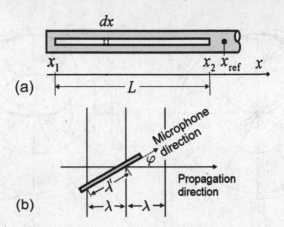

Fig. 15.17 Slit-microphone coordinates and orientation (**a**) Coordinates of microphone, with the reference point at x_{ref}, the length of the slit opening, L (**b**) Orientation relative to the direction of plane-wave propagation

The sound pressure at the reference point is determined by integration of the elementary contribution over the slit length as follows,

$$\underline{p}_{\text{ref}} = C\,\mathrm{e}^{-\mathrm{j}2\pi\frac{x_{\text{ref}}}{\lambda}} \int_{x_1}^{x_1+L} \underline{p}(x)\,\mathrm{e}^{\mathrm{j}2\pi\frac{x}{\lambda}}\mathrm{d}x\,. \tag{15.161}$$

Considering (15.160), the sound-pressure function is determined by

$$\underline{p}(x) = \underline{p}_0\,\mathrm{e}^{-\mathrm{j}2\pi\frac{x\cos\varphi}{\lambda}}\,, \tag{15.162}$$

substituting (15.162) into (15.161) leads to

$$\underline{p}_{\text{ref}} = C\,\underline{p}_0\,\mathrm{e}^{-\mathrm{j}2\pi\frac{x_{\text{ref}}}{\lambda}} \int_{x_1}^{x_1+L} \underline{p}(x)\,\mathrm{e}^{\mathrm{j}2\pi\frac{x(1-\cos\varphi)}{\lambda}}\mathrm{d}x$$

$$= C\,\underline{p}_0\,\mathrm{e}^{-\mathrm{j}2\pi\frac{x_{\text{ref}}}{\lambda}}\,\mathrm{e}^{\mathrm{j}2\pi\frac{x_1(1-\cos\varphi)}{\lambda}}\,\frac{\mathrm{e}^{\mathrm{j}2\pi\frac{L(1-\cos\varphi)}{\lambda}}-1}{\mathrm{j}2\pi(1-\cos\varphi)/\lambda}\,. \tag{15.163}$$

Taking the magnitude and applying Euler's formula, (15.163) results in

$$\left| \underline{p}_{\text{ref}} \right| = C \, \underline{p}_0 \frac{\left| \cos\left[\frac{2\pi L}{\lambda}(1 - \cos\varphi)\right] - 1 + j\sin\left[\frac{2\pi L}{\lambda}(1 - \cos\varphi)\right] \right|}{2\pi(1 - \cos\varphi)/\lambda}$$

$$= C \, \underline{p}_0 L \frac{\sqrt{\cos^2[\ ] + 1 - 2\cos[\ ] + \sin^2[\ ]}}{2\pi L\,(1 - \cos\varphi)/\lambda}$$

$$= C \, \underline{p}_0 L \frac{\sqrt{2 - 2\cos[2\pi L\,(1 - \cos\varphi)/\lambda]}}{2\pi L\,(1 - \cos\varphi)/\lambda}$$

$$= C \, \underline{p}_0 L \frac{2\sqrt{\{1 - \cos[2\pi L\,(1 - \cos\varphi)/\lambda]\}/2}}{2\pi L\,(1 - \cos\varphi)/\lambda} . \tag{15.164}$$

Applying $2\sin^2\alpha = 1 - \cos 2\alpha$ to (15.164) yields

$$\left| \underline{p}_{\text{ref}} \right| = C \, \underline{p}_0 L \left| \frac{\sin[\pi L\,(1 - \cos\varphi)/\lambda]}{\pi L\,(1 - \cos\varphi)/\lambda} \right| , \tag{15.165}$$

or

$$\Gamma(\varphi) = \left| \frac{\sin[\pi L\,(1 - \cos\varphi)/\lambda]}{\pi L\,(1 - \cos\varphi)/\lambda} \right| = \left| \frac{\sin[\pi\gamma\,(1 - \cos\varphi)]}{\pi\gamma\,(1 - \cos\varphi)} \right| . \tag{15.166}$$

The *length-to-wavelength ratio*, $\gamma = L/\lambda$, turns out to be a key quantity for controlling the directional characteristic. The directional characteristic of a slit microphone are of si–function type.

Figure 15.18 illustrates the directional characteristic of a slit microphone for length-to-wavelength ratios of $\gamma = 0.5, 1, 2, 4$, respectively.

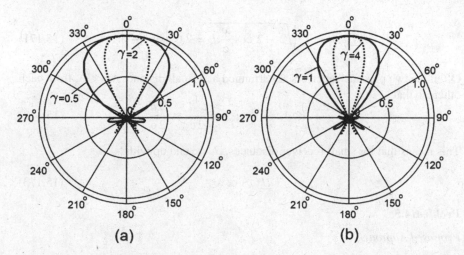

(a)   **(b)**

Fig. 15.18 Directional characteristic of a slit microphone **(a)** Length-to-wavelength ratios, $\gamma = 0.5$ and 2 **(b)** Length-to-wavelength ratios, $\gamma = 1$ and 4

Problem 4.4.

Proposed Approach

Pressure receiver (microphones) react to the sound pressure in front of the membrane as follows,

$$\underline{p}_{\mathrm{m}}(x) = \underline{p}_0 \mathrm{e}^{\mathrm{j}2\pi\frac{x}{\lambda}} , \tag{15.167}$$

where $\beta = 2\pi/\lambda = \omega/c$ with λ, ω, and c, are wavelength, angular frequency, and sound speed, respectively.

Pressure-gradient receivers are frequently used for measuring/receiving the particle velocity in sound waves. Using (15.167), their outputs are signals as follows.

$$\underline{F}_{\mathrm{g}}(x) \propto \underline{p}_{\mathrm{m}}(x) - \underline{p}_{\mathrm{m}}(x+d) = \underline{p}_0 \mathrm{e}^{\mathrm{j}2\pi\frac{x}{\lambda}} - \underline{p}_0 \mathrm{e}^{\mathrm{j}2\pi\frac{x+d}{\lambda}}$$
$$= \underline{p}_{\mathrm{m}}(x)\left(1 - \mathrm{e}^{\mathrm{j}\beta d}\right) , \tag{15.168}$$

with

$$\underline{p}_{\mathrm{m}}(x) = \underline{p}_0\, \mathrm{e}^{\mathrm{j}2\pi\frac{x}{\lambda}} = \underline{p}_0\, \mathrm{e}^{\mathrm{j}\beta x} , \tag{15.169}$$

Expressing (15.168) as a function of frequency results in

$$\underline{H}(\omega) = \underline{C}\left(1 - \mathrm{e}^{\mathrm{j}\beta d}\right) = \left[1 - \cos(\beta d) - \mathrm{j}\sin(\beta d)\right] , \tag{15.170}$$

with $\underline{C} = \underline{p}_0(x)$ being a complex-valued coefficient with a magnitude of

$$\left|\underline{H}(\omega)\right| = \underline{C}\sqrt{2 - 2\cos\frac{\omega}{c}d} = 2\underline{C}\left|\sin\frac{\omega d}{2c}\right| . \tag{15.171}$$

Often the two pressure receivers are arranged at a small distance, $d \ll \lambda$, from each other, so that

$$\frac{2\pi d}{2\lambda} \ll 1, \quad \rightarrow \quad \sin\frac{\omega d}{2c} \approx \frac{\omega d}{2c} . \tag{15.172}$$

This means that for small receiver distances, d, we end up with

$$\underline{H}(\omega) \propto \omega . \tag{15.173}$$

Problem 4.5.

Proposed Approach

For measuring the sensitivity coefficient of transducers of the magnetic-field type, for example, to be applied to structure-borne-sound, the following procedure is common practice.

The magnetic-field transducer under test is rigidly connected to two other reversible transducers. Mass and compliance of the rigid-connection elements are given. Consequently, the determination of the transducers' sensitivity coefficients follows solely from electric measurements, whereby the reversible transducers are first used as transmitters and then as receivers. Since the mass and the compliance of the rigid-connection elements are given, the mechanic impedance is determined by

$$\underline{Z} = \mathrm{j}\,\omega\,m + \frac{1}{\mathrm{j}\,\omega\,n}\,. \tag{15.174}$$

The transducers are used well above their resonance frequency, so that

$$\underline{v} \approx \frac{\underline{F}}{\mathrm{j}\,\omega\,m}\,. \tag{15.175}$$

In Fig. 15.19, the reversible transducers are labeled by subscript 1 (M_1, $\underline{F}_1$, $\underline{u}_1$, $\underline{i}_1$) and subscript 2 (M_2, $\underline{F}_2$, $\underline{u}_2$, $\underline{i}_2$), respectively. The transducer to be determined is labeled by subscript x—it is a receiving transducer. At node, k, in Fig. 15.19 the forces follow

$$\underline{F}_1 + \underline{F}_2 + \underline{F}_x + \mathrm{j}\,\omega\,m\,\underline{v} = 0\,, \tag{15.176}$$

where the compliances are negligible because the operating frequency is much higher than the resonance frequency.

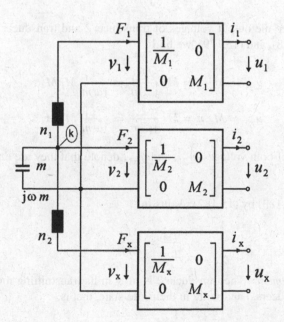

Fig. 15.19 Electromechanic circuit of magnetic-field transducers rigidly connected together (analogy #2). Transducers 1, and 2 are reversible

The analogy#2 is applicable to solving this problem by following (15.151), namely,

$$\begin{bmatrix} \underline{v}_i \\ \underline{F}_i \end{bmatrix} = \begin{bmatrix} 1/M_i & 0 \\ 0 & M_i \end{bmatrix} \begin{bmatrix} \underline{u}_i \\ \underline{i}_i \end{bmatrix}, \tag{15.177}$$

also expressed as

$$\underline{u}_i = M_i \underline{v}_i \quad \text{and} \quad \underline{i}_i = \frac{\underline{F}_i}{M_i}. \tag{15.178}$$

The rigid connections imply

$$\underline{v}_1 = \underline{v}_2 = \underline{v}_x = \underline{v}, \tag{15.179}$$

as is also illustrated in Fig. 15.19. Here the transducer to be determined has a sensitivity coefficient, M_x, and is rigidly connected to two other reversible transducers. *First measurement:* The reversible transducer, 1, is in its transmitting mode, while the receiving transducers, 2 and x, are in their idle state, that is,

$$\underline{F}_2 = 0 = \underline{F}_x. \tag{15.180}$$

Substituting (15.180) into (15.176) yields

$$\underline{v} = \frac{-\underline{F}_1}{j\omega m}. \tag{15.181}$$

Now we measure the output voltages of transducer 2 and transducer x. Employing (15.181), (15.178), and (15.179), we find

$$\underline{u}_{21} = M_2 \underline{v} = M_2 \frac{-\underline{F}_1}{j\omega m} = \frac{-\underline{i}_1}{j\omega m} M_1 M_2, \tag{15.182}$$

$$\underline{u}_{x1} = M_x \underline{v} = M_x \frac{-\underline{F}_1}{j\omega m} = \frac{-\underline{i}_1}{j\omega m} M_1 M_x. \tag{15.183}$$

The subscripts of both voltages, $\underline{u}_{21}$, and $\underline{u}_{x1}$, denote that they are driven by transducer 1.

Dividing (15.183) by (15.182) results in

$$\frac{\underline{u}_{x1}}{\underline{u}_{21}} = \frac{M_x}{M_2}. \tag{15.184}$$

Second measurement: The transducer 2 is now in its transmitting mode, while the receiving transducers, 1 and x, are in their idle state, that is,

$$\underline{F}_1 = 0 = \underline{F}_x. \tag{15.185}$$

Substituting (15.185) into (15.176) gives

$$\underline{v} = \frac{-\underline{F}_2}{j\,\omega\,m}. \tag{15.186}$$

Here, only the output voltage of transducer x needs to be measured. Using (15.186), (15.178), and (15.179), we get

$$\underline{u}_{x2} = M_x\,\underline{v} = M_x\,\frac{-\underline{F}_2}{j\,\omega\,m} = \frac{-\underline{i}_2}{j\,\omega\,m}\,M_x\,M_2, \tag{15.187}$$

where the subscript of voltage $\underline{u}_{x2}$, denotes that it is driven by the transducer 2.

Re-arranging (15.187) yields

$$\left|\frac{\underline{u}_{x2}}{\underline{i}_2}\right| = \left|\frac{M_x\,M_2}{j\,\omega\,m}\right|, \tag{15.188}$$

and multiplying both sides of (15.184) with those of (15.188) leads to

$$\left|\frac{\underline{u}_{x2}}{\underline{i}_2}\right|\left|\frac{\underline{u}_{x1}}{\underline{u}_{21}}\right| = \frac{M_x\,M_2}{\omega\,m}\,\frac{M_x}{M_2} = \frac{M_x^2}{\omega\,m}. \tag{15.189}$$

Solving for $\dot{M}_x$ results in

$$M_x = \sqrt{\omega\,m\,\left|\frac{\underline{u}_{x2}}{\underline{i}_2}\right|\left|\frac{\underline{u}_{x1}}{\underline{u}_{21}}\right|}. \tag{15.190}$$

The result in (15.190) indicates that only electric measurements, that is, measurements of $\underline{u}_{x1}$, $\underline{u}_{x2}$, $\underline{u}_{21}$, and $\underline{i}_2$, are necessary to determine the sensitivity coefficient, M_x, given that the operating frequency and the mass are known or measurable.

Problem 4.6.

Proposed Approach

Task (**a**) If a transmission system consists of subsystems connected in series, and each of them is represented by a chain matrix, the entire system results from multiplying the individual chain matrices.

Task (**b**) Figure 15.20a illustrates a transmission system as a "black box". The input and output variables are $\underline{u}_1$, $\underline{i}_1$, and $\underline{u}_2$, $\underline{i}_2$, respectively. The system is represented by a matrix as follows,

Fig. 15.20 Transmission systems **(a)** "Black box" with input and output ports **(b)** and **(c)** Two-port representation of acoustic and/or electric elements

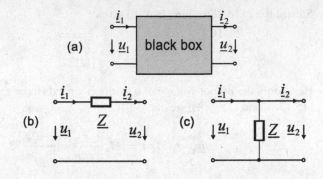

$$\begin{bmatrix} \underline{u}_1 \\ \underline{i}_1 \end{bmatrix} = \begin{bmatrix} \underline{A}_{11} & \underline{A}_{12} \\ \underline{A}_{21} & \underline{A}_{22} \end{bmatrix} \begin{bmatrix} \underline{u}_2 \\ \underline{i}_2 \end{bmatrix}, \tag{15.191}$$

or, respectively,

$$\underline{u}_1 = \underline{A}_{11}\underline{u}_2 + \underline{A}_{12}\underline{i}_2$$
$$\underline{i}_1 = \underline{A}_{21}\underline{u}_2 + \underline{A}_{22}\underline{i}_2 \tag{15.192}$$

The four matrix elements are defined as follows,

$$\underline{A}_{11} = \left.\frac{\underline{u}_1}{\underline{u}_2}\right|_{\underline{i}_2=0} \qquad \underline{A}_{12} = \left.\frac{\underline{u}_1}{\underline{i}_2}\right|_{\underline{u}_2=0}$$

$$\underline{A}_{21} = \left.\frac{\underline{i}_1}{\underline{u}_2}\right|_{\underline{i}_2=0} \qquad \underline{A}_{22} = \left.\frac{\underline{i}_1}{\underline{i}_2}\right|_{\underline{u}_2=0} \tag{15.193}$$

Task **(c)** For a network as illustrated in Fig. 15.20b, the matrix elements are given by

$$\underline{A}_{11} = 1 \qquad \underline{A}_{12} = \underline{Z}$$
$$\underline{A}_{21} = 0 \qquad \underline{A}_{22} = 1 \tag{15.194}$$

Similarly, for a network as illustrated in Fig. 15.20c, the matrix elements read as follows,

$$\underline{A}_{11} = 1 \qquad \underline{A}_{12} = 0$$
$$\underline{A}_{21} = 1/\underline{Z} \qquad \underline{A}_{22} = 1 \tag{15.195}$$

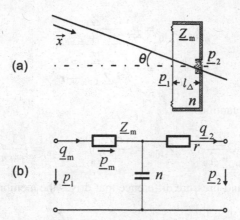

Fig. 15.21 Microphone in the sound field and its analog circuit (a) Plane-wave incident upon the microphone membrane (b) Electroacoustic analog circuit

Problem 4.7.

Proposed Approach

A basic assumption here is that the microphone's linear dimensions are small enough to not significantly disturb the sound field around it. Further, we assume plane-wave propagation since the wavelengths under which the device is operating are much larger than its dimension. The microphone is orientated arbitrarily at an angle, θ, relative to the direction of plane-wave propagation, $\vec{x}$ —as depicted in Fig. 15.21a.

As illustrated in Figs. 4.23 and 15.21a, the microphone membrane possesses an impedance, $\underline{Z}_m$, toward the sound field. The volume behind the microphone membrane exhibits a compliance, n. A damped opening at the back side represents an acoustic resistance, r. Figure 15.21b shows a respective analog circuit. The solution needs to focus on the sound pressure on the membrane, $\underline{p}_m$.

Under the plane-wave assumption, the sound pressure on the front side of the microphone membrane—according to Fig. 15.21a—is

$$\underline{p}(x) = \underline{p}_0 \, e^{-j\beta x}, \tag{15.196}$$

where $\beta = 2\pi/\lambda = \omega/c$ is the propagation coefficient. λ, ω, and c are wavelength, angular frequency, and sound speed, respectively.

The sound pressure at the back side of the microphone is given by

$$\underline{p}_2 = \underline{p}_1 + \frac{\partial \underline{p}_1}{\partial x} L \cos\theta,$$
$$= \underline{p}_1 (1 - j\beta L \cos\theta). \tag{15.197}$$

Figure 15.21b also shows that

$$\underline{p}_m\Big|_{\underline{p}_2=0} = \underline{p}_1 \frac{\underline{Z}_m}{\underline{Z}_m + \underline{Z}_{r\|n}},$$

$$\underline{p}_m\Big|_{\underline{p}_1=0} = -\underline{p}_2 \frac{\underline{Z}_{m\|n}}{\underline{Z}_{m\|n} + r}, \tag{15.198}$$

with the following quantities,

$$\underline{Z}_{m\|n} = \frac{\underline{Z}_m \underline{Z}_n}{\underline{Z}_m + \underline{Z}_n}, \qquad \underline{Z}_{r\|n} = \frac{r \underline{Z}_n}{r + \underline{Z}_n}, \qquad \underline{Z}_n = \frac{1}{j\omega n}. \tag{15.199}$$

This leads to the sound-pressure difference that drives the membrane, namely,

$$\underline{p}_m = \underline{p}_1 \frac{\underline{Z}_m}{\underline{Z}_m + \underline{Z}_{r\|n}} - \underline{p}_2 \frac{\underline{Z}_{m\|n}}{\underline{Z}_{m\|n} + r}. \tag{15.200}$$

Substituting (15.197) into (15.200) provides

$$\underline{p}_m = \underline{p}_1 \left[\frac{\underline{Z}_m}{\underline{Z}_m + \underline{Z}_{r\|n}} - \frac{\underline{Z}_{m\|n}(1 - j\beta L_\delta \cos\theta)}{\underline{Z}_{m\|n} + r} \right], \tag{15.201}$$

and substituting (15.199) into (15.201) leads to

$$\begin{aligned}
\underline{p}_m &= \underline{p}_1 \left[\frac{\underline{Z}_m(r + \underline{Z}_n)}{r\underline{Z}_m + \underline{Z}_m\underline{Z}_n + r\underline{Z}_n} - \frac{\underline{Z}_m\underline{Z}_n(1 - j\beta L_\delta \cos\theta)}{r\underline{Z}_m + \underline{Z}_m\underline{Z}_n + r\underline{Z}_n} \right] \\
&= \underline{p}_1 \left[\frac{\underline{Z}_m(r + \underline{Z}_n) - \underline{Z}_m\underline{Z}_n(1 - j\beta L_\delta \cos\theta)}{r\underline{Z}_m + \underline{Z}_m\underline{Z}_n + r\underline{Z}_n} \right], \\
&= \underline{p}_1 \left[\frac{\underline{Z}_m r + \underline{Z}_m\underline{Z}_n j\beta L_\delta \cos\theta}{r\underline{Z}_m + \underline{Z}_m\underline{Z}_n + r\underline{Z}_n} \right].
\end{aligned} \tag{15.202}$$

Finally, substituting $\underline{Z}_n$ in (15.199) into (15.202) produces

$$\begin{aligned}
\underline{p}_m &= \underline{p}_1 \underline{Z}_m r \left[\frac{1 + \frac{L_\delta}{cnr}\cos\theta}{r\underline{Z}_m + \underline{Z}_m\underline{Z}_n + r\underline{Z}_n} \right] \\
&= \underline{p}_1 \underline{A}(1 + B\cos\theta),
\end{aligned} \tag{15.203}$$

with

$$\underline{A} = \frac{\underline{Z}_m r}{r\underline{Z}_m + \underline{Z}_m\underline{Z}_n + r\underline{Z}_n}, \tag{15.204}$$

and

$$B = \frac{L}{cnr}. \tag{15.205}$$

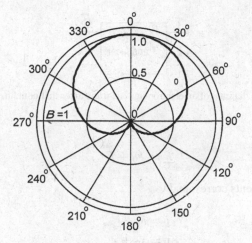

Fig. 15.22 Directional characteristic of the microphone as shown in Fig. 4.23

Consequently, the directional characteristic results in

$$\Gamma = \frac{|\underline{p}_m|}{|\underline{p}_{m,\max}|} = 1/2 + B/2 \cos\theta, \tag{15.206}$$

where B is real-valued and often equals $B \le 1$, depending on the sound speed, the cavity volume behind the membrane, and the resistive damper. Figure 15.22 illustrates, as an example, the directional characteristic for $B = 1$.

15.5   Chapter 5

Problem 5.1.

Proposed Approach

Figure. 5.12 shows the inner transducer of an electromagnetic transducer as a two-port element. The functional relationship of the electric and mechanic sides are illustrated in Fig. 15.23 and represented in the subsequent matrix,

$$\begin{pmatrix} \underline{u} \\ \underline{i} \end{pmatrix} = \begin{pmatrix} A_{11} & A_{12} \\ A_{21} & A_{22} \end{pmatrix} \begin{pmatrix} \underline{v} \\ \underline{F} \end{pmatrix}. \tag{15.207}$$

as well as in the equations,

$$\underline{u} = A_{11}\,\underline{v} + A_{12}\,\underline{F}$$
$$\underline{i} = A_{21}\,\underline{v} + A_{22}\,\underline{F}. \tag{15.208}$$

where the force consists of a constant (bias) force, F_0 and an alternating force,

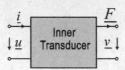

Fig. 15.23 Schematic diagram of an electromagnetic transducer that contains an inner transducer as shown in Fig. 5.12

$$\underline{F} = F_0 + \underline{F}_\sim.$$

These two components correspond to

$$\underline{i} = i_0 + \underline{i}_\sim,$$
$$\underline{x} = x_0 + \underline{x}_\sim, \tag{15.209}$$
$$\underline{\Phi} = \Phi_0 + \underline{\Phi}_\sim.$$

The force is given by (5.15), that is,

$$\underline{F} = \frac{\nu^2 \mu_0 A}{4\, x^2}\, i^2 = \frac{\Phi^2}{\mu_0 A}\,. \tag{15.210}$$

whereby,

$$|\,\underline{\Phi}\,| = L \cdot i\,. \tag{15.211}$$

From either (15.207) or (15.208) it follows,

$$A_{11} = \left.\frac{\underline{u}}{\underline{v}}\right|_{\underline{F}_\sim = 0;\ \underline{F} = F_0}. \tag{15.212}$$

This condition leads to

$$\underline{\Phi} = \Phi_0, \quad \underline{u} = -\frac{d\underline{\Phi}}{dt} \quad \Rightarrow \quad \underline{u} = 0, \quad \text{and} \quad A_{11} = 0. \tag{15.213}$$

Also, again from (15.207) or (15.208), we write

$$A_{12} = \left.\frac{\underline{u}}{\underline{F}}\right|_{\underline{v} = 0;\ L = \text{const.}}. \tag{15.214}$$

Note that the magnetic flux, Φ, consists of two components, a constant one, Φ_0, and an alternating one, $\underline{\Phi}_\sim$, aiming at $\Phi_0 \gg \underline{\Phi}_\sim$ to achieve linearization.

Substituting $\underline{\Phi} = \Phi_0 + \underline{\Phi}_\sim$ with $\Phi_0 \gg \underline{\Phi}_\sim$ into (15.210) leads to

$$\underline{F} = \frac{(\Phi_0 + \underline{\Phi}_\sim)^2}{\mu_0 A} = \frac{1}{\mu_0 A}\left(\Phi_0^2 + 2\Phi_0\underline{\Phi}_\sim + \underline{\Phi}_\sim^2\right)$$

$$\approx \frac{1}{\mu_0 A}\left(\Phi_0^2 + 2\Phi_0\underline{\Phi}_\sim\right) = F_0 + \underline{F}_\sim. \tag{15.215}$$

This results in

$$\underline{F}_\sim = \frac{2\Phi_0}{\mu_0 A}\underline{\Phi}_\sim = \frac{2x_0\Phi_0}{x_0\mu_0 A}\underline{\Phi}_\sim = \frac{\Phi_0}{x_0 L_0}\underline{\Phi}_\sim, \tag{15.216}$$

where $L_0 = \mu_0 A/(2x_0)$ comes into use.

Substituting $M = \Phi_0/x_0 = (L_0 i_0)/x_0$ into (15.216) produces

$$\underline{F}_\sim = \frac{M}{L_0}\underline{\Phi}_\sim, \tag{15.217}$$

and multiplication of $j\omega$ on both sides of (15.217) yields

$$j\omega\underline{F}_\sim = \frac{M}{L_0}j\omega\underline{\Phi}_\sim = \frac{M}{L_0}\underline{u}, \tag{15.218}$$

with $\underline{u} = j\omega\underline{\Phi}_\sim$.

Now, substitution of (15.218) into (15.214) results in

$$A_{12} = \frac{\underline{u}}{\underline{F}} = \frac{\underline{u}}{\underline{F}_\sim} = \frac{j\omega L_0}{M}. \tag{15.219}$$

Further, from (15.207) or (15.208) we get

$$A_{21} = \frac{\underline{i}}{\underline{v}}\bigg|_{\underline{F}_\sim=0;\ \underline{F}=F_0;\ \underline{\Phi}=\Phi_0}. \tag{15.220}$$

and, using (15.211)

$$\underline{\Phi} = \Phi_0 = L\,i = \frac{\mu_0 A}{2(x_0 + \underline{x}_\sim)}(i_0 + \underline{i}_\sim), \tag{15.221}$$

we get

$$(i_0 + \underline{i}_\sim) = \frac{2\Phi_0}{\mu_0 A}(x_0 + \underline{x}_\sim), \quad \text{or} \quad \underline{i}_\sim = \frac{2\Phi_0}{\mu_0 A}\underline{x}_\sim = \frac{M}{L_0}\underline{x}_\sim. \tag{15.222}$$

Multiplication with $j\omega$ on both sides of (15.222) brings us to

$$j\omega\,\underline{i}_{\sim} = \frac{M}{L_0}\,j\omega\,\underline{x}_{\sim} = \frac{M}{L_0}\,(-\underline{v}_{\sim}). \tag{15.223}$$

Note the negative sign in Fig. 15.23!

From here we proceed as follows,

$$A_{21} = \frac{\underline{i}}{\underline{v}} = \frac{\underline{i}_{\sim}}{\underline{v}_{\sim}} = \frac{-M}{j\,\omega\,L_0}, \tag{15.224}$$

and, again from (15.207) or (15.208), we obtain

$$A_{22} = \frac{\underline{i}}{\underline{F}}\bigg|_{\underline{v}=0;\ L=L_0;\ \underline{x}=x_0}. \tag{15.225}$$

Using (15.210) generates

$$
\begin{aligned}
\underline{F} = F_0 + \underline{F}_{\sim} &= \frac{\nu^2\mu_0 A}{4\,x_0^2}\,\underline{i}^2 = \frac{\nu^2\mu_0 A}{4\,x_0^2}\,(i_0 + \underline{i}_{\sim})^2 \\
&= \frac{\nu^2\mu_0 A}{4\,x_0^2}\,(i_0^2 + 2\,i_0\,\underline{i}_{\sim} + \underline{i}_{\sim}^2) \\
&\approx \frac{\nu^2\mu_0 A}{4\,x_0^2}\,(i_0^2 + 2\,i_0\,\underline{i}_{\sim}),
\end{aligned}
\tag{15.226}
$$

and

$$\underline{F}_{\sim} = \frac{\nu^2\mu_0 A}{2\,x_0^2}\,i_0\,\underline{i}_{\sim} = \frac{L_0\,i_0}{x_0}\,\underline{i}_{\sim} = M\,\underline{i}_{\sim}, \tag{15.227}$$

prompting

$$A_{22} = \frac{\underline{i}}{\underline{F}} = \frac{\underline{i}_{\sim}}{\underline{F}_{\sim}} = \frac{1}{M}. \tag{15.228}$$

Substituting (15.214), (15.219), (15.224), and (15.228) into the square matrix in (15.207) results in

$$
\begin{pmatrix} A_{11} & A_{12} \\ A_{21} & A_{22} \end{pmatrix} =
\begin{pmatrix} 0 & \dfrac{j\omega L_0}{M} \\[2mm] \dfrac{-M}{j\omega L_0} & \dfrac{1}{M} \end{pmatrix}.
\tag{15.229}
$$

Centered around the electromagnetic transducer in the middle between the electric input side and the mechanic output side, the right-hand side of the matrix above is separable into three sub-matrices in chain, namely,

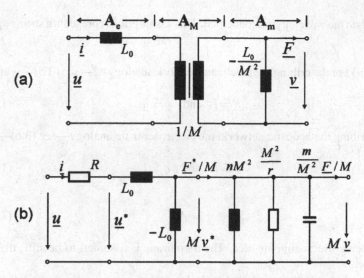

Fig. 15.24 Equivalent circuit of an electromagnetic transducer as shown in Fig. 5.12 according to the analogy #2

$$\begin{pmatrix} 0 & \frac{j\omega L_0}{M} \\ \frac{-M}{j\omega L_0} & \frac{1}{M} \end{pmatrix} = \begin{pmatrix} 1 & j\omega L_0 \\ 0 & 1 \end{pmatrix} \begin{pmatrix} M & 0 \\ 0 & \frac{1}{M} \end{pmatrix} \begin{pmatrix} 1 & 0 \\ \frac{-M^2}{j\omega L_0} & 1 \end{pmatrix} \tag{15.230}$$

$$= \quad \mathbf{A}_e \qquad \mathbf{A}_M \qquad \mathbf{A}_m .$$

The inner transducer is described as an ideal transformer with the transformation coefficient, M, as follows,

$$\mathbf{A}_M = \begin{pmatrix} M & 0 \\ 0 & \frac{1}{M} \end{pmatrix}, \tag{15.231}$$

while

$$\mathbf{A}_e = \begin{pmatrix} 1 & j\omega L_0 \\ 0 & 1 \end{pmatrix}, \quad \text{and} \quad \mathbf{A}_m = \begin{pmatrix} 1 & 0 \\ \frac{-M^2}{j\omega L_0} & 1 \end{pmatrix}. \tag{15.232}$$

The sub-matrices in (15.230), (15.231) and (15.232) facilitate to draw up an equivalent circuit as shown in Fig. 15.24a. To consider the loss, we introduce a resistance, R, at the electric side, while at the mechanic side we add a finite mass, a compliance, and a damper. The compliance includes the field-compliance. In analogy #2, the transformer converts the mechanic elements into electrical ones correspondingly—as shown in Fig. 15.24b.

Problem 5.2.

Proposed Approach

Preliminary remark: In the following, we assume that all acoustically effective elements are lumped ones because the linear dimensions of the elements are much smaller than the wavelengths concerned. This assumption represents a strong approximation.

Task (**a**) For describing the mechanic network, analogy #2—see (3.5)—is applied,

$$\underline{F} \hat{=} \underline{i} \quad \text{and} \quad \underline{v} \hat{=} \underline{u}. \tag{15.233}$$

For describing the acoustic network, the electroacoustic analogy—see (3.6)—is utilized, that is,

$$\underline{p} \hat{=} \underline{u} \quad \text{and} \quad \underline{q} \hat{=} \underline{i}, \tag{15.234}$$

where

$$\underline{q} = A\,\underline{v}, \tag{15.235}$$

with A being the membrane area. The membrane is assumed to be stiff, massless, and plane.

The mechanic-force/sound-pressure relationship is

$$\underline{p}_f\, A = \underline{F}_m + \underline{p}_b\, A, \tag{15.236}$$

where $\underline{p}_f$, and $\underline{p}_b$ represent the sound pressures in front of and at the back of the membrane, respectively. $\underline{F}_m$ represents the total force imposed on the membrane,

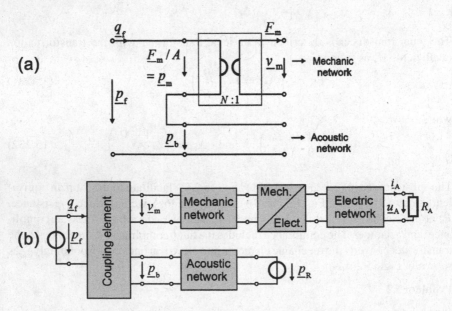

Fig. 15.25 Concerning the electromagnetic transducer shown in Fig. 5.16

including the force component from elastic tensions of the spring-type support, and the force component from the coil attached to it.

Figure 15.25a illustrates an equivalent circuit. Functionally, the circuit that is equivalent to the block diagram shown in Fig. 15.25b. $\underline{u}_A$ and R_A are the voltage and the inner resistance of a driving amplifier.

Mechanical components: The mechanic network features a damper, r_M, a mass, m_M, and a compliance, n_M—see Fig. 15.26a. The mechanic-electric converter is performed by an ideal electromagnetic transformer—compare (5.10) for the transformer equations.

The equivalent circuit is depicted in Fig. 15.26c. On the electric side, a resistance, R, in series with an inductor, L, is connected to a suitable driver (amplifier). The driving amplifier is specified by its current, voltage, and inner resistance, noted by $\underline{i}_A, \underline{u}_A$, and R_A, respectively. Figure 15.26d shows the the electric side. When putting these three components together—(a–c) in Fig. 15.26—the ideal transformer converts the values of the electric components according to (5.8) as follows,

$$\underline{v}_m = \frac{1}{M}\underline{u}_A, \qquad \underline{F}_m = M\underline{i}_A. \tag{15.237}$$

Task, (b) The velocity at the mechanic side is given by

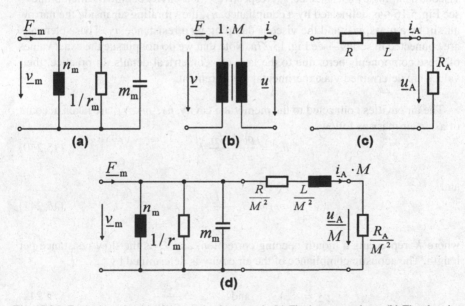

Fig. 15.26 Equivalent circuits of the mechanic network. (a) The inner transducer (b) The electric network (c) The electromagnetic transducer as shown in Fig. 5.16 (d) Circuits (a)–(c) combined

Fig. 15.27 Equivalent
circuits **(a)** Back cavity right
behind the membrane as
shown in Fig. 5.16
(b) Cavity 2 (m_2, r_2, n_2) and
cavity 3 (m_3, r_3, n_3)—in the
same circuit structure

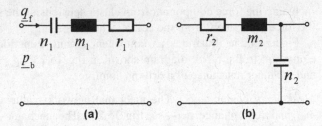

$$\underline{v}_m = \frac{1}{M}\,\underline{u}_A = \frac{1}{M}\,\underline{i}_A(R + j\omega L + R_A)$$

$$= \frac{1}{M^2}\,\underline{F}_m(R + j\omega L + R_A), \qquad (15.238)$$

resulting in

$$\underline{v}_m = \underline{F}_m\left(\frac{R}{M^2} + j\omega\,\frac{L}{M^2} + \frac{R_A}{M^2}\right). \qquad (15.239)$$

Figure 15.26d combines the equivalent circuit of the converted components at the electric side with those at the mechanic side.

Acoustic components: Section 2.6 dealt with acoustic elements. A cavity is described by a compliance, n_A, vibrating air inside a narrow duct/tube by a mass, m_A, and the viscous damping by a resistance, r_A, respectively. The cavity behind the membrane— see Fig. 5.16—is delineated by a compliance, n_1, the vibrating air inside the narrow air slit by a mass, m_1, and the viscous damping by a resistance, r_1. These elements are connected in series—see Fig. 15.27a. Note that we do not pursue the exact values of these components here, due to the lack of geometrical details. In practice, their values are determined via experimental measurement.

The air cavities connected to the membrane cavity, n_1, m_1, r_1, are taken account of as an air mass as follows,

$$m_2 = \frac{\varrho_0\,A_2\,l_{21}}{A_2^2}\,K, \qquad (15.240)$$

and

$$r_2 = \Xi\,\frac{A_2\,l_{22}}{A_2^2}, \qquad (15.241)$$

where K represents a mouth-opening correction, and Ξ is the flow resistance per length. The acoustic compliance of the air cavity is determined by

$$n_2 = \kappa_-V_2, \quad \text{and} \quad \kappa_- = \frac{c_v}{c_p\,p_0}. \qquad (15.242)$$

Since the air cavities are coupled to thin air ducts/tubes on two opposite sides—see Sect. 8.6 for more details—the air cavity, m_2, r_2, n_2, is described as an equivalent circuit—see Fig. 15.27b. Another cavity, m_3, r_3, n_3, is described in a similar way. Further, there is an additional cavity that has an air-slit opening to the rear part of the transducer, r_R, m_R—Fig. 5.16.

Using the equivalent-circuit components of Fig. 15.27, we merge the relevant components into a combined network—see Fig. 15.28.

As indicated in Fig. 15.25, both the mechanic and the electric network are coupled as a whole, whereby the mechanic network is connected with the acoustic network through a gyrator—see Fig. 15.29.

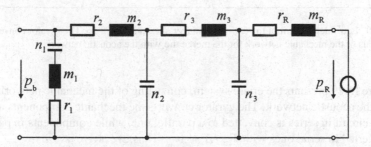

Fig. 15.28 Equivalent circuit of the acoustic network with all the cavity components together

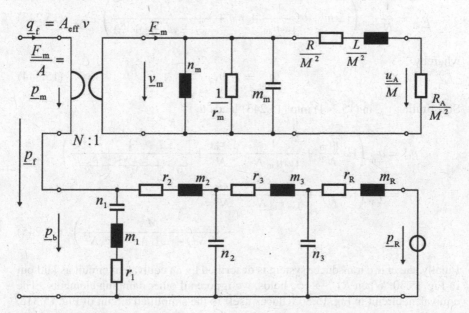

Fig. 15.29 Equivalent circuits of the entire transducer system with the mechanic network and the acoustic network coupled

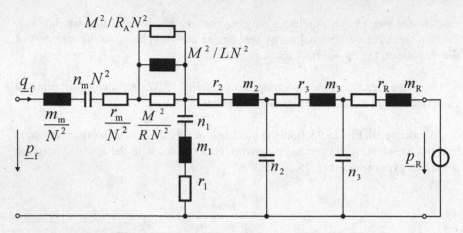

Fig. 15.30 Equivalent circuit of the entire transducer shown in Fig. 5.16. The gyrator converts the components of the mechanic network for the merge the with the acoustic network

Figure 15.29 presents the entire system, consisting of the mechanic network coupled to the acoustic network. The gyrator converts the mechanic components as follows. A circuit in series is converted to a parallel one, while components in parallel are converted to serial ones with $\underline{F}_m/A = \underline{p}_m$. Thus, we arrive at

$$\underline{p}_m = \frac{1}{N}\underline{v}_m\left(j\omega m_m + \frac{1}{j\omega n_m} + r_m + \frac{1}{\frac{R}{M^2} + j\omega\frac{L}{M^2} + \frac{R_A}{M^2}}\right), \qquad (15.243)$$

whereby

$$\underline{q}_m = N\underline{v}_m. \qquad (15.244)$$

Substituting $\underline{v}_m$ in (15.244) into (15.243) leads to

$$\begin{aligned}
p_m &= \underline{q}_m\left(j\omega\frac{m_m}{N^2} + \frac{1}{j\omega n_m N^2} + \frac{r_m}{N^2} + \frac{1}{\frac{RN^2}{M^2} + \frac{j\omega LN^2}{M^2} + \frac{R_A N^2}{M^2}}\right) \\
&= \underline{q}_m\left(j\omega\frac{m_m}{N^2} + \frac{1}{j\omega n_m N^2} + \frac{r_m}{N^2}\right.\\
&\qquad\qquad\left. + \frac{M^2}{RN^2 + j\omega LN^2 + R_A N^2}\right). \quad (15.245)
\end{aligned}$$

Finally, the entire transducer system is described by an equivalent circuit as laid out in Fig. 15.30. When $R_A \rightarrow \infty$ holds, we ignore all other damping elements. The equivalent circuit in Fig. 15.30 reduces itself to the simplified circuit of Fig. 15.31a.

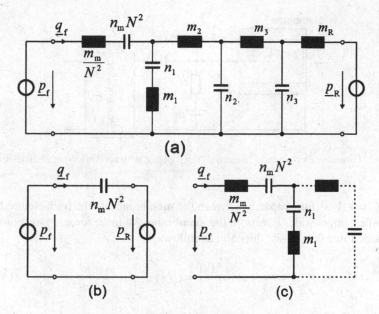

Fig. 15.31 Equivalent circuits with simplifications (a) For $R_A \rightarrow \infty$ (b) For the low-frequency range, derived from (a) (c) For the low-frequency to the middle-frequency range, derived and simplified from (a)

Further, due to this simplification, all inductances in the circuits are negligible in the case of very low operating frequencies—see Fig. 15.31b. This is the case where the transducer (microphone) acts as a gradient receiver.

From the low-frequency range up to the middle frequency range, the simplified circuit takes on the structure shown in Fig. 15.31c, indicating a resonant circuit with inductive and capacitive elements.

15.6 Chapter 6

Problem 6.1.

Proposed Approach

Figure 15.32 illustrates the arrangement of a dielectric microphone. The electric field energy between the membrane and the back electrode is given—see (6.10)—by

$$W_C = \frac{1}{2} C u^2 = \frac{1}{2} \frac{\varepsilon A}{x} u^2, \tag{15.246}$$

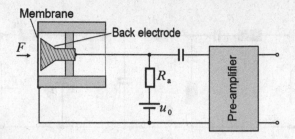

Fig. 15.32 Diagram of a condenser (dielectric) microphone with accompanying electric circuit

where $C = \varepsilon A/x$ is the capacity between the membrane and the back electrode.

A force from outside, F, acts on the membrane. An inner force, F_i, responses to the outside force in opposite direction as follows,

$$F_i = -F(x) = -\frac{\mathrm{d}}{\mathrm{d}x}\left[\frac{1}{2}\,C(x)\,u^2\right] = \frac{A\,\varepsilon}{2}\frac{u^2}{x^2}. \tag{15.247}$$

The electric field reacts with an increase in voltage, $u = u_0 + \mu$, due to a decrease of electrode distance, $x = x_0 - \delta$. This leads to

$$F_i = \frac{A\,\varepsilon}{2}\frac{(u_0 + \mu)^2}{(x_0 - \delta)^2}. \tag{15.248}$$

Differentiating the inner force in (15.248) concerning μ and δ brings

$$\mathrm{d}F_i = \frac{\partial F_i}{\partial \mu}\,\mathrm{d}\mu + \frac{\partial F_i}{\partial \delta}\,\mathrm{d}\delta = A\,\varepsilon\,\frac{u_0 + \mu}{(x_0 - \delta)^2}\,\mathrm{d}\mu + A\,\varepsilon\,\frac{(u_0 + \mu)^2}{(x_0 - \delta)^3}\,\mathrm{d}\delta. \tag{15.249}$$

For a linearized operation, the microphone is usually offset by a high DC voltage with $u_0 \gg \mu$, and $x_0 \gg \delta$,

$$\mathrm{d}F_i \approx A\,\varepsilon\,\frac{u_0}{x_0^2}\,\mathrm{d}\mu + A\,\varepsilon\,\frac{u_0^2}{x_0^3}\,\mathrm{d}\delta. \tag{15.250}$$

Integration of both sides of (15.250) leads to

$$F_i \approx A\,\varepsilon\,\frac{u_0}{x_0^2}\,\mu + A\,\varepsilon\,\frac{u_0^2}{x_0^3}\,\delta. \tag{15.251}$$

Task(**a**) With a static capacity, $C_0 = A\,\varepsilon/x_0$, and a transducer constant, $N = C_0\,u_0/x_0$, which is essentially a transformation factor of the inner transducer (gyrator), the above equation becomes

$$F_i \approx N\mu + \frac{N^2}{C_0}\delta$$
$$= \mu + \frac{1}{j\omega\, C_0/N^2}\, \underline{v} \quad \text{with} \quad \delta = \frac{\underline{v}}{j\omega}, \tag{15.252}$$

leading to

$$\mu \approx \frac{1}{N}\left[F_i - \underbrace{\frac{1}{j\omega\, C_0/N^2}}_{\text{inductance}}\, \underline{v} \right], \tag{15.253}$$

where the second term inside the bracket, which according to analogy #2 (… inductance $\hat{=}$ spring) is of inductive nature. The force and velocity have the same direction/polarity. Therefore, the second term indicates a negative compliance of the electric field between the membrane and the back electrode, that is,

$$-n_f = -\frac{C_0}{N^2}. \tag{15.254}$$

This negative compliance, also called *spring-impedance*, is indicated in Fig. 15.33a.

At the gyrator, we have

$$v = \frac{1}{N}\, i, \quad F = N\, u. \tag{15.255}$$

The analogy #2 with $\underline{v} \hat{=} u$ and $F \hat{=} i$ leads to the electric admittance, Y_{elec}, as

$$\frac{v}{F} = \frac{1}{N^2}\frac{i}{u}, \quad \Leftrightarrow \quad Y_{\text{elec}} = \frac{i}{u} = N^2\frac{v}{F}. \tag{15.256}$$

In particular, we have

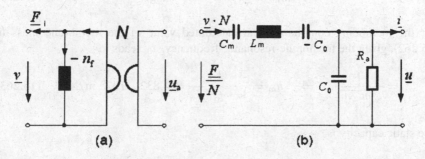

Fig. 15.33 Equivalent circuit of a condenser (dielectric) microphone (**a**) Inner transducer (gyrator) (**b**) Equivalent circuit of the microphone membrane

$$j\omega\, n_m\, N^2 = \underline{Y}_{n,\,elec} \quad\Rightarrow\quad \underline{Z}_{n,\,elec} = \frac{1}{j\omega\, C_m}, \quad \text{with } C_m = n_m\, N^2 \qquad (15.257)$$

and

$$\frac{N^2}{j\omega\, m_m} = \underline{Y}_{m,\,elec} \quad\Rightarrow\quad \underline{Z}_{m,\,elec} = j\omega\, L_m, \quad \text{with } L_m = \frac{m_m}{N^2}. \qquad (15.258)$$

The mechanic "voltage", v, transfers into the electric current, i, and the mechanic "current", F, transfer into the electric voltage, u,—in terms of $v \,\hat{=}\, u,\ F \,\hat{=}\, i$.

After the conversion by the gyrator, we see circuit-equivalence of the mechanic and the analog electric sides. Figure 15.33b shows the equivalent circuit, where

$$N = \frac{C_0\, u_0}{x_0}, \qquad C_0 = \varepsilon_0\, \varepsilon_r\, \frac{A}{x_0}, \qquad (15.259)$$

$$C_m = n_m\, N^2, \quad L_m = \frac{m_m}{N^2}. \qquad (15.260)$$

The total capacity along the series part in Fig. 15.33b becomes

$$C_{sum} = \frac{-C_0\, C_m}{C_m - C_0}. \qquad (15.261)$$

For $C_m \geq C_0$, $C_{sum} < 0$, the "effective" spring is negative. A negative spring has the property that it does not oppose the force from outside but enhances it until, finally, the membrane clings to the back electrode. Figure 15.34 illustrates the complete equivalent circuit.

The effective membrane mass is

$$m_m = \frac{1}{2}\, A\, h\, \varrho = 36.4 \text{ mg}. \qquad (15.262)$$

For the membrane compliance, the calculated value of the membrane mass in (15.262), given the membrane-resonance frequency, f_0, leads to

$$\omega_0 = \frac{1}{\sqrt{m_m\, n_m}}, \quad\Rightarrow\quad n_m = \frac{1}{4\pi^2\, f_0^2\, m_m} = 4.832 \cdot 10^{-6} \text{ m/N}. \qquad (15.263)$$

The static capacity is

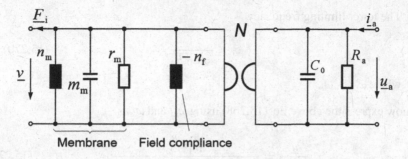

Fig. 15.34 Complete equivalent circuit of a condenser (dielectric) microphone

$$C_0 = \frac{\varepsilon_0 A}{x_0} = 403 \text{ pF}. \tag{15.264}$$

The transformation factor is a

$$N = \frac{u_0 C_0}{x_0} = 3.02 \cdot 10^{-3} \text{ N/V}. \tag{15.265}$$

Further analog electric quantities are

$$C_m = n_m N^2 = 44.1 \text{ pF}, \quad L_m = \frac{m_m}{N^2} = 3.98 \text{ H}. \tag{15.266}$$

Task (b) According to Fig. 15.33b, the output, $\underline{u}$, versus the input, $\underline{F}/N$, is

$$\frac{\underline{u}}{\underline{F}/N} = \frac{\frac{R_a \cdot 1/(j\,\omega\,C_0)}{R_a + 1/(j\,\omega\,C_0)}}{j\,\omega\,L_m - \frac{C_m - C_0}{j\,\omega\,C_m C_0} + \frac{R_a \cdot 1/(j\,\omega\,C_0)}{R_a + 1/(j\,\omega\,C_0)}}$$

$$= \frac{\frac{j\,\omega\,C_m C_0 R_a}{1 + j\,\omega\,R_a C_0}}{C_0 - C_m - \omega^2 L_m C_m C_0 + \frac{j\,\omega\,C_m C_0 R_a}{1 + j\,\omega\,R_a C_0}}, \tag{15.267}$$

$$\frac{\underline{u}}{\underline{F}/N} = \frac{\frac{j\,\omega\,C_m R_a}{1 + j\,\omega\,C_0 R_a}}{1 - \frac{C_m}{C_0} - \omega^2 L_m C_m + \frac{j\,\omega\,C_m R_a}{1 + j\,\omega\,C_0 R_a}}. \tag{15.268}$$

Our further discussion benefits from the following definitions of two angular frequencies, namely,

– The membrane-resonance frequency

$$\omega_m = \frac{1}{\sqrt{L_m C_m}}, \tag{15.269}$$

– The lower limiting frequency

$$\omega_l = \frac{1}{\sqrt{C_0\, R_a}}, \tag{15.270}$$

where $\omega_m > \omega_l$.

We now express the above Eq. (15.268) using ω_m and ω_l as

$$\frac{u}{F/N} = \frac{\frac{j\omega\, C_m\, R_a}{1 + j\omega/\omega_l}}{1 - \frac{C_m}{C_0} - \left(\frac{\omega}{\omega_m}\right)^2 + \frac{j\omega\, C_m\, R_a}{1 + j\omega/\omega_l}}. \tag{15.271}$$

This result facilitates our discussions concerning the transfer function in the three frequency regions listed below.

(1) $\omega \gg \omega_m$

From (15.270)

$$C_m\, R_a\, \omega_l = \frac{C_m\, R_a}{C_0\, R_a} = \frac{C_m}{C_0} \quad\Rightarrow\quad \frac{j\omega\, C_m\, R_a}{1 + j\omega/\omega_l} \approx \frac{C_m}{C_0}, \tag{15.272}$$

substitution of (15.272) into (15.271) leads to

$$\frac{u}{F/N} = \frac{C_m/C_0}{1 - (\omega/\omega_m)^2} \approx \frac{C_m/C_0}{(\omega/\omega_m)^2}, \tag{15.273}$$

indicating a decrease rate of 40 dB/decade—see Fig. 15.35.

(2) $\omega \ll \omega_l$

From (15.271) it follows

$$\frac{j\omega\, C_m\, R_a}{1 + j\omega/\omega_l} \approx j\omega\, C_m\, R_a, \tag{15.274}$$

leading to

$$\left| \frac{u}{F/N} \right| \approx \left| \frac{j\omega\, C_m\, R_a}{1 - \frac{C_m}{C_0} + j\omega\, C_m\, R_a} \right| = \left| \frac{j\omega\, C_0\, R_a}{\frac{C_0}{C_m} - 1 + j\omega\, C_0\, R_a} \right|$$

$$\approx \left| \frac{j\omega\, C_0\, R_a}{C_0/C_m - 1} \right|, \tag{15.275}$$

indicating an increase rate of 20 dB/decade—see Fig. 15.35.

(3) $\omega_l \ll \omega \ll \omega_m$

For this frequency range, we obtain

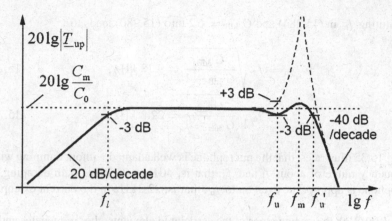

Fig. 15.35 Magnitude spectrum of the condenser (dielectric) microphone

$$\frac{j\omega\, C_{\mathrm{m}}\, R_{\mathrm{a}}}{1 + j\omega/\omega_l} \approx \omega_l\, C_{\mathrm{m}}\, R_{\mathrm{a}} = \frac{C_{\mathrm{m}}}{C_0}. \tag{15.276}$$

The transfer function reduces to

$$\left| \frac{u}{\underline{F}/N} \right| \approx \left| \frac{C_{\mathrm{m}}/C_0}{1 - C_{\mathrm{m}}/C_0 + C_{\mathrm{m}}/C_0} \right| = \frac{C_{\mathrm{m}}}{C_0} \neq f(\omega), \tag{15.277}$$

indicating a frequency independence in this frequency range. The sensitivity coefficient of the microphone results as

$$\underline{T}_{\mathrm{up}} = \left| \frac{u}{\underline{p}} \right| = \left| \frac{u}{\underline{F}} \right| A = \left| \frac{u}{\underline{F}/N} \right| \frac{A}{N}$$

$$= \frac{C_{\mathrm{m}}}{C_0} \frac{A}{N} = \frac{n_{\mathrm{m}} N^2 A}{C_0 N} = \frac{n_{\mathrm{m}} u_0 C_0 A}{C_0 x_0}, \tag{15.278}$$

$$\underline{T}_{\mathrm{up}} = \frac{n_{\mathrm{m}} u_0 A}{x_0} = n_{\mathrm{m}} u_0 \frac{C_0}{\varepsilon}. \tag{15.279}$$

Figure 15.35 plots the magnitude spectrum of the condenser microphone with the limiting frequencies, f_l and f_{m}. Without membrane damping, the transfer function's magnitude (sensitivity coefficient) would become singular around the membrane resonance at f_{m}—indicated by dot-line curves. If the microphone is properly damped the singularity is adequately reduced. In case of sufficient damping, there are two frequencies about ω_{m} or f_{m} at which the normalized transfer function takes on a value of $-3\,\mathrm{dB}$—equivalent to $G_{\mathrm{3dB}} = \sqrt{2}$—that is,

$$f_{\mathrm{u/u'}} = f_{\mathrm{m}} \sqrt{\frac{G_{\mathrm{3dB}}}{G_{\mathrm{3dB}} \pm 1}}. \tag{15.280}$$

Substituting f_m in (15.284) and $G_{3dB} = \sqrt{2}$ into (15.280) leads to

$$f_u = f_m \sqrt{\frac{G_{3dB}}{G_{3dB} + 1}} = 9.18 \text{ kHz},\qquad(15.281)$$

$$f_{u'} = f_m \sqrt{\frac{G_{3dB}}{G_{3dB} - 1}} = 22.2 \text{ kHz}.\qquad(15.282)$$

As Fig. 15.35 illustrates, that the microphone is well adapted without damping within a frequency range of about f_l and f_u, that is, 40 Hz—9.2 kHz. With damping, the microphone is appropriate even up to $f_{u'}$, that is, 22.2 kHz in the current example.

Task (c) With the given values of the associated elements, the frequencies and the sensitivity coefficient, we calculated as follows,

$$f_l = \frac{\omega_l}{2\pi} = \frac{1}{2\pi C_0 R_m} = \frac{1}{2\pi \cdot (403 \text{ /pF}) \cdot (10 \text{ /M}\Omega)} \approx 40 \text{ Hz},\qquad(15.283)$$

$$f_m = \frac{1}{2\pi \sqrt{L_m C_m}} = \frac{1}{2\pi \sqrt{(44.1 \text{ /pF}) \cdot (3.98 \text{ /H})}} \approx 12 \text{ kHz},\qquad(15.284)$$

and

$$\underline{T}_{up} = \frac{n_m u_0 A}{x_0} = \frac{(4.8 \cdot 10^{-6} \text{/m/N}) \cdot (150\text{/V}) \cdot (9.1 \cdot 10^{-4} \text{/m}^2)}{2 \cdot 10^{-5} \text{/m}}$$

$$\approx 33 \frac{\text{m V}}{\text{Pa}}.\qquad(15.285)$$

Problem 6.2.

Proposed Approach

Figure 15.36 shows an equivalent circuit of the electret microphone. Using (6.11), capacities in Fig. 15.36 for each layer in Fig. 6.23 we write

$$C_2 = \frac{\varepsilon_2 A}{d_2}, \quad C_3 = \frac{\varepsilon_3 A}{d_3}, \quad \text{and} \quad C_4 = \frac{\varepsilon_4 A}{d_4},\qquad(15.286)$$

where capacitor, C_3, represents the electret. On its surface, it generates an electric charge, $Q_3 = \sigma A$, thus, that

$$u_3 = \frac{Q_3}{C_3} = \frac{\sigma A}{\varepsilon_3 A/d_3} = \frac{\sigma d_3}{\varepsilon_3},\qquad(15.287)$$

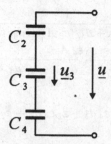

Fig. 15.36 Equivalent circuit of an electric microphone as illustrated in Fig. 6.23

where σ is the electric-charge density. The total capacity, C, is determined by a series of three capacitors as

$$\frac{1}{C} = \frac{1}{C_2} + \frac{1}{C_3} + \frac{1}{C_4}. \tag{15.288}$$

and

$$Q(t) = C(t)\, u(t). \tag{15.289}$$

Note that the capacity is time-dependent!

Figure 15.36 shows an equivalent circuit of the electret microphone. According to (6.11), we find the capacities in Fig. 15.36 for each layer of Fig. 6.23.

Task **(a)** When C changes, the electric charges cannot flow away, in other words,

$$\frac{\mathrm{d}Q(t)}{\mathrm{d}t} = \frac{\mathrm{d}\,[C(t)\,u(t)]}{\mathrm{d}t} = C(t)\frac{\mathrm{d}u(t)}{\mathrm{d}t} + u(t)\frac{\mathrm{d}C(t)}{\mathrm{d}t} \doteq 0, \tag{15.290}$$

leading to

$$\frac{\mathrm{d}u(t)}{\mathrm{d}t} = -\frac{u(t)}{C(t)}\frac{\mathrm{d}C(t)}{\mathrm{d}t}. \tag{15.291}$$

Introducing (6.16) with $d_{41} \ll d_{40}$, and

$$
\begin{aligned}
C &= C_0 + C_\sim, \quad \text{with} \quad C_\sim \ll C_0, \\
u &= u_0 + u_\sim, \quad \text{with} \quad u_\sim \ll u_0,
\end{aligned} \tag{15.292}
$$

into (15.291) leads to

$$\frac{\mathrm{d}u(t)}{\mathrm{d}t} \approx -\frac{u_0}{C_0}\frac{\mathrm{d}C(t)}{\mathrm{d}t} = -\frac{\sigma\, d_3}{\varepsilon_3\, C_0}\frac{\mathrm{d}C(t)}{\mathrm{d}t}, \tag{15.293}$$

where (15.287) comes into use. From (15.288) we obtain

$$C = \frac{1}{d_2/(\varepsilon_2\,A) + d_3/(\varepsilon_3\,A) + d_4/(\varepsilon_4\,A)}$$

$$\doteq \frac{\varepsilon_2\,\varepsilon_3\,\varepsilon_4\,A}{\varepsilon_3\,\varepsilon_4\,d_2 + \varepsilon_2\,\varepsilon_4\,d_3 + \varepsilon_2\,\varepsilon_3\,d_4} = \frac{\varepsilon_2\,\varepsilon_3\,\varepsilon_4\,A}{\Gamma}, \qquad (15.294)$$

with

$$\Gamma = \varepsilon_3\,\varepsilon_4\,d_2 + \varepsilon_2\,\varepsilon_4\,d_3 + \varepsilon_2\,\varepsilon_3\,d_4, \qquad (15.295)$$

$$\frac{dC(t)}{dt} = \varepsilon_2\,\varepsilon_3\,\varepsilon_4\,A\,\frac{d}{dt}\left(\frac{1}{\Gamma}\right) = -\frac{\varepsilon_2\,\varepsilon_3\,\varepsilon_4\,A}{\Gamma^2}\,\frac{d\Gamma}{dt}$$

$$= -\frac{\varepsilon_2\,\varepsilon_3\,\varepsilon_4\,A}{\Gamma^2}\,\varepsilon_2\,\varepsilon_3\,d_{41}\,\omega\,\cos(\omega\,t). \qquad (15.296)$$

Substituting (15.296) into (15.293) results in

$$\frac{du(t)}{dt} \approx -\frac{u_0}{C_0}\,\frac{dC(t)}{dt}$$

$$= \left(-\frac{\sigma\,d_3}{\varepsilon_3}\,\frac{\Gamma}{\varepsilon_2\,\varepsilon_3\,\varepsilon_4\,A}\right)\left[\frac{\varepsilon_2\,\varepsilon_3\,\varepsilon_4\,A}{\Gamma^2}\,\varepsilon_2\,\varepsilon_3\,d_{41}\,\omega\,\cos(\omega\,t)\right]$$

$$= \frac{\sigma\,\varepsilon_2\,d_3}{\Gamma}\,d_{41}\,\omega\,\cos(\omega\,t). \qquad (15.297)$$

Integrating (15.297) renders

$$u_\sim(t) = \frac{\sigma\,\varepsilon_2\,d_3}{\Gamma}\,d_{41}\,\sin(\omega\,t) = \frac{\sigma\,\varepsilon_2\,d_3\,d_{41}\,\sin(\omega\,t)}{\varepsilon_3\,\varepsilon_4\,d_2 + \varepsilon_2\,\varepsilon_4\,d_3 + \varepsilon_2\,\varepsilon_3\,d_4}. \qquad (15.298)$$

Task (**b**) Given $\sigma = 10^{-8}\,\mathrm{C/cm^2}, d_3 = 1.3\,\cdot\,10^{-3}\,\mathrm{cm}$ and $\varepsilon_2 = \varepsilon_3 = 3\,\varepsilon_0$ the bias voltage is

$$u_0 = \frac{\sigma\,d_3}{\varepsilon_3} = 48.9/\mathrm{V}. \qquad (15.299)$$

Task (**c**) The sound pressure is determined by

$$\underline{p} = \frac{1}{j\,\omega\,n_a}\,\underline{q} = \frac{\underline{v}\,A}{j\,\omega\,n_a} = \frac{j\,\omega\,A\,d_{41}}{j\,\omega\,n_a} = \frac{A}{n_a}\,d_{41}, \qquad (15.300)$$

with the volume velocity, $\underline{q} = \underline{v}\,A$, and the velocity, $\underline{v} = j\,\omega\,d_{41}$. This leads to the sensitivity coefficient (transfer function),

$$\underline{T}_{\mathrm{up}} = \frac{\underline{u}_\sim}{\underline{p}} = \frac{\sigma\,\varepsilon_2\,d_3\,d_{41}}{\Gamma}\,\frac{n_a}{A\,d_{41}} = \frac{\sigma\,\varepsilon_2\,d_3\,n_a}{\Gamma\,A}. \qquad (15.301)$$

Problem 6.3.

Proposed Approach

Task **(a)** The series branch in Fig. 6.24 possesses an admittance of

$$\underline{Y}_1 = 1/\underline{Z}_1, \tag{15.302}$$

and a respective impedance of

$$\underline{Z}_1 = R_1 + j\omega L_1 + \frac{1}{j\omega C_2}. \tag{15.303}$$

The angular resonance frequency is

$$\omega_s = \frac{1}{\sqrt{L_1 C_2}}. \tag{15.304}$$

Task **(b)** When the parallel branches are considered together, the total admittance becomes

$$\underline{Y} = \underline{Y}_1 + j\omega C_3. \tag{15.305}$$

According to (6.9), the angular resonance frequency of parallel circuits is determined by

$$\omega_p = \frac{1}{\sqrt{L_1\left(\frac{C_2 C_3}{C_2 + c_3}\right)}}. \tag{15.306}$$

Since $C_3 \gg C_2$, the paralleled capacitance in (15.306) is slightly smaller than C_2, and the two resonance frequencies are, therefore, close together, with $\omega_p > \omega_s$. Figure 15.37a, b illustrate this impedance and its corresponding admittance as functions of frequency in the complex plane.

The relative frequency interval,

$$\frac{\omega_p - \omega_s}{\omega_s} = \frac{\omega_\Delta}{\omega_s} = \frac{\omega_p}{\omega_s} - 1$$

$$= \sqrt{\frac{(C_2 - C_3)(1 - R_1^2 C_3/L_1)}{C_3(1 + R_1^2 C_3/L_1)}} - 1. \tag{15.307}$$

Using an approximation

$$\sqrt{\frac{1-x}{1+x}} \approx 1 - x, \tag{15.308}$$

the above Eq. (15.307) reduces to

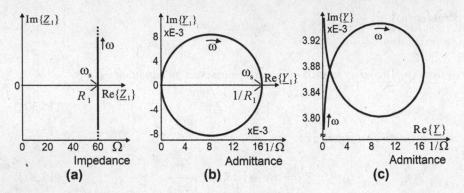

Fig. 15.37 Impedance and admittance of the quartz in Fig. 6.24 as frequency function (**a**) Trajectory of the impedance of the series branch (**b**) Trajectory of the admittance of the series branch, $\underline{Y}_1$ (**c**) Trajectory of the total admittance, $\underline{Y}$

$$\frac{\omega_\Delta}{\omega_s} \approx \sqrt{\frac{C_2 + C_3}{C_3}\left(1 - \frac{R_1^2 C_3}{L_1}\right)} - 1. \tag{15.309}$$

A further approximation renders

$$\sqrt{\frac{C_2 + C_3}{C_3}} \approx 1 + \frac{C_2}{2\,C_3}, \tag{15.310}$$

and the expression (15.309) thus reduces to

$$\frac{\omega_\Delta}{\omega_s} \approx \left(1 + \frac{C_2}{2\,C_3}\right)\left(1 - \frac{R_1^2 C_3}{L_1}\right) - 1$$

$$= \frac{C_2}{2\,C_3} - \frac{R_1^2 C_3}{L_1} - \frac{R_1^2 C_2}{2\,L_1} = \frac{C_2}{2\,C_3}\left[1 - \frac{R_1^2}{L_1}\left(\frac{C_3^2}{C_2} - C_3\right)\right], \tag{15.311}$$

and, further, to

$$\frac{\omega_\Delta}{\omega_s} \approx \frac{C_2}{2\,C_3} = 1.33 \cdot 10^{-4}. \tag{15.312}$$

15.7   Chapter 7

Problem 7.1.

Proposed Approach

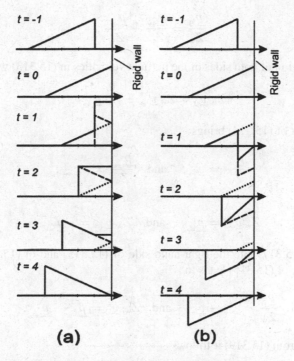

Fig. 15.38 Superposition of an incident sawtooth wave and a reflected wave in a tube with a wall being rigidly terminated (**a**) Pressure components (**b**) Velocity components

Task (**a**) Figure 15.38a, illustrates the wave progression in five steps.

Task (**b**) The progression of the wavefront is outlined in Fig.15.38.

Problem 7.2.

Proposed Approach

At the termination, $x = l = 0$, the sound pressure in (7.40) and the particle velocity (7.41) at the termination become

$$\underline{p}_0 = \underline{p}_\rightarrow + \underline{p}_\leftarrow , \quad \underline{v}_0 = \underline{v}_\rightarrow + \underline{v}_\leftarrow . \tag{15.313}$$

Substitution of $e^{\pm j\beta l}|_{l=0} = 1$ into (7.45) results in

$$Z_w = \frac{\underline{p}_\rightarrow}{\underline{v}_\rightarrow} = -\frac{\underline{p}_\leftarrow}{\underline{v}_\leftarrow} . \tag{15.314}$$

Division of both sides of the pressures in (15.313) by Z_w leads to

$$\frac{\underline{p}_0}{Z_w} = \frac{\underline{p}_\rightarrow}{Z_w} + \frac{\underline{p}_\leftarrow}{Z_w}, \tag{15.315}$$

and multiplication of both sides of the particle velocities in (15.313) with Z_w yields

$$Z_w \underline{v}_0 = Z_w \underline{v}_\rightarrow + Z_w \underline{v}_\leftarrow. \tag{15.316}$$

Rearrangement of (15.314) brings

$$\frac{\underline{p}_\rightarrow}{Z_w} = \underline{v}_\rightarrow, \quad \text{and} \quad \frac{\underline{p}_\leftarrow}{Z_w} = -\underline{v}_\leftarrow, \tag{15.317}$$

and

$$Z_w \underline{v}_\rightarrow = \underline{p}_\rightarrow, \quad \text{and} \quad Z_w \underline{v}_\leftarrow = -\underline{p}_\leftarrow. \tag{15.318}$$

Substituting (15.317) into the right-hand side of (15.315) and of (15.318) into the right-hand side of (15.316) leads to

$$\frac{\underline{p}_0}{Z_w} = \underline{v}_\rightarrow - \underline{v}_\leftarrow \quad \text{and} \quad Z_w \underline{v}_0 = \underline{p}_\rightarrow - \underline{p}_\leftarrow. \tag{15.319}$$

Furthermore, from (15.319) it follows

$$\underline{p}_\rightarrow = Z_w \underline{v}_0 + \underline{p}_\leftarrow \quad \text{and} \quad \underline{p}_\leftarrow = \underline{p}_\rightarrow - Z_w \underline{v}_0, \tag{15.320}$$

and

$$\underline{v}_\rightarrow = \frac{\underline{p}_0}{Z_w} + \underline{v}_\leftarrow \quad \text{and} \quad \underline{v}_\leftarrow = \underline{v}_\rightarrow - \frac{\underline{p}_0}{Z_w}. \tag{15.321}$$

Thereafter, a substitution of $\underline{p}_0 = \underline{p}_\rightarrow + \underline{p}_\leftarrow$ in (15.313) into (15.320) yields

$$\underline{p}_\rightarrow = Z_w \underline{v}_0 + (\underline{p}_0 - \underline{p}_\rightarrow) \quad \text{and} \quad \underline{p}_\leftarrow = (\underline{p}_0 - \underline{p}_\leftarrow) - Z_w \underline{v}_0. \tag{15.322}$$

This results in

$$\underline{p}_\rightarrow = \frac{1}{2}\left(\underline{p}_0 + Z_w \underline{v}_0\right) \quad \text{and} \quad \underline{p}_\leftarrow = \frac{1}{2}\left(\underline{p}_0 - Z_w \underline{v}_0\right). \tag{15.323}$$

as stated in (7.48).

In similar fashion, substitution of $\underline{v}_0 = \underline{v}_\rightarrow + \underline{v}_\leftarrow$ of (15.313) into (15.321) leads to

$$\underline{v}_\rightarrow = \frac{\underline{p}_0}{Z_w} + (\underline{v}_0 - \underline{v}_\rightarrow) \quad \text{and} \quad \underline{v}_\leftarrow = (\underline{v}_0 - \underline{v}_\leftarrow) - \frac{\underline{p}_0}{Z_w}, \tag{15.324}$$

which results in

$$\underline{v}_\rightarrow = \frac{1}{2}\left(\frac{\underline{p}_0}{Z_w} + \underline{v}_0\right), \quad \text{and} \quad \underline{v}_\leftarrow = -\frac{1}{2}\left(\frac{\underline{p}_0}{Z_w} - \underline{v}_0\right), \tag{15.325}$$

as stated in (7.49).

Problem 7.3.

Proposed Approach

The transmission-line equations in a tube at a distance, l, to the termination are given by (7.52) and (7.53) as

$$\underline{p}(l) = \underline{p}_0 \cos{(\beta l)} + j\, Z_w\, \underline{v}_0 \sin{(\beta l)}, \tag{15.326}$$

and

$$\underline{v}(l) = j\frac{\underline{p}_0}{Z_w} \sin{(\beta l)} + \underline{v}_0 \cos{(\beta l)}, \tag{15.327}$$

which is equivalent to

$$\frac{\underline{p}(l)}{\underline{p}_0} = \cos{(\beta l)} + j\frac{Z_w}{Z_0} \sin{(\beta l)}, \tag{15.328}$$

and

$$\frac{\underline{v}(l)}{\underline{v}_0} = \cos{(\beta l)} + j\frac{Z_0}{Z_w} \sin{(\beta l)}. \tag{15.329}$$

Task **(a)** We obtain the input impedance by substituting L for l, and Z_L for Z_w, and then dividing the above two equations by each other, such getting to

$$Z_{\text{input}} = \frac{\underline{p}(L)\underline{v}_0}{\underline{v}(L)\underline{p}_0} = \frac{1}{Z_0}\frac{\cos{(\beta L)} + j\frac{Z_L}{Z_0}\sin{(\beta L)}}{\cos{(\beta L)} + j\frac{Z_0}{Z_L}\sin{(\beta L)}}. \tag{15.330}$$

Task **(b)** For the case $Z_0 = 0$, that is, *soft termination*, $\underline{p}_0 = 0$, substitution of $\underline{p}_0 = 0$, and L for l into (15.326) provides

$$\underline{p}(L) = (0)\cos{(\beta L)} + j\, Z_L\, \underline{v}_0 \sin{(\beta L)} = j\, Z_L\, \underline{v}_0 \sin{(\beta L)}. \tag{15.331}$$

For the case $Z_0 = \infty$, that is, *hard (rigid) termination*, also meaning that $\underline{v}_0 = 0$, substitution of $\underline{v}_0 = 0$ and L for l into (15.327) delivers

$$\underline{p}(L) = \underline{p}_0 \cos{(\beta L)} + j\, Z_L \cdot (0) = \underline{p}_0 \cos{(\beta L)}. \tag{15.332}$$

For $Z_0 = Z_L$, there is no reflection at the terminal. Consequently, from (15.328) we end with the following equation,

$$\frac{\underline{p}(L)}{\underline{p}_0} = \cos{(\beta L)} + j\frac{Z_L}{Z_L}\sin{(\beta L)} = e^{j\beta L}. \tag{15.333}$$

Problem 7.4.

Proposed Approach

Task **(a)** The sound pressure inside the tube is given as follows,

$$\begin{aligned}
\underline{p}(l) &= \hat{p}\left[e^{+j\beta l} + \underline{R}\,e^{-j\beta l}\right] = \hat{p}\left[e^{+j\beta l} + |\underline{R}|\,e^{-j(\beta l - \phi_R)}\right]\\
&= \hat{p}\left[e^{+j\beta l} + |\underline{R}|\,e^{j(\phi_R - \beta l)}\right], \tag{15.334}
\end{aligned}$$

where $\underline{R} = |\underline{R}|e^{-j\phi_R}$.

Firstly, we calculate the absolute value of the pressure field by using $|a+jb| = \sqrt{a^2 + b^2}$,

$$|\underline{p}(l)| = \hat{p}\,|\,[\cos\beta l + j\sin\beta l] + |\underline{R}|\,[\cos(\phi_R - \beta l) \\ + j\sin(\phi_R - \beta l)]\,|. \tag{15.335}$$

After regrouping and taking the square root of the square of the real part and the imaginary part, with $b = \phi_R - \beta l$, we have

$$\begin{aligned}
|\underline{p}(l)| &= \hat{p}\,\sqrt{[\cos(\beta l) + |\underline{R}|\cos b]^2 + [\sin(\beta l) + |\underline{R}|\sin b]^2},\\
&= \hat{p}\,\big[\cos^2(\beta l) + |\underline{R}|^2\cos^2 b + 2|\underline{R}|\cos(\beta l)\cos b\\
&\quad + \sin^2(\beta l) + |\underline{R}|^2\sin b + 2|\underline{R}|\sin(\beta l)\sin b\big]^{1/2}. \tag{15.336}
\end{aligned}$$

This expression turns simpler by applying the trigonometric identity $\sin^2\theta + \cos^2\theta = 1$, namely,

$$|\underline{p}(l)| = \hat{p}\,\sqrt{1 + |\underline{R}|^2 + 2|\underline{R}|\,[\cos(\beta l)\cos b + \sin(\beta l)\sin b]}, \tag{15.337}$$

which we be further reduce with the identity $\cos(a \pm b) = \cos a\,\cos b \mp \sin a\,\sin b$, where $a = \beta l$ and $b = \phi_R - \beta l$.

Therewith it follows

$$|\underline{p}(l)| = \hat{p}\,\sqrt{1 + |\underline{R}|^2 + 2|\underline{R}|\cos(2\beta l - \phi_R)}. \tag{15.338}$$

Because of $\beta = \frac{\omega}{c} = \frac{2\pi f}{c} = \frac{2\pi}{\lambda}$, the substitution leads to

$$|\underline{p}(l)| = \hat{p}\sqrt{1 + |\underline{R}|^2 + 2|\underline{R}|\cos\left(\frac{4\pi}{\lambda}l - \phi_R\right)}. \tag{15.339}$$

Task (**b**) Nodes are located at pressure minima, while anti-nodes are at pressure maxima. The cosine term in the above equation has the value of −1 at the minima and of +1 at the maxima.

Task (**c**) At the minima, $\cos(\frac{4\pi}{\lambda}l - \phi_R) = -1$, we have

$$|\underline{p}(l_{\min})| = \hat{p}\sqrt{1 + |\underline{R}|^2 - 2|\underline{R}|} = \hat{p}\sqrt{(1 - |\underline{R}|)^2} = \hat{p}(1 - |\underline{R}|), \tag{15.340}$$

and

$$\cos\left(\frac{4\pi}{\lambda}l_{\min} - \phi_R\right) = -1 \rightarrow \frac{4\pi}{\lambda}l_{\min} - \phi_R = n\pi, \tag{15.341}$$

where n consists of the set containing all odd integers, that is,

$$l_{\min} = \frac{\lambda}{4\pi}(n\pi + \phi_R). \tag{15.342}$$

In the case of the cosine term, $\cos(\frac{4\pi}{\lambda}l - \phi_R) = +1$, we get

$$|\underline{p}(l_{\max})| = \hat{p}\sqrt{1 + |\underline{R}|^2 + 2|\underline{R}|} = \hat{p}\sqrt{(1 + |\underline{R}|)^2} = \hat{p}(1 + |\underline{R}|), \tag{15.343}$$

and

$$\cos\left(\frac{4\pi}{\lambda}l_{\max} - \phi_R\right) = 1 \rightarrow \frac{4\pi}{\lambda}l_{\max} - \phi_R = n\pi, \tag{15.344}$$

where n consists of the set containing all even integers. Thus follows

$$l_{\max} = \frac{\lambda}{4\pi}(n\pi + \phi_R). \tag{15.345}$$

The standing wave ratio from (15.340) and (15.343) is given by

$$S = \frac{|\underline{p}(l_{\max})|}{|\underline{p}(l_{\min})|} = \frac{1 + |\underline{R}|}{1 - |\underline{R}|}, \tag{15.346}$$

leading to the solution for $|\underline{R}|$,

$$|\underline{R}| = \frac{S - 1}{S + 1}. \tag{15.347}$$

Task **(d)** For the reflectance, $\underline{R}$, at the position $l = 0$, as in (15.313), we find

$$\underline{p}_0 = \underline{p}_{\rightarrow} + \underline{p}_{\leftarrow}, \quad \underline{v}_0 = \underline{v}_{\rightarrow} + \underline{v}_{\leftarrow}, \tag{15.348}$$

and

$$Z_{\mathrm{w}} = \frac{\underline{p}_{\rightarrow}}{\underline{v}_{\rightarrow}} = -\frac{\underline{p}_{\leftarrow}}{\underline{v}_{\leftarrow}}, \tag{15.349}$$

so that

$$\underline{p}_0 / Z_{\mathrm{w}} = \underline{v}_{\rightarrow} - \underline{v}_{\leftarrow}, \quad Z_{\mathrm{w}} \underline{v}_0 = \underline{p}_{\rightarrow} - \underline{p}_{\leftarrow}, \tag{15.350}$$

as also stated in (15.319). This leads to

$$\underline{p}_{\rightarrow} = \frac{1}{2} \left(\underline{p}_0 + Z_{\mathrm{w}} \underline{v}_0 \right), \quad \underline{p}_{\leftarrow} = \frac{1}{2} \left(\underline{p}_0 - Z_{\mathrm{w}} \underline{v}_0 \right). \tag{15.351}$$

Consequently, the reflectance $\underline{R}$, follows as

$$\underline{R} = \frac{\underline{p}_{\leftarrow}}{\underline{p}_{\rightarrow}} = \frac{\underline{p}_0 - Z_{\mathrm{w}} \underline{v}_0}{\underline{p}_0 + Z_{\mathrm{w}} \underline{v}_0} = \frac{\underline{Z}_0 - Z_{\mathrm{w}}}{\underline{Z}_0 + Z_{\mathrm{w}}}, \tag{15.352}$$

since $\underline{Z}_0 = \underline{p}_0 / \underline{v}_0$.

Substituting for $\underline{R}$ leads to

$$\frac{1 + \underline{R}}{1 - \underline{R}} = \frac{1 + \frac{\underline{Z}_0 - Z_{\mathrm{w}}}{\underline{Z}_0 + Z_{\mathrm{w}}}}{1 - \frac{\underline{Z}_0 - Z_{\mathrm{w}}}{\underline{Z}_0 + Z_{\mathrm{w}}}} = \frac{\frac{2\underline{Z}_0}{\underline{Z}_0 + Z_{\mathrm{w}}}}{\frac{2 Z_{\mathrm{w}}}{\underline{Z}_0 + Z_{\mathrm{w}}}} = \frac{\underline{Z}_0}{Z_{\mathrm{w}}}, \tag{15.353}$$

with the final result

$$\underline{Z}_0 = Z_{\mathrm{w}} \frac{1 + \underline{R}}{1 - \underline{R}}. \tag{15.354}$$

Problem 7.5.

Proposed Approach

Assume that the string is stretched along the x–direction—as sketched in Fig. 15.39—and the string's displacement is so small that we consider F_x to be constant. The force component along y–axis, F_y, causes the string to oscillate. The oscillations follow *Newton*'s law

$$F_y = m \ddot{y}. \tag{15.355}$$

Figure 15.39 describes the string along the x–axis in a discrete, elementary way. Considering point k on the string, the forces along the y–axis about point k in Fig. 15.39b is expressible as

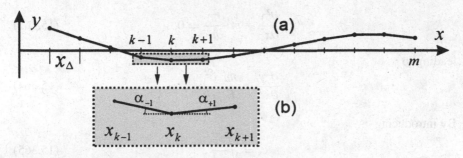

Fig. 15.39 String stretched along the x–direction (**a**) Discretizing the string elements along the x–axis (**b**) Elementary segment about point k

$$F_{y_{-1}} = F_x \sin \alpha_{-1} \approx F_x \frac{y_{k-1} - y_k}{x_\Delta}. \tag{15.356}$$

$$F_{y_{+1}} = F_x \sin \alpha_{+1} \approx F_x \frac{y_{k+1} - y_k}{x_\Delta}, \tag{15.357}$$

where $x_\Delta = x_k - x_{k-1} = x_{k+1} - x_k$. This leads to a pull-back force formed by the sum of $F_{y_{-1}}$ and $F_{y_{+1}}$, namely,

$$F_{y_k} = F_{y_{-1}} + F_{y_{+1}} = F_x \frac{y_{k+1} - 2y_k + y_{k-1}}{x_\Delta}. \tag{15.358}$$

Substituting the force in (15.358) into (15.355)—according to *Newton*'s law—and assuming that the differential string element concerned possesses an elementary mass, m, we obtain

$$F_x \frac{y_{k+1} - 2y_k + y_{k-1}}{x_\Delta} - m \ddot{y}_k = 0, \tag{15.359}$$

and

$$F_x \frac{y_{k+1} - 2y_k + y_{k-1}}{x_\Delta^2} - \frac{m}{x_\Delta} \ddot{y}_k = 0. \tag{15.360}$$

Now let the differential element go to zero, keeping in mind that

$$\lim_{x_\Delta \to 0} \frac{y_{k+1} - 2y_k + y_{k-1}}{x_\Delta^2} = \frac{\partial^2 y}{\partial x^2}, \tag{15.361}$$

and, that for $x_\Delta \to 0$ we have a *mass load*, that is, the following mass distribution along the string,

$$m' = \lim_{x_\Delta \to 0} \frac{m}{x_\Delta}. \tag{15.362}$$

From this we obtain

$$F_x \frac{\partial^2 y}{\partial x^2} - m' \frac{\partial^2 y}{\partial t^2} = 0, \tag{15.363}$$

leading to

$$\frac{\partial^2 y}{\partial x^2} = \frac{m'}{F_x} \frac{\partial^2 y}{\partial t^2}. \tag{15.364}$$

By introducing

$$c_s = \sqrt{\frac{F_x}{m'}}, \tag{15.365}$$

the string oscillation equation results as

$$\frac{\partial^2 y}{\partial x^2} = \frac{1}{c_s^2} \frac{\partial^2 y}{\partial t^2}. \tag{15.366}$$

Note that this equation describes how the string oscillates vertically to the traveling direction of the wave—with a traveling-wave speed of c_s. The expression represents the fundamental equation for transversal waves in one dimension.

15.8   Chapter 8

Problem 8.1.

Proposed Approach

The the acoustic power radiated by a horn, $\overline{Z}$, is proportional to the real part of its radiation impedance, $\underline{Z}_{\mathrm{rad}}$. For *conical horns*, this radiation resistance, r_{rad}, is given according to (8.36), as

$$r_{\mathrm{rad}} = A_0 \varrho c \frac{\left(\frac{\omega}{c} x_0\right)^2}{1 + \left(\frac{\omega}{c} x_0\right)^2}, \tag{15.367}$$

where x_0 is the position of the mouth of the horn, and A_0 the cross-sectional area at the mouth position.

For *exponential horns*, the radiation impedance, according to (8.31) and (8.34), is given as

$$\underline{Z}_{\mathrm{rad}} = A_0 \varrho c \left[\sqrt{1 - \left(\frac{\omega_l}{\omega}\right)^2} + j\left(\frac{\omega_l}{\omega}\right) \right], \tag{15.368}$$

where $\omega_l = \epsilon \cdot c$ is the limiting frequency, with ϵ being the flare coefficient of the horn.

For exponential horns, the radiation impedance becomes imaginary when $\omega < \omega_1$, indicating that the radiation resistance becomes zero. This means that there is no radiation power available. Yet, for $\omega > \omega_1$, the radiation resistance gets

$$r_{\text{rad}} = A_0\,\varrho\,c\,\sqrt{1 - \left(\frac{\omega_1}{\omega}\right)^2}, \tag{15.369}$$

indicating that the horn radiates sound power.

In contrast to exponential horns, the radiation resistance of conical horns is nonzero even at low frequencies. This means that the horn still radiates some sound power even when, according to $\omega \to 0$, r_{rad} is decreasing. With increasing frequency, $\omega \to \infty$, the radiations resistance, r_{rad}, approaches $A_0\,\varrho\,c$. Conical horns then radiate power like tubes with a uniform cross area, A_0.

Problem 8.2.

Proposed Approach

Task (a) The field impedance of an exponential horn is

$$\underline{Z}_f = \varrho\,c\left[\sqrt{1 - \left(\frac{\omega_1}{\omega}\right)^2} + j\left(\frac{\omega_1}{\omega}\right)\right]. \tag{15.370}$$

With the cross-sectional area, A_0, at the mouth of the horn, the mechanical input impedance, (8.34), is

$$\underline{Z}_{\text{mech}} = A_0\,\varrho\,c\left[\sqrt{1 - \left(\frac{\omega_1}{\omega}\right)^2} + j\left(\frac{\omega_1}{\omega}\right)\right]. \tag{15.371}$$

For frequencies $\omega \gg \omega_l$, the term ω_1/ω inside (15.371) becomes negligible, and the radiation (mechanic) impedance of the horn approaches

$$\underline{Z}_{\text{mech}} \to A_0\,\varrho\,c. \tag{15.372}$$

The radiated power of the exponential horn is

$$\overline{P} = \frac{1}{2}\,\text{Re}\{\underline{Z}_{\text{rad}}\}\,|\underline{v}|^2, \tag{15.373}$$

where the real part of the radiation impedance, according to (15.371), is

$$\text{Re}\{\underline{Z}_{\text{rad}}\} = \text{Re}\{\underline{Z}_{\text{mech}}\} = A_0\,\varrho\,c\,\sqrt{1 - \left(\frac{\omega_1}{\omega}\right)^2}. \tag{15.374}$$

For frequencies $\omega \gg \omega_1$, the radiation power becomes

$$\overline{P}_\infty = \frac{1}{2} A_0 \varrho c |\underline{v}|^2, \tag{15.375}$$

with the velocity, $\underline{v}$, being constant.

For the particular frequency $f = 500\,\text{Hz}$, the radiated power gets

$$\overline{P}_{500\text{Hz}} = \frac{1}{2} A_0 \varrho c |\underline{v}|^2 \sqrt{1 - \left(\frac{f_1}{f_{500\,\text{Hz}}}\right)^2}, \tag{15.376}$$

and the ratio of the two powers results in

$$\overline{P}_{500\,\text{Hz}} / \overline{P}_\infty = \sqrt{1 - \left(\frac{f_1}{f_{500\,\text{Hz}}}\right)^2}. \tag{15.377}$$

Task (**b**) At the frequency of 500 Hz, we obtain

$$10 \log_{10} \left[\overline{P}_{500\,\text{Hz}} / \overline{P}_\infty \right] =$$

$$10 \log_{10} \left[\sqrt{1 - \left(\frac{f_1}{500/\text{Hz}}\right)^2} \right] = -3 \text{ dB}. \tag{15.378}$$

Substitution of $2\pi f_1 = \omega_1 = \varepsilon \cdot c$ and $c = 344\,\text{m/s}$ yields

$$\sqrt{1 - \left(\frac{\omega_1}{2\pi \cdot 500/\text{Hz}}\right)^2} = \frac{1}{2}, \tag{15.379}$$

and

$$\frac{\varepsilon \cdot 344/(\text{m/s})}{2\pi \cdot 500/\text{Hz}} = \sqrt{\frac{3}{4}}. \tag{15.380}$$

Hence, the flare coefficient, ε, is found to be

$$\varepsilon = \frac{\sqrt{3} \cdot 500/\text{Hz} \cdot \pi}{344/(\text{m/s})} \, \text{m}^{-1} = 7.91\,\text{m}^{-1}. \tag{15.381}$$

Task (**c**) Figure 15.40a illustrates a sketch of the pressure chamber. A_1 and A_2 are cross-sectional areas of the input and output sides of the pressure chamber. Thereby we assume $l \ll \lambda$.

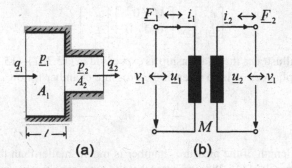

Fig. 15.40 (a) Pressure chamber (b) Equivalent circuit

We express the boundary conditions for the sound pressures and the volume velocities in the pressure chamber as follows,

$$\underline{p}_1 = \underline{p}_2 \qquad \underline{q}_1 = \underline{q}_2 \qquad (15.382)$$

To develop an equivalent mechanic circuit of the pressure chamber and the horn, we consider the following variables,

$$\underline{F}_1 = \underline{p}_1 A_1 \qquad \underline{F}_2 = \underline{p}_2 A_2 \qquad (15.383)$$

$$\underline{q}_1 = \underline{v}_1 A_1 \qquad \underline{q}_2 = \underline{v}_2 A_2 \qquad (15.384)$$

The boundary conditions in (15.382) lead to

$$\frac{\underline{F}_1}{A_1} = \frac{\underline{F}_2}{A_2} \qquad \underline{v}_1 A_1 = \underline{v}_2 A_2 \qquad (15.385)$$

We rewrite this in matrix form as

$$\begin{bmatrix} \underline{F}_1 \\ \underline{v}_1 \end{bmatrix} = \begin{bmatrix} 1/M & 0 \\ 0 & M \end{bmatrix} \begin{bmatrix} \underline{F}_2 \\ \underline{v}_2 \end{bmatrix}, \qquad (15.386)$$

with $M = A_2/A_1$.

Taking analogy #2 as in (3.5) leads to

$$\underline{F} \,\hat{=}\, \underline{i} \qquad \underline{v} \,\hat{=}\, \underline{u} \qquad (15.387)$$

from which follows the matrix

$$\begin{bmatrix} \underline{u}_1 \\ \underline{i}_1 \end{bmatrix} = \begin{bmatrix} M & 0 \\ 0 & 1/M \end{bmatrix} \begin{bmatrix} \underline{u}_2 \\ \underline{i}_2 \end{bmatrix}. \tag{15.388}$$

Figure 15.40b illustrates the relationship as expressed in (15.387)–(15.388). Furthermore, the compliances on both sides of the pressure chamber follow

$$n_1 \propto \frac{1}{A_1} \qquad n_2 \propto \frac{1}{A_2} \tag{15.389}$$

Given that the length of the pressure chamber is much smaller than the wavelength of interest, the result of (15.388) indicates that the pressure chamber in the mechanic subsystem is equivalent to an ideal transformer with the transformation coefficient,

$$M = \frac{n_1}{n_2} = \frac{A_2}{A_1}. \tag{15.390}$$

Task **(d)** To develop an equivalent circuit for the entire system with $\omega \gg \omega_1$, we simplify the equivalent circuit for the frequency range where the frequency course remains constant.

The moving coil inside the air-gap of a magnet builds up an electromagnetic transducer that links the force and the velocity of the moving coil to the electrical current flowing through it and to the electrical voltage across it, namely,

$$\underline{F} = B l \, \underline{i} \quad \text{and} \quad \underline{v} = \frac{\underline{u}}{B l}, \tag{15.391}$$

with B being the magnetic flux.

The quantity, $B \cdot l$, is the transformation coefficient of the electromagnetic transducer. The electric side also contains an inductance, L, and a resistance, R_l. At the mechanic side, we have a mass, m, a spring, n, and some losses expressed as damping, r—compare Fig. 5.3.

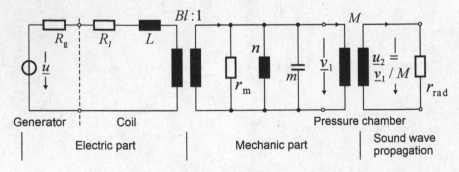

Generator | Coil

Electric part

Pressure chamber

Mechanic part

Sound wave propagation

Fig. 15.41 Equivalent circuit of the complete loudspeaker system as in Fig. 8.12

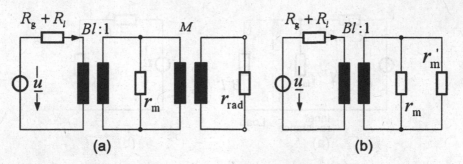

Fig. 15.42 Simplified equivalent circuits of the horn loudspeaker system as in Fig. 8.12 **(a)** The circuit consists of electric, mechanic, and acoustic parts, which are connected by two transformers (moving coils in the static magnet field and the pressure chamber) **(b)** The radiation resistance at the acoustic side is transformed to the mechanic side, there denoted r'_m

Figure 15.41 illustrates the complete circuit of the horn loudspeaker, including the pressure-chamber. The entire circuit consists of three parts, an electric, a mechanic, and an acoustic one. When a signal source feeds electric voltage signals into the horn loudspeaker's electric input, this causes an electric current to flow through the moving coil inside the air-gap of the magnet. This current induces a force in the mechanic part of the horn loudspeaker. The moving coil in the static magnetic field acts as an ideal transformer with a transformation coefficient, $B \cdot l$, where l is the effective coil length in the magnet field.

Further, the equivalent circuit reduces to $\omega \gg \omega_1$ because the inductive and the capacitive elements are negligible. This simplification leads to the circuit shown in Fig. 15.42a. A further assumption is that $\mathrm{Re}\{\underline{Z}_m\}$ is dominated by

$$r_{rad} = A_2 \varrho c. \tag{15.392}$$

According to the analog #2 in (15.387), we then have

$$r'_m = \frac{F_1}{\underline{v}_1} \hat{=} \frac{i_1}{\underline{u}_1} = \frac{1}{M^2} \frac{i_2}{\underline{u}_2} \hat{=} \frac{1}{M^2} \frac{F_2}{\underline{v}_2} = \frac{1}{M^2} r_{rad}, \tag{15.393}$$

which means that the pressure chamber transforms the radiation resistance from the acoustic side into the mechanic side through the transformation coefficient, M, and by substituting r_{rad} in (15.392),

$$r'_m = \frac{A_1^2}{A_2^2} r_{rad} = \frac{1}{M^2} A_2 \varrho c, \tag{15.394}$$

Figure 15.42b illustrates this transformation. The two resistances are transformed to the electric side as follows. The transformation coefficient, $B\,l$, applies to transform-

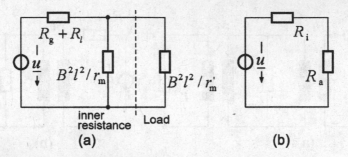

Fig. 15.43 Further simplification of the equivalent circuit of Fig. 15.42

ing the resistances, r_m and r'_m, to the electric ones, namely, $B_2 \, l_2/r_\mathrm{m}$ and $B_2 \, l_2/r'_m$. Figure 15.43a depicts this transformation.

  Task (**e**) The inner source contains a resistive sum, R_g and R_l, paralleled with a transformed resistance, $B^2 \, l^2/r_\mathrm{m}$, while the radiation resistance is transformed to the mechanic side. Further transformation to the electric side renders $B^2 \, l^2/r'_m$ as the load of the equivalent circuit.

  The inner resistance, R_i, of the equivalent source—as shown in Fig. 15.43a— comes out as

$$R_\mathrm{i} = \frac{R_\mathrm{s} \, B^2 \, l^2/r_\mathrm{m}}{R_\mathrm{s} + B^2 \, l^2/r_\mathrm{m}}, \quad \text{with} \quad R_\mathrm{s} = R_\mathrm{g} + R_l \,. \tag{15.395}$$

and the load resistance as

$$R_\mathrm{a} = \frac{B^2 \, l^2}{r'_\mathrm{m}}, \tag{15.396}$$

These two terms should be equal to match the impedances. This match is indeed the condition for achieving maximal radiation power, namely,

$$R_\mathrm{i} = \frac{R_\mathrm{s} \, B^2 \, l^2/r_\mathrm{m}}{R_\mathrm{s} + B^2 \, l^2/r_\mathrm{m}} = \frac{B^2 \, l^2}{r'_\mathrm{m}} = R_\mathrm{a} \,. \tag{15.397}$$

Substitution of r'_m in (15.394) into (15.397) leads to

$$\frac{1}{M^2} = \frac{r_\mathrm{m}}{A_2 \, \varrho \, c} + \frac{B^2 \, l^2}{(R_\mathrm{g} + R_l) \, A_2 \, \varrho \, c} \,. \tag{15.398}$$

The diameter of the loudspeaker membrane, $d = 6\,\mathrm{cm}$, and the specific impedance of air, $\varrho \, c$, are given, so that

$$r_\mathrm{rad} = A_2 \, \varrho \, c = \pi \frac{d^2}{4} \varrho \, c = \frac{2.83 \cdot 10^{-3}}{\mathrm{m}^2} \frac{406}{(\mathrm{Ns/m}^3)} = 1.148 \, \mathrm{Ns/m} \,. \tag{15.399}$$

Substituting (15.399) into (15.398) with given values of r_m [in Ns/m], $B^2 l^2$ [in Vs^2/m^2], and $R_g + R_l$ [in Ω = Vs/Nm] leads to

$$\frac{1}{M^2} = \frac{1.2(\text{Ns/m})}{1.148(\text{Ns/m})} + \frac{44\,(\text{Vs}^2/\text{m}^2)}{(24+24)\,(\text{Vs/Nm}) \cdot 1.148\,(\text{Ns/m})}$$
$$= 1.045 + 0.798, \tag{15.400}$$

resulting in

$$M = \frac{1}{\sqrt{1.045 + 0.798}} = 0.736. \tag{15.401}$$

The pressure chamber ensures that the radiation power becomes maximal. The chamber acts as an ideal transformer, converting the electromechanic transducer's inner resistance to equal the radiation resistance of the horn loudspeaker.

Problem 8.3.

Proposed Approach

The expressions for the radiation resistances in the three different configurations are given by (8.35)–(8.37). The resulting graph looks like Fig. 15.44.

Among the horns that are governed by *Webster*'s equation—that is, by assuming plane waves inside the horn—the exponential horn is the one with the steepest increase of Re$\{\underline{Z}_{rad}\} = r_{rad}$ as a function of frequency. Below $\omega = \omega_1$, there is no wave propagation in exponential horns and, thus, no radiation.

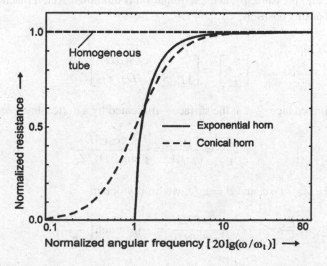

Fig. 15.44 Radiation resistances of a conical and an exponential horn, normalized to $A_0 \varrho c$, and that of a tube with a uniform cross-sectional area

Problem 8.4.

Proposed Approach

Assuming that the tube diameter is much smaller than the wavelength of the rightward propagating wave, we suppose plane waves. From the dotted-line to the rigid termination, the tube length is D—refer to Fig. 8.13. Because the termination is rigid it follows that both the volume velocity, $\underline{q}_t$, and the velocity, $\underline{v}_t$, are zero at the termination.

Written as a matrix—compare(8.48)—we have

$$\begin{bmatrix} \underline{p}_s \\ \underline{q}_s \end{bmatrix} = \begin{bmatrix} \cos(\beta D) & j\, Z_D\, \sin(\beta D) \\ j\, \sin(\beta D)/Z_D & \cos(\beta D) \end{bmatrix} \begin{bmatrix} \underline{p}_t \\ \underline{q}_t \end{bmatrix}, \tag{15.402}$$

where the sound pressure and the volume velocity are noted by $\underline{p}_s$ and $\underline{q}_s$ at the surface indicated by the dotted-line. The tube is assumed to have a cross-sectional area of A and a segment length of D, with the segment impedance $Z_D = Z_w/A = \varrho c/A$ and the volume velocity $\underline{q}_s = \underline{v}_s A$.

Because of $\underline{q}_t = 0$ and $\underline{v}_t = 0$ at the rigid termination the matrix above becomes

$$\begin{bmatrix} \underline{p}_s \\ \underline{q}_s \end{bmatrix} = \begin{bmatrix} \cos(\beta D) & j\, Z_D\, \sin(\beta D) \\ j\, \sin(\beta D)/Z_D & \cos(\beta D) \end{bmatrix} \begin{bmatrix} \underline{p}_t \\ 0 \end{bmatrix}, \tag{15.403}$$

where the sound pressure, $\underline{p}_t$, at the termination is unknown. Yet, it reaches its maximum value at this position.

This leads to

$$\begin{bmatrix} \underline{p}_s \\ \underline{q}_s \end{bmatrix} = \begin{bmatrix} \underline{p}_t\, \cos(\beta D) \\ j\, \underline{p}_t\, \sin(\beta D)/Z_D \end{bmatrix}. \tag{15.404}$$

The (field) impedance, $\underline{Z}_s$, at the surface—indicated by a dotted-line—is

$$\underline{Z}_s = \frac{\underline{p}_s}{\underline{v}_s} = \frac{\underline{p}_s}{\underline{q}_s/A} = \frac{A\, \cos(\beta D)}{j\, \sin(\beta D)/Z_D}. \tag{15.405}$$

With $Z_D A = Z_w = \varrho c$, and $\beta = \omega/c$, we finally obtain

$$\underline{Z}_s = \frac{\underline{p}_s}{\underline{v}_s} = -j Z_D A \frac{\cos(\beta D)}{\sin(\beta D)} = -j\, \varrho c\, \cot\left(\frac{\omega D}{c}\right). \tag{15.406}$$

Problem 8.5.

Proposed Approach

The right end of the tube, A_2, is open, so that $p_m = 0$.[2] As in (8.48), the transmission line equation in matrix form is then given by

$$\begin{bmatrix} \underline{p}_1 \\ \underline{q}_1 \end{bmatrix} = \begin{bmatrix} \cos(\beta l) & j Z_L \sin(\beta l) \\ j \sin(\beta l)/Z_L & \cos(\beta l) \end{bmatrix} \begin{bmatrix} \underline{p}_2 \\ \underline{q}_2 \end{bmatrix}. \tag{15.407}$$

Task (a) By employing the transmission-line equation matrix, we get

$$\begin{bmatrix} \underline{p}_g \\ \underline{q}_g \end{bmatrix} = [M_1][M_2] \begin{bmatrix} \underline{p}_m \\ \underline{q}_m \end{bmatrix}, \tag{15.408}$$

where $[M_1]$ is the transmission line matrix for the first tube segment, and $[M_2]$ the one for the second tube segment.

The equations for the first tube segment are

$$\begin{bmatrix} \underline{p}_g \\ \underline{q}_g \end{bmatrix} = \begin{bmatrix} \cos(\beta l_1) & j Z_{l_1} \sin(\beta l_1) \\ j \sin(\beta l_1)/Z_{l_1} & \cos(\beta l_1) \end{bmatrix} \begin{bmatrix} \underline{p}_0 \\ \underline{q}_0 \end{bmatrix}, \tag{15.409}$$

where the output of the first tube is $\underline{p}_0$ and $\underline{q}_0$.

Note that we assume the termination impedance to be very small compared to the tube resistance. This assumption means that the tube segment, l_2, is open, with $\underline{p}_m \approx 0$. Thus, the matrix for the second duct becomes

$$\begin{bmatrix} \underline{p}_0 \\ \underline{q}_0 \end{bmatrix} = \begin{bmatrix} \cos(\beta l_2) & j Z_{l_2} \sin(\beta l_2) \\ j \sin(\beta l_2)/Z_{l_2} & \cos(\beta l_2) \end{bmatrix} \begin{bmatrix} 0 \\ \underline{q}_m \end{bmatrix}. \tag{15.410}$$

Substitution of (15.410) into (15.409) yields

$$\begin{bmatrix} \underline{p}_g \\ \underline{q}_g \end{bmatrix} = [M_1] \cdot [M_2] \begin{bmatrix} 0 \\ \underline{q}_m \end{bmatrix}, \tag{15.411}$$

with

---

[2] Which means that no substantial radiation is assumed.

$$[M_1][M_2] = \begin{bmatrix} N_{11} & N_{12} \\ N_{21} & N_{22} \end{bmatrix} =$$

$$\begin{bmatrix} \cos(\beta l_1) & j\, Z_{l_1} \sin(\beta l_1) \\ j\, \sin(\beta l_1)/Z_{l_1} & \cos(\beta l_1) \end{bmatrix} \begin{bmatrix} \cos(\beta l_2) & j\, Z_{l_2} \sin(\beta l_2) \\ j\, \sin(\beta l_2)/Z_{l_2} & \cos(\beta l_2) \end{bmatrix}, \quad (15.412)$$

whereby N_{11}, N_{12}, N_{21}, and N_{22} are elements of the resulting matrix. They are, however, less relevant for the current problem due to $\underline{p}_m = 0$. Therefore, we write

$$\underline{q}_g = 0 \cdot N_{21} + \left[\cos(\beta l_1) \cos(\beta l_2) - \frac{Z_{l_2}}{Z_{l_1}} \sin(\beta l_1) \sin(\beta l_2) \right] \underline{q}_m, \quad (15.413)$$

which leads to

$$H(\omega) = \frac{\underline{q}_m}{\underline{q}_g} = \frac{1}{\cos(\beta l_1) \cos(\beta l_2) - Z_{l_2} \sin(\beta l_1) \sin(\beta l_2)/Z_{l_1}}. \quad (15.414)$$

Task (**b**) The poles are apparent when the denominator of the above equation assumes zero, that is,

$$\cos(\beta l_1) \cos(\beta l_2) \left[1 - \frac{Z_{l_2}}{Z_{l_1}} \tan(\beta l_1) \tan(\beta l_2) \right] \overset{!}{=} 0, \quad (15.415)$$

This expression needs two conditions to be fulfilled. The first one is βl_1 or $\beta l_2 = \pi/2$. The second one requires that

$$\frac{Z_{l_2}}{Z_{l_1}} \tan(\beta l_1) \tan(\beta l_2) = 1, \quad \text{or} \quad \frac{A_1}{A_2} \tan(\beta l_1) \tan(\beta l_2) = 1. \quad (15.416)$$

Taken that $l_1, l_2 \ll \lambda$, $\tan(\beta l_1) \approx \beta l_1$ and $\tan(\beta l_2) \approx \beta l_2$, we arrive at

$$\frac{A_2}{A_1} = \beta^2 l_1 l_2, \quad (15.417)$$

and, with $\beta = \omega/c$, the second condition for the poles results as

$$\omega^2 = \frac{c^2}{l_1 l_2} \frac{A_2}{A_1}, \quad \text{or} \quad \omega = c \sqrt{\frac{A_2}{l_1 l_2 A_1}}. \quad (15.418)$$

15.9 Chapter 9

Problem 9.1.

Proposed Approach

The product rule of differentiation of the product of two functions is

$$(f \cdot g)' = f'g + fg', \tag{15.419}$$

and *Euler*'s equation in spherical coordinates is

$$-\frac{\partial p}{\partial r} = \varrho \frac{\partial v}{\partial t}. \tag{15.420}$$

The field impedance given by

$$\underline{Z}_{\mathrm{f}} = \frac{\underline{p}_{\rightarrow}(r)}{\underline{v}_{\rightarrow}(r)}. \tag{15.421}$$

Task (a) Firstly, we apply the product rule to expand $\frac{1}{r}\frac{\partial^2(rp)}{\partial r^2}$ as follows,

$$\begin{aligned}
\frac{1}{r}\frac{\partial^2(rp)}{\partial r^2} &= \frac{1}{r}\frac{\partial}{\partial r}\left[\frac{\partial}{\partial r}(rp)\right] = \frac{1}{r}\frac{\partial}{\partial r}\left(r\frac{\partial p}{\partial r} + p\right) \\
&= \frac{1}{r}\left(r\frac{\partial^2 p}{\partial r} + 2\frac{\partial p}{\partial r}\right) = \frac{\partial^2 p}{\partial r} + \frac{2}{r}\frac{\partial p}{\partial r}.
\end{aligned} \tag{15.422}$$

Usage of the product rule renders

$$\frac{1}{r^2}\frac{\partial}{\partial r}\left(r^2\frac{\partial}{\partial r}p\right) = \frac{1}{r^2}\left(r^2\frac{\partial^2 p}{\partial r^2} + 2r\frac{\partial p}{\partial r}\right) = \frac{\partial^2 p}{\partial r^2} + \frac{2}{r}\frac{\partial p}{\partial r}, \tag{15.423}$$

and applying (15.422), (15.423) and the right-hand side of (9.3), the spherical wave equation finally results in

$$\frac{1}{r^2}\frac{\partial}{\partial r}\left(r^2\frac{\partial}{\partial r}p\right) = \frac{\partial^2 p}{\partial r^2} + \frac{2}{r}\frac{\partial p}{\partial r} = \frac{1}{r}\frac{\partial^2(rp)}{\partial r^2} = \frac{1}{c^2}\frac{\partial^2 p}{\partial t^2}. \tag{15.424}$$

Task (b) According to *Euler*'s equation (15.420), the derivative of $\underline{p}$ with respect to r in (9.67), using the product rule, yields

$$-\frac{\partial \underline{p}_{\rightarrow}}{\partial r} = \frac{\partial}{\partial r}\left(\frac{\underline{g}_{\rightarrow}}{r}\mathrm{e}^{-\mathrm{j}\beta r}\right) = -\underline{g}_{\rightarrow}\left(\frac{-\mathrm{j}\,\beta}{c\,r} + \frac{-1}{r^2}\right)\mathrm{e}^{-\mathrm{j}\beta r}. \tag{15.425}$$

Now assume that $\underline{v}(r, t) = \underline{v}(r)\underline{v}(t)$ and $\underline{v}(t) = \mathrm{e}^{\mathrm{j}\omega t}$. Using the right-hand side of (15.420), we get

$$\varrho\frac{\partial \underline{v}(r, t)}{\partial t} = \mathrm{j}\,\varrho\,\omega\,\underline{v}(r, t). \tag{15.426}$$

By considering *Euler*'s equation, equating (15.425) to (15.426), and dividing both sides by $\mathrm{j}\omega\,\varrho$, we obtain

$$\underline{v}_\rightarrow(r) = \underline{g}_\rightarrow \left(\frac{1}{\varrho c\, r} + \frac{1}{j\omega\, \varrho\, r^2} \right) e^{-j\beta r}. \tag{15.427}$$

Task (c) Using (9.67) and (9.68), the field impedance, $\underline{Z}_f$, becomes

$$\underline{Z}_f = \frac{\underline{p}_\rightarrow(r)}{\underline{v}_\rightarrow(r)} = \frac{1}{\frac{1}{\varrho c} + \frac{1}{j\omega\, \varrho}} = \varrho c\, \frac{j\frac{2\pi r}{\lambda}}{1 + j\frac{2\pi r}{\lambda}}. \tag{15.428}$$

This equation contains two parts—compare (8.16) in Sect. 8.2. Let the absolute values of these two parts be equal, so that we get

$$\left| \frac{1}{\varrho c\, r_{ff}} \right| = \left| \frac{1}{j\omega\, \varrho\, c\, r_{ff}^2} \right|, \tag{15.429}$$

For a discrimination of *near-field* and *far-field*, we use

$$r_{ff} = \frac{\lambda}{2\pi} = \frac{c}{\omega_{ff}}. \tag{15.430}$$

For $r \gg r_{ff} = \lambda/2\pi$, we talk of the far-field of (15.428), which possesses the field impedance $\underline{Z}_f \rightarrow \varrho c$. This is the field resistance of fluid media.

For $r \ll r_{ff} = \lambda/2\pi$, we speak of the near-field, which possesses a field impedance of $\underline{Z}_f \rightarrow j\omega\, \varrho\, r$, denoting an imaginary, mass-like impedance.

Problem 9.2.

Proposed Approach

The general expression for the sound pressure of a 0th-order sphere is

$$\underline{p}_\rightarrow(r) = \underline{g}_\rightarrow \frac{e^{-j\beta r}}{r}, \tag{15.431}$$

and the particle velocity, determined by *Euler*'s equation is

$$\underline{v}_\rightarrow(r) = \underline{g}_\rightarrow \left(\frac{1}{\varrho c\, r} + \frac{1}{j\omega\, \varrho\, r^2} \right) e^{-j\beta r}. \tag{15.432}$$

The general form of the volume-velocity is a function of the radius/distance, expressly,

$$\underline{q}_r = 4\pi r^2 \underline{v}_\rightarrow(r). \tag{15.433}$$

Task (**a**) At a distance of $r = a$, the volume velocity in (15.433) via (15.432) becomes

$$\underline{q}_a = 4\pi a^2 \underline{v}_\to(a) = 4\pi a^2 \underline{g}_\to \left(\frac{1}{\varrho c a} + \frac{1}{j\omega \varrho a^2} \right) e^{-j\beta a}, \qquad (15.434)$$

and, after a rearrangement of (15.434), we obtain the amplitude, $\underline{g}_\to$,

$$\underline{g}_\to = \frac{\underline{q}_a}{4\pi a^2} \frac{e^{+j\beta a}}{\frac{1}{\varrho c a} + \frac{1}{j\omega \varrho a^2}} = \frac{\underline{q}_a}{4\pi} \frac{e^{+j\beta a}}{\frac{a}{\varrho c} + \frac{1}{j\omega \varrho}}. \qquad (15.435)$$

Substitution of (15.435) into (15.431) provides the pressure distribution in the space outside the sphere as

$$\underline{p}_{s\to}(r) = \frac{\underline{q}_a}{4\pi r} \frac{e^{-j\beta(r-a)}}{\frac{a}{\varrho c} + \frac{1}{j\omega \varrho}}, \qquad (15.436)$$

and by inserting this into (15.432), we arrive at the velocity distribution in the space outside the sphere, to wit,

$$\underline{v}_{s\to}(r) = \frac{\underline{q}_a}{4\pi r^2} \frac{\left(\frac{r}{\varrho c} + \frac{1}{j\omega \varrho} \right)}{\frac{a}{\varrho c} + \frac{1}{j\omega \varrho}} e^{-j\beta(r-a)}. \qquad (15.437)$$

Task (**b**) With the volume velocity, $\underline{q}_a$, given, to produce the same sound field by a point source at a distance r, the sound pressures generated by the sphere, $\underline{p}_{s\to}(r)$, and that of the point source, $\underline{p}_{0\to}(r)$, have to be equal. Thereupon, we obtain

$$\underline{p}_{0\to}(r) = \underline{p}_{s\to}(r). \qquad (15.438)$$

For a point source, that is, a *monopole*, the sound pressure, $\underline{p}_{0\to}(r)$, at a distance r, is

$$\underline{p}_{0\to}(r) = \frac{j\omega \varrho \underline{q}_0}{4\pi} \frac{e^{-j\beta r}}{r}, \qquad (15.439)$$

and for a spherical source of radius a, the sound pressure, $\underline{p}_{s\to}(r)$, at a distance r is expressed in (15.436). According to (15.438), the following holds,

$$\underline{p}_{0\to}(r) = \frac{j\omega \varrho \underline{q}_0}{4\pi} \frac{e^{-j\beta r}}{r} = \frac{\underline{q}_a}{4\pi r} \frac{e^{-j\beta(r-a)}}{\frac{a}{\varrho c} + \frac{1}{j\omega \varrho}} = \underline{p}_{s\to}(r), \qquad (15.440)$$

and, hence,

$$\underline{q}_0 = \frac{\underline{q}_a}{j\omega \varrho} \frac{e^{+j\beta a}}{\frac{a}{\varrho c} + \frac{1}{j\omega \varrho}} = \frac{\underline{q}_a e^{j\beta a}}{\frac{j\omega a}{c} + 1} = \frac{\underline{q}_a e^{j\beta a}}{1 + \frac{j2\pi a}{\lambda}}, \qquad (15.441)$$

with $\omega/c = 2\pi/\lambda$.

Problem 9.3.

Proposed Approach

In general, the radiated power is

$$P_w = \frac{1}{2}\mathrm{Re}\{\underline{Z}_{\text{mech}}\}\,|\underline{v}(r)|^2,\qquad(15.442)$$

and for a monopole,

$$\mathrm{Re}\{\underline{Z}_{\text{mech}}\} = A(r)\mathrm{Re}\{\underline{Z}_f\} = 4\pi r^2\,\mathrm{Re}\{\underline{Z}_f\}.\qquad(15.443)$$

Task **(a)** For a monopole ($0th$-order) source, the following applies,

$$\mathrm{Re}\{\underline{Z}_f\} = \mathrm{Re}\left\{\frac{1}{\frac{1}{\varrho c}+\frac{1}{j\omega\varrho r}}\right\} = \varrho c\,\frac{\left(\frac{\omega r}{c}\right)^2}{1+\left(\frac{\omega r}{c}\right)^2},\qquad(15.444)$$

and the velocity is

$$\begin{aligned}\underline{v}_\to(r) &= \frac{j\omega\varrho\underline{q}_0}{4\pi}\left(\frac{1}{\varrho c r}+\frac{1}{j\omega\varrho r^2}\right)e^{-j\beta r}\\ &= \frac{\underline{q}_0}{4\pi r^2}\left(\frac{j\omega r}{c}+1\right)e^{-j\beta r}.\end{aligned}\qquad(15.445)$$

Thus, this leads to

$$|\underline{v}(r)|^2 = \frac{q_0^2}{(4\pi r^2)^2}\left[1+\left(\frac{\omega r}{c}\right)^2\right].\qquad(15.446)$$

Substitution of (15.444) and (15.446) into (15.442) and (15.443) renders

$$P_w = \frac{1}{2}\left(\frac{\varrho c q_0^2}{4\pi r^2}\right)\frac{\left(\frac{\omega r}{c}\right)^2}{1+\left(\frac{\omega r}{c}\right)^2}\left[1+\left(\frac{\omega r}{c}\right)^2\right] = \frac{1}{2}\frac{\varrho c\beta^2}{4\pi}q_0^2.\qquad(15.447)$$

Consequently, it turns out that the radiated power of a monopole source is independent of the distance, r.

Task **(b)** For a dipole, that is, a 1st-order source, the sound field is no longer point-symmetric. This means that the shells around the sphere do not represent areas of equal phase. Unlike in (15.443), the mechanic impedance cannot be determined

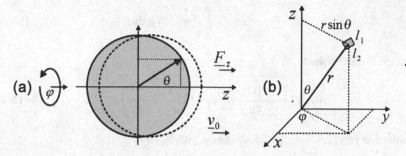

Fig. 15.45 Notations of the spherical source moving along the z-direction (**a**) Basic definitions of quantities and the moving direction (**b**) Spherical co-ordinates for determining the elementary area, $dA_s = l_1 \cdot l_2$

straightly from the field impedance, $\underline{Z}_f$, but rather needs to be expressed by the original definition, viz,

$$\underline{Z}_{\text{mech}} = \frac{\underline{F}_z}{\underline{v}}, \tag{15.448}$$

where $\underline{F}_z$ represents the force along the z-direction—see Fig. 15.45. We express this force via the sound pressure, $\underline{p}(r, \delta)$, as

$$\underline{F}_z = \int_{A_s} \underline{p}(r, \theta) \cos\theta \, dA_s, \tag{15.449}$$

where A_s is the surface of the sphere. Note that $dA_s = l_1 \cdot l_2$—see Fig. 15.45b for the notations. Expressly, we write

$$\frac{d\theta}{2\pi} = \frac{l_1}{2\pi r} \quad \rightarrow \quad l_1 = r \, d\theta, \tag{15.450}$$

and

$$\frac{d\phi}{2\pi} = \frac{l_2}{2\pi r \sin\theta} \quad \rightarrow \quad l_2 = r \sin\theta \, d\phi, \tag{15.451}$$

which leads to

$$dA_s = r^2 \sin\theta \, d\phi \, d\theta, \tag{15.452}$$

with $0 \leq \phi \leq 2\pi$ and $0 \leq \theta \leq \pi$, so that (15.449) becomes

$$\underline{F}_z = \int_0^{2\pi} d\phi \int_0^\pi \underline{p}(r, \theta) \cos\theta \, r^2 \sin\theta \, d\theta. \tag{15.453}$$

Substitution of $u = \cos\theta$ provides $du = -\sin\theta \, d\theta$. When $\theta = 0 \rightarrow u = 1$ and $\theta = \pi$, $u = -1$, the force in (15.453) results as

$$\underline{F}_z = 2\pi r^2 \int_{-1}^{1} \underline{p}(r, \theta)\, u \, du. \tag{15.454}$$

From (9.23), we obtain

$$\underline{p}(r, \theta) = \frac{-j\omega\varrho}{4\pi}\, \underline{\mu}_d \cos\theta \left(\frac{1}{r^2} + \frac{j\beta}{r} \right) e^{-j\beta r}. \tag{15.455}$$

Substitution of (15.455) into (15.454) now yields

$$\underline{F}_z = 2\pi r^2 \frac{-j\omega\varrho}{4\pi}\, \underline{\mu}_d \left(\frac{1}{r^2} + \frac{j\beta}{r} \right) e^{-j\beta r} \underbrace{\int_{-1}^{1} u^2 \, du}_{=2/3}, \tag{15.456}$$

$$\underline{F}_z = \frac{-j\omega\varrho r^2}{3}\, \underline{\mu}_d \left(\frac{1}{r^2} + \frac{j\beta}{r} \right) e^{-j\beta r}. \tag{15.457}$$

From (9.24) we bring in

$$\underline{v}_\to (r, \theta) = \frac{\underline{\mu}_d}{4\pi} \cos\theta \left(\frac{\beta^2}{r} - \frac{2}{r^3} - \frac{2j\beta}{r^2} \right) e^{-j\beta r}. \tag{15.458}$$

For the direction of radiation, $\theta = 0$, $\cos\theta = 1$, subsequently follows

$$\underline{v}_\to (r, \theta = 0) = \frac{\underline{\mu}_d}{4\pi} \left(\frac{\beta^2}{r} - \frac{2}{r^3} - \frac{2j\beta}{r^2} \right) e^{-j\beta r}. \tag{15.459}$$

Since the mechanical impedance becomes

$$\underline{Z}_{\text{mech}} = \frac{\underline{F}_z}{\underline{v}_\to (r, 0)}, \tag{15.460}$$

substitution of (15.457) and (15.459) into (15.460) provides

$$\begin{aligned}
\underline{Z}_{\text{mech}} &= \frac{\frac{-j\omega\varrho r^2}{3}\, \underline{\mu}_d \left(\frac{1}{r^2} + \frac{j\beta}{r} \right) e^{-j\beta r}}{\frac{\underline{\mu}_d}{4\pi} \left(\frac{\beta^2}{r} - \frac{2}{r^3} - \frac{2j\beta}{r^2} \right) e^{-j\beta r}} \\
&= \frac{4}{3}\pi r^2 \varrho c\, \frac{\left(\frac{\beta^2}{r} - \frac{j\beta}{r^2} \right)}{\left[\left(\frac{\beta^2}{r} - \frac{2}{r^3} \right) - \frac{2j\beta}{r^2} \right]} \\
&= \frac{4}{3}\pi r^2 \varrho c\, \frac{\beta^2 + j \left(\frac{\beta}{r} + \frac{2}{\beta r^3} \right)}{\beta^2 \left[1 + \frac{4}{(\beta r)^4} \right]}. \tag{15.461}
\end{aligned}$$

and, hence,

$$\mathrm{Re}\{\underline{Z}_{\mathrm{mech}}\} = \frac{4}{3}\pi r^2 \varrho c \frac{(\beta r)^4}{4+(\beta r)^4}. \tag{15.462}$$

The radiated power—compare (15.442)—results in

$$P_{\mathrm{w}} = \frac{2}{3}\pi r^2 \varrho c \frac{(\beta r)^4}{4+(\beta r)^4}|\underline{v}|^2, \tag{15.463}$$

and, from (15.459) follows

$$|\underline{v}|^2 = \left(\frac{\underline{\mu}_{\mathrm{d}}}{4\pi r^2}\right)^2 \left[(\beta^2 r)^2 + \frac{4}{r^2}\right], \tag{15.464}$$

so that

$$P_{\mathrm{w}} = \frac{2}{3}\pi r^2 \varrho c \left(\frac{\underline{\mu}_{\mathrm{d}}}{4\pi r^2}\right)^2 \left[\frac{(\beta r)^4}{4+(\beta r)^4}\right]\left[(\beta^2 r)^2 + \frac{4}{r^2}\right]. \tag{15.465}$$

For the near-field, we approximate this expression by

$$P_{\mathrm{w}} \approx \frac{1}{6}\left(\frac{\varrho c}{4\pi}\right)\underline{\mu}_{\mathrm{d}}^2 \beta^4. \tag{15.466}$$

Task (c) To find the dependency of the sound intensity on the distance to the sound source, we start with the general expression for the sound intensity as derived in (16.16), namely,

$$I(r) = \frac{1}{2}\underline{p}(r)\underline{v}^*(r), \tag{15.467}$$

where $\underline{v}^*(r)$ is the complex conjugate of $\underline{v}(r)$.

For a 0th-order spherical sound source—compare (9.18)—the sound pressure is equal to

$$\underline{p}_{\rightarrow}(r) = \frac{\mathrm{j}\omega\varrho}{4\pi}\underline{q}_0 \frac{e^{-\mathrm{j}\beta r}}{r}, \tag{15.468}$$

and the velocity follows by applying *Euler*'s equation to (15.468), namely,

$$\underline{v}_{\rightarrow}(r) = \frac{\mathrm{j}\omega\varrho}{4\pi}\underline{q}_0\left(\frac{1}{\varrho c}+\frac{1}{\mathrm{j}\omega\varrho r}\right)\frac{e^{-\mathrm{j}\beta r}}{r}. \tag{15.469}$$

Substitution of (15.468) and (15.469) into (15.467) delivers

$$I_0(r) = \frac{1}{2} \frac{j\omega\varrho}{4\pi} \underline{q}_0 \frac{e^{-j\beta r}}{r} \left[\frac{j\omega\varrho}{4\pi} \underline{q}_0 \left(\frac{1}{\varrho c} + \frac{1}{j\omega\varrho r} \right) \frac{e^{-j\beta r}}{r} \right]^*, \qquad (15.470)$$

prompting

$$I_0(r) = \frac{1}{2} \frac{\omega^2 \varrho}{16\pi^2 r^2} \underline{q}_0^2 \left(\frac{1}{c} - \frac{1}{j\omega r} \right). \qquad (15.471)$$

Its real part is equal to

$$I_{w_0} = \text{Re}\{I_0(r)\} = \frac{\omega^2 \varrho}{32\pi^2 c r^2} \underline{q}_0^2. \qquad (15.472)$$

For a 1st-order spherical sound source, the sound pressure and the particle velocity—see (9.23) and (9.24)—are equal to

$$\underline{p}_\to(r, \theta) = \frac{-j\omega\varrho}{4\pi r} \underline{\mu}_d \cos\theta \left(\frac{1}{r} + j\beta \right) e^{-j\beta r}, \qquad (15.473)$$

and

$$\underline{v}_\to(r, \theta) = \frac{\underline{\mu}_d}{4\pi r} \cos\theta \left[\left(\beta^2 - \frac{2}{r^2} \right) - j \frac{2\beta}{r} \right] e^{-j\beta r}, \qquad (15.474)$$

so that

$$\begin{aligned}
\underline{I}_1 &= \frac{1}{2} |\underline{\mu}_d|^2 \frac{-j\omega\varrho}{(4\pi r)^2} \cos^2\theta \left(\frac{1}{r} + j\beta \right) \left[\left(\beta^2 - \frac{2}{r^2} \right) + j \frac{2\beta}{r} \right] \\
&= \frac{1}{2} |\underline{\mu}_d|^2 \frac{\omega\varrho}{(4\pi r)^2} \cos^2\theta \left[\beta^3 + j \left(\frac{\beta^2}{r} + \frac{2}{r^3} \right) \right],
\end{aligned} \qquad (15.475)$$

and

$$I_{w_1} = \text{Re}\{\underline{I}_1\} = \frac{\omega^4 \varrho}{32\pi^2 r^2 c^3} |\underline{\mu}_d|^2 \cos^2\theta. \qquad (15.476)$$

Problem 9.4.

Proposed Approach

Let us begin with the free-field velocity, $\underline{v}_f$, and the velocity in the horn, $\underline{v}_h$. The volume-velocity, $\underline{q}_0$, completely feeds into the horn, such that

$$\underline{q}_0 = \lim_{r \to 0} \left[\underline{v}_h(r) A_h(r) \right] = \lim_{r \to 0} \left[\underline{v}_f(r) 4\pi r^2 \right], \qquad (15.477)$$

where $A_h(r) = \Omega r^2$.

This leads to

$$\underline{v}_h(r) = \underline{v}_f(r)\frac{4\pi}{\Omega} = \underline{v}_f(r)\,R, \tag{15.478}$$

where $R = 4\pi/\Omega$.

Because of $\underline{g}_h = \underline{g}_f\,R$ we also find

$$\underline{p}_h(r) = \underline{p}_f(r)\,R. \tag{15.479}$$

and, following $I = \frac{1}{2}\text{Re}[\,\underline{p}\cdot\underline{v}^*]$, we get

$$I_h = I_f\,R^2. \tag{15.480}$$

The power is related to the intensity through $P = \int I\,dA$. Therefore, according to (15.480), we end with

$$I_h = \frac{P_h}{A_h} = \frac{P_h}{\Omega r^2} = R^2\,I_f = \left(\frac{4\pi}{\Omega}\right)^2\frac{P_f}{4\pi r^2}, \tag{15.481}$$

which leads to

$$P_h = \frac{4\pi}{\Omega}P_f = P_f\,R. \tag{15.482}$$

Problem 9.5.

Proposed Approach

Task (**a**) The directional characteristic is of the figure-of-eight type—see Fig. 15.46.

Task (**b**) The sound pressures of a monopole and a dipole are equal to

$$\underline{p}_m = \frac{j\,\omega\,\varrho}{4\pi}\underline{q}_0\frac{e^{-j\beta r}}{r}, \tag{15.483}$$

and

$$\underline{p}_d = \frac{j\,\omega\,\varrho}{4\pi}\underline{M}\cos\theta\left(\frac{1}{r^2}+\frac{j\beta}{r}\right)\frac{e^{-j\beta r}}{r}, \tag{15.484}$$

respectively.

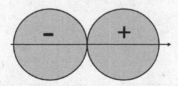

Fig. 15.46 Directional characteristic of a dipole spherical source

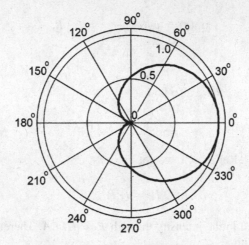

Fig. 15.47 Directional characteristic of two superposed spherical sound sources—a monopole, and a dipole

In the far-field, the two sources superpose as follows,

$$p = \frac{j\omega \varrho}{4\pi} \frac{e^{-j\beta r}}{r} \left(\underline{M} \, j\beta \cos\theta + \underline{q}_0 \right). \tag{15.485}$$

With the directional characteristic being defined as

$$\Gamma = \frac{\underline{p}(r, \theta)}{\underline{p}(r, 0)} = \frac{1 + \cos\theta}{2} = \frac{(\underline{M}/\underline{q}_0)\, j\beta \cos\theta + 1}{(\underline{M}/\underline{q}_0)\, j\beta + 1}, \tag{15.486}$$

this expression prompts

$$\frac{\underline{M}}{\underline{q}_0} \, j\beta = 1, \tag{15.487}$$

which implies

$$|\underline{q}_0| = \underline{M}\,\beta \quad \text{and} \quad \beta(\underline{q}_0) = \frac{\pi}{2}. \tag{15.488}$$

Task (**c**) The directional characteristic is of the cardioid type—as depicted in Fig. 15.47

Problem 9.6.

Proposed Approach

The three-dimensional wave equation in general form reads as follows, (7.34) and (7.36),

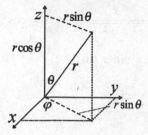

Fig. 15.48 *Cartesian* and spherical coordinates

$$\nabla^2 p = \frac{1}{c^2} \frac{\partial^2 p}{\partial t^2}. \tag{15.489}$$

The expressions of this notation in *Cartesian* coordinates are treated in Chap. 7. To transform them into spherical coordinates, we use the following operators as given in (7.30)–(7.31),

– *Gradient*, $\overrightarrow{\nabla}$
– *Divergence*, ∇
– *Laplacian operator*, ∇^2

For the transformation, we employ the following equations—see Fig. 15.48,

$$x = (r \sin \theta) \cos \varphi, \quad y = (r \sin \theta) \sin \varphi, \quad z = r \cos \theta, \tag{15.490}$$

$$r = \sqrt{x^2 + y^2 + z^2}, \quad \varphi = \arctan \frac{y}{x}, \quad \theta = \arctan \frac{\sqrt{x^2 + y^2}}{z}, \tag{15.491}$$

with r being the radial variable, θ the elevation variable, and φ the azimuth variable.

In each *Cartesian* direction, we express the partial derivatives of a function, that is, p, in terms of spherical-coordinate variables as follows,

$$\begin{aligned}
\frac{\partial p}{\partial x} &= \frac{\partial p}{\partial r} \frac{\partial r}{\partial x} + \frac{\partial p}{\partial \varphi} \frac{\partial \varphi}{\partial x} + \frac{\partial p}{\partial \theta} \frac{\partial \theta}{\partial x}, \\
\frac{\partial p}{\partial y} &= \frac{\partial p}{\partial r} \frac{\partial r}{\partial y} + \frac{\partial p}{\partial \varphi} \frac{\partial \varphi}{\partial y} + \frac{\partial p}{\partial \theta} \frac{\partial \theta}{\partial y}, \\
\frac{\partial p}{\partial z} &= \frac{\partial p}{\partial r} \frac{\partial r}{\partial z} + \frac{\partial p}{\partial \varphi} \frac{\partial \varphi}{\partial z} + \frac{\partial p}{\partial \theta} \frac{\partial \theta}{\partial z}.
\end{aligned} \tag{15.492}$$

From (15.491) it follows that

$$\frac{\partial r}{\partial x} = \frac{x}{\sqrt{x^2 + y^2 + z^2}} = \frac{x}{r} = \cos\varphi \, \sin\theta,$$

$$\frac{\partial r}{\partial y} = \frac{y}{r} = \sin\varphi \, \sin\theta, \tag{15.493}$$

$$\frac{\partial r}{\partial z} = \frac{z}{r} = \cos\theta,$$

and

$$\frac{\partial\varphi}{\partial x} = \frac{-y}{x^2 + y^2} = \frac{-\sin\varphi}{r\cos\theta},$$

$$\frac{\partial\varphi}{\partial y} = \frac{x}{x^2 + y^2} = \frac{\cos\varphi}{r\cos\theta}, \tag{15.494}$$

$$\frac{\partial\varphi}{\partial z} = 0,$$

and

$$\frac{\partial\theta}{\partial x} = \frac{xz}{r^2\sqrt{x^2 + y^2}} = \frac{\cos\varphi\cos\theta}{r},$$

$$\frac{\partial\theta}{\partial y} = \frac{yz}{r^2\sqrt{x^2 + y^2}} = \frac{\sin\varphi\cos\theta}{r}, \tag{15.495}$$

$$\frac{\partial\theta}{\partial z} = \frac{-\sqrt{x^2 + y^2}}{r^2} = \frac{-\sin\theta}{r}.$$

Substituting (15.493)–(15.495) into (15.492) yields three gradient (unit) vectors, viz,

$$\frac{\partial p}{\partial x} = \frac{\partial p}{\partial r}(\cos\varphi \, \sin\theta) + \frac{\partial p}{\partial\varphi}\left(\frac{-\sin\varphi}{r\sin\theta}\right) + \frac{\partial p}{\partial\theta}\left(\frac{\cos\varphi\cos\theta}{r}\right),$$

$$\frac{\partial p}{\partial y} = \frac{\partial p}{\partial r}(\sin\varphi \, \sin\theta) + \frac{\partial p}{\partial\varphi}\left(\frac{\cos\varphi}{r\sin\theta}\right) + \frac{\partial p}{\partial\theta}\left(\frac{\sin\varphi\cos\theta}{r}\right), \tag{15.496}$$

$$\frac{\partial p}{\partial z} = \frac{\partial p}{\partial r}(\cos\theta) + \frac{\partial p}{\partial\theta}\left(\frac{-\sin\theta}{r}\right).$$

The *first linear* 1st-*order equation*, namely, the *Euler* equation in (7.32), is obtained as follows.

The three unit vectors $\vec{r}$, $\vec{\varphi}$, and $\vec{\theta}$ are represented in spherical coordinates by

$$\vec{r} = \cos\varphi \sin\theta \, \frac{\partial p}{\partial x} + \sin\varphi \, \sin\theta \, \frac{\partial p}{\partial y} + \cos\theta \, \frac{\partial p}{\partial z},$$

$$\vec{\varphi} = -\sin\varphi \, \frac{\partial p}{\partial x} + \cos\varphi \, \frac{\partial p}{\partial y}, \tag{15.497}$$

$$\vec{\theta} = \cos\varphi \cos\theta \, \frac{\partial p}{\partial x} + \sin\varphi \, \cos\theta \, \frac{\partial p}{\partial y} - \sin\theta \, \frac{\partial p}{\partial z}.$$

Now, substituting each gradient component in (15.492) into (15.497), after regrouping them with respect to their partial derivatives $\frac{\partial}{\partial r}, \frac{\partial}{\partial \varphi}, \frac{\partial}{\partial \theta}$, and simplifying with the trigonometric identity, $\sin^2 \alpha + \cos^2 \alpha = 1$, leads to

$$\vec{r} = \frac{\partial p}{\partial r}, \qquad \vec{\varphi} = \frac{1}{r \sin\theta} \frac{\partial p}{\partial \varphi}, \qquad \vec{\theta} = \frac{1}{r} \frac{\partial p}{\partial \theta}. \tag{15.498}$$

We express the *Euler*-equation in (7.32), in the general form using (15.498) for 1st-order spherical waves, as

$$-\left(\frac{\partial}{\partial r} \vec{r} + \frac{1}{r \sin\theta} \frac{\partial}{\partial \varphi} \vec{\varphi} + \frac{1}{r} \frac{\partial}{\partial \theta} \vec{\theta} \right) p = \varrho \, \frac{\partial v}{\partial t}. \tag{15.499}$$

We now generate the *second linear* 1st-*order equation*—the *continuity* equation in (7.33). First, we express the divergence of a velocity vector in spherical coordinates as

$$\nabla \cdot \vec{v} = \nabla \cdot (v_r \, \vec{r} + v_\varphi \, \vec{\varphi} + v_\theta \, \vec{\theta}) \tag{15.500}$$

$$= (\nabla v_r)\vec{r} + v_r \nabla \cdot \vec{r} + (\nabla v_\varphi)\vec{\varphi} + v_\varphi \nabla \cdot \vec{\varphi} + (\nabla v_\theta)\vec{\theta} + v_\theta \nabla \cdot \vec{\theta}$$

$$= \left(\frac{\partial v}{\partial r} + \frac{1}{r \sin\theta} \frac{\partial v}{\partial \varphi} + \frac{1}{r} \frac{\partial v}{\partial \theta} \right) + v_r \nabla \cdot \vec{r} + v_\varphi \nabla \cdot \vec{\varphi} + v_\theta \nabla \cdot \vec{\theta}.$$

Thereby, it is convenient to calculate the divergence of the unit vectors in *Cartesian* coordinates and then convert them to spherical coordinates afterward, namely,

$$\nabla \cdot \vec{r} = \frac{\partial}{\partial x} \left(\frac{\partial r}{\partial x} \right) + \frac{\partial}{\partial y} \left(\frac{\partial r}{\partial y} \right) + \frac{\partial}{\partial z} \left(\frac{\partial r}{\partial z} \right)$$

$$= \frac{\partial}{\partial x} \left(\frac{x}{\sqrt{x^2 + y^2 + z^2}} \right) + \frac{\partial}{\partial y} \left(\frac{y}{\sqrt{x^2 + y^2 + z^2}} \right) + \frac{\partial}{\partial z} \left(\frac{z}{\sqrt{x^2 + y^2 + z^2}} \right)$$

$$= \frac{3}{\sqrt{x^2 + y^2 + z^2}} - \frac{x^2 + y^2 + z^2}{\left(\sqrt{x^2 + y^2 + z^2} \right)^3} = \frac{2}{r}. \tag{15.501}$$

$$\nabla \cdot \vec{\varphi} = \frac{\partial}{\partial x}\left(\frac{\partial \varphi}{\partial x}\right) + \frac{\partial}{\partial y}\left(\frac{\partial \varphi}{\partial y}\right) + \frac{\partial}{\partial z}\left(\frac{\partial \varphi}{\partial z}\right)$$

$$= \frac{\partial}{\partial x}\left(\frac{-y}{\sqrt{x^2 + y^2}}\right) + \frac{\partial}{\partial y}\left(\frac{x}{\sqrt{x^2 + y^2}}\right) = 0. \qquad (15.502)$$

$$\nabla \cdot \vec{\theta} = \frac{\partial}{\partial x}\left(\frac{\partial \theta}{\partial x}\right) + \frac{\partial}{\partial y}\left(\frac{\partial \theta}{\partial y}\right) + \frac{\partial}{\partial z}\left(\frac{\partial \theta}{\partial z}\right)$$

$$= \frac{\partial}{\partial x}\left(\frac{zx}{\sqrt{x^2 + y^2 + z^2}\sqrt{x^2 + y^2}}\right) + \frac{\partial}{\partial y}\left(\frac{zy}{\sqrt{x^2 + y^2 + z^2}\sqrt{x^2 + y^2}}\right)$$

$$+ \frac{\partial}{\partial z}\left(\frac{\sqrt{x^2 + y^2}}{\sqrt{x^2 + y^2 + z^2}}\right) = \frac{z}{\sqrt{x^2 + y^2}} = \frac{\cos\theta}{r\sin\theta}. \qquad (15.503)$$

Substituting (15.501)–(15.503) into (15.500) yields

$$\nabla \cdot \vec{v} = \frac{\partial v_r}{\partial r} + \frac{2}{r}v_r + \frac{1}{r\sin\theta}\frac{\partial v_\varphi}{\partial \varphi} + \frac{1}{r}\frac{\partial v_\theta}{\partial \theta} + v_\theta\frac{\cos\theta}{r\sin\theta}. \qquad (15.504)$$

From (7.33) and (15.504), the second linear 1st-order wave equation in spherical coordinates is expressible as

$$-\left[\frac{1}{r^2}\frac{\partial}{\partial r}(r^2 v_r) + \frac{1}{r\sin\theta}\frac{\partial v_\varphi}{\partial \varphi} + \frac{1}{\sin\theta}\frac{\partial}{\partial \theta}(\sin\theta\, v_\theta)\right] = \kappa\frac{\partial p}{\partial t}. \qquad (15.505)$$

The 2nd-*order wave equation* is derived by combining the two 1st-order equations in the following way. Apply the 1st-order derivatives a second time concerning x, y, and z, using (15.496), that is,

$$\frac{\partial^2 p}{\partial x^2} = \frac{\partial}{\partial r}\left(\frac{\partial p}{\partial x}\right)(\cos\varphi\sin\theta) + \frac{\partial}{\partial \varphi}\left(\frac{\partial p}{\partial x}\right)\left(\frac{-\sin\varphi}{r\sin\theta}\right) + \frac{\partial}{\partial \theta}\left(\frac{\partial p}{\partial x}\right)\left(\frac{\cos\varphi\cos\theta}{r}\right)$$

$$\frac{\partial^2 p}{\partial y^2} = \frac{\partial}{\partial r}\left(\frac{\partial p}{\partial y}\right)(\sin\varphi\sin\theta) + \frac{\partial}{\partial \varphi}\left(\frac{\partial p}{\partial y}\right)\left(\frac{\cos\varphi}{r\sin\theta}\right) + \frac{\partial}{\partial \theta}\left(\frac{\partial p}{\partial y}\right)\left(\frac{\sin\varphi\cos\theta}{r}\right)$$

$$\frac{\partial^2 p}{\partial z^2} = \frac{\partial}{\partial r}\left(\frac{\partial p}{\partial z}\right)(\cos\theta) + \frac{\partial}{\partial \theta}\left(\frac{\partial p}{\partial z}\right)\left(\frac{-\sin\theta}{r}\right). \qquad (15.506)$$

Finally, substituting the first equation in (15.496) into the first equation in (15.506) renders

$$\frac{\partial^2 p}{\partial x^2} = \frac{\partial}{\partial r}\left[\frac{\partial p}{\partial r}(\cos\varphi\sin\theta) + \frac{\partial p}{\partial\varphi}\left(\frac{-\sin\varphi}{r\sin\theta}\right) + \frac{\partial p}{\partial\theta}\left(\frac{\cos\varphi\cos\theta}{r}\right)\right](\cos\varphi\sin\theta)$$

$$+ \frac{\partial}{\partial\varphi}\left[\frac{\partial p}{\partial r}(\cos\varphi\sin\theta) + \frac{\partial p}{\partial\varphi}\left(\frac{-\sin\varphi}{r\sin\theta}\right) + \frac{\partial p}{\partial\theta}\left(\frac{\cos\varphi\cos\theta}{r}\right)\right]\left(\frac{-\sin\varphi}{r\sin\theta}\right)$$

$$+ \frac{\partial}{\partial\theta}\left[\frac{\partial p}{\partial r}(\cos\varphi\sin\theta) + \frac{\partial p}{\partial\varphi}\left(\frac{-\sin\varphi}{r\sin\theta}\right) + \frac{\partial p}{\partial\theta}\left(\frac{\cos\varphi\cos\theta}{r}\right)\right]\left(\frac{\cos\varphi\cos\theta}{r}\right)$$

$$= \left[\frac{\partial^2 p}{\partial r^2}(\cos^2\varphi\sin^2\theta) + \frac{\partial p}{\partial\varphi}\left(\frac{\sin\varphi\cos\varphi}{r^2}\right) - \frac{\partial^2 p}{\partial r\partial\varphi}\left(\frac{\sin\varphi\cos\varphi}{r}\right)\right.$$

$$\left. - \frac{\partial p}{\partial\theta}\left(\frac{\cos^2\varphi\sin\theta\cos\theta}{r^2}\right) + \frac{\partial^2 p}{\partial r\partial\theta}\left(\frac{\cos^2\varphi\sin\theta\cos\theta}{r}\right)\right] \qquad (15.507)$$

$$+ \left[\frac{\partial p}{\partial r}\left(\frac{\sin^2\varphi}{r}\right) - \frac{\partial^2 p}{\partial r\partial\varphi}\left(\frac{\sin\varphi\cos\varphi}{r}\right) + \frac{\partial p}{\partial\varphi}\left(\frac{\sin\varphi\cos\varphi}{r^2\sin^2\theta}\right) + \frac{\partial^2 p}{\partial\varphi^2}\left(\frac{\sin^2\varphi}{r^2\sin^2\theta}\right)\right.$$

$$\left. + \frac{\partial p}{\partial\theta}\left(\frac{\sin^2\varphi\cos\theta}{r^2\sin\theta}\right) - \frac{\partial^2 p}{\partial\varphi\partial\theta}\left(\frac{\sin\varphi\cos\varphi\cos\theta}{r^2\sin\theta}\right)\right]$$

$$+ \left[\frac{\partial p}{\partial r}\left(\frac{\cos^2\varphi\cos^2\theta}{r}\right) + \frac{\partial^2 p}{\partial r\partial\theta}\left(\frac{\cos^2\varphi\sin\theta\cos\theta}{r}\right) - \frac{\partial p}{\partial\varphi}\left(\frac{\cos^2\varphi\cos\theta}{r^2\sin\theta}\right)\right.$$

$$\left. - \frac{\partial^2 p}{\partial\varphi\partial\theta}\left(\frac{\sin\varphi\cos\varphi\cos\theta}{r^2\sin\theta}\right) - \frac{\partial p}{\partial\theta}\left(\frac{\cos^2\varphi\sin\theta\cos\theta}{r^2}\right) + \frac{\partial^2 p}{\partial\theta^2}\left(\frac{\cos^2\varphi\cos^2\theta}{r^2}\right)\right].$$

Two further 2nd-order derivatives are obtained similarly, that is, by adding up the three 2nd-order derivatives for the extraneous terms to cancel, thus leaving

$$\frac{\partial^2 p}{\partial x^2} + \frac{\partial^2 p}{\partial y^2} + \frac{\partial^2 p}{\partial z^2} = \frac{\partial^2 p}{\partial r^2} + \frac{1}{r^2\sin^2\theta}\frac{\partial^2 p}{\partial\varphi^2} + \frac{1}{r^2}\frac{\partial^2 p}{\partial\theta^2} + \frac{2}{r}\frac{\partial p}{\partial r} + \frac{\cos\theta}{r^2\sin\theta}\frac{\partial p}{\partial\theta}$$

$$= \frac{1}{r^2}\frac{\partial}{\partial r}\left(r^2\frac{\partial p}{\partial r}\right) - \frac{1}{r^2\sin\theta}\frac{\partial}{\partial\theta}\left(\sin\theta\frac{\partial p}{\partial\theta}\right) + \frac{1}{r^2\sin^2\theta}\frac{\partial^2 p}{\partial\varphi^2}. \quad (15.508)$$

In consequence, we express the 2nd-order wave equation in spherical coordinates as

$$\left[\frac{1}{r^2}\frac{\partial}{\partial r}\left(r^2\frac{\partial}{\partial r}\right) + \frac{1}{r^2\sin\theta}\frac{\partial}{\partial\theta}\left(\sin\theta\frac{\partial}{\partial\theta}\right) + \frac{1}{r^2\sin^2\theta}\frac{\partial^2}{\partial\varphi^2}\right]p = \frac{1}{c^2}\frac{\partial^2 p}{\partial t^2}. \quad (15.509)$$

Problem 9.7.

Proposed Approach

Regroup (9.71) into the following three terms,

$$-\frac{n(n+1)}{r^2} + \left[\frac{n(n+1)}{r^2} - \frac{m^2}{r^2\sin^2\theta}\right] + \frac{m^2}{r^2\sin^2\theta} = 0, \qquad (15.510)$$

Add the three terms to the three terms contained in the brackets of (9.70), as

$$
\beta^2 + \frac{1}{r^2} \frac{1}{R} \frac{\partial}{\partial r} \left(r^2 \frac{\partial R}{\partial r} \right) - \frac{n(n+1)}{r^2}
$$

$$
+ \frac{1}{r^2} \frac{1}{\Theta \sin \theta} \frac{\partial}{\partial \theta} \left(\sin \theta \frac{\partial \Theta}{\partial \theta} \right) + \left[\frac{n(n+1)}{r^2} - \frac{m^2}{r^2 \sin^2 \theta} \right]
$$

$$
+ \frac{1}{r^2} \frac{1}{\Phi \sin^2 \theta} \frac{\partial^2 \Phi}{\partial \varphi^2} + \frac{m^2}{r^2 \sin^2 \theta} = 0. \tag{15.511}
$$

Equation (15.511) still holds if each of its three lines equals zero, that is,

$$
\frac{1}{r^2} \frac{1}{R} \frac{\partial}{\partial r} \left(r^2 \frac{\partial R}{\partial r} \right) + \beta^2 - \frac{n(n+1)}{r^2} = 0, \tag{15.512}
$$

$$
\frac{1}{r^2} \frac{1}{\Theta \sin \theta} \frac{\partial}{\partial \theta} \left(\sin \theta \frac{\partial \Theta}{\partial \theta} \right) + \left[\frac{n(n+1)}{r^2} - \frac{m^2}{r^2 \sin^2 \theta} \right] = 0, \tag{15.513}
$$

and

$$
\frac{1}{r^2} \frac{1}{\Phi \sin^2 \theta} \frac{\partial^2 \Phi}{\partial \varphi^2} + \frac{m^2}{r^2 \sin^2 \theta} = 0. \tag{15.514}
$$

For the radial component, multiplication with $\underline{R}$ on both sides of (15.512) leads to

$$
\frac{1}{r^2} \frac{\partial}{\partial r} \left(r^2 \frac{\partial R}{\partial r} \right) + \underline{R} \beta^2 - \frac{n(n+1)}{r^2} \underline{R} = 0, \tag{15.515}
$$

as expressed in (9.30).

For the elevation component, multiplication of $r^2 \underline{\Theta}$ on both sides of (15.513) results in

$$
\frac{1}{\sin \theta} \frac{\partial}{\partial \theta} \left(\sin \theta \frac{\partial \Theta}{\partial \theta} \right) + \left[n(n+1) - \frac{m^2}{\sin^2 \theta} \right] \underline{\Theta} = 0, \tag{15.516}
$$

as noted in (9.31).

For the azimuth component, multiplication of $r^2 \underline{\Phi} \sin^2 \theta$ on both sides of (15.514) yields

$$
\frac{\partial^2 \Phi}{\partial \varphi^2} + m^2 \underline{\Phi} = 0, \tag{15.517}
$$

as already stated in (9.32).

Problem 9.8.

Proposed Approach

The spherical harmonics are given in (9.42) as

$$\underline{Y}_n^m(\theta, \varphi) = \sqrt{\frac{2n+1}{4\pi} \frac{(n-m)!}{(n+m)!}} \, P_n^m(\cos \theta) \, e^{jm\varphi}, \qquad (15.518)$$

where $P_n^m(\eta)$ is the *Legendre* function with $\eta = \cos \theta$. We get

$$P_n^m(\eta) = (-1)^m (1 - \eta^2)^{m/2} \frac{d^m}{d\eta^m} P_n(\eta), \qquad (15.519)$$

with $P_n(\eta)$ being the *Legendre* polynomial

$$P_n(\eta) = \frac{1}{2^n \, n!} \frac{d^n}{d\eta^n} (\eta^2 - 1)^n. \qquad (15.520)$$

Thereby, mind that $0! = 1$, and $1! = 1$.

Substituting $n = 0, 1, 2$ into (15.520) leads to

$$P_0(\eta) = 1, \quad P_1(\eta) = \eta, \qquad (15.521)$$

and

$$P_2(\eta) = \frac{1}{8} \frac{d^2}{d\eta^2} (\eta^2 - 1)^2 = \frac{1}{8} \frac{d^2}{d\eta^2} (\eta^4 - 2\eta^2 + 1) = \frac{1}{2}(3\eta^2 - 1). \qquad (15.522)$$

The following derivations use the trigonometric relation $\cos^2 \theta + \sin^2 \theta = 1$.

We begin with the order $n = 0$ and the degree $m = 0$. Since $\eta = \cos \theta$, $P_0^0(\eta) = P_0(\eta) = 1$, according to (15.518) and (15.521), and write

$$\underline{Y}_0^0(\theta, \varphi) = \sqrt{\frac{1}{4\pi}}. \qquad (15.523)$$

For $n = 1$, and $m = 0$, substituting n, and m into (15.518) yields

$$\underline{Y}_1^0(\theta, \varphi) = \sqrt{\frac{3}{4\pi}} \, P_1^0(\cos \theta), \qquad (15.524)$$

with

$$P_1^0(\cos \theta) = P_1(\cos \theta) = \frac{1}{2} \frac{d}{d \cos \theta} (\cos^2 \theta - 1) = \cos \theta. \qquad (15.525)$$

Consequently, it follows that

$$\underline{Y}_1^0(\theta, \varphi) = \sqrt{\frac{3}{4\pi}} \cos \theta. \tag{15.526}$$

For $n = 1$ and $m = 1$, substituting n and m into (15.518) brings

$$\underline{Y}_1^1(\theta, \varphi) = \sqrt{\frac{3}{4\pi}} P_1^1(\cos \theta) \, \mathrm{e}^{\mathrm{j}\varphi}, \tag{15.527}$$

with

$$P_1^1(\eta) = -\sqrt{1 - \eta^2} \frac{\mathrm{d}}{\mathrm{d}\eta} P_1(\eta) = -\sin\theta \frac{\mathrm{d}}{\mathrm{d}\eta} \eta = -\sin\theta, \tag{15.528}$$

since (15.521) comes into use.

Substituting (15.528) into (15.527) results in

$$\underline{Y}_1^1(\theta, \varphi) = -\sqrt{\frac{3}{4\pi}} \sin\theta \, \mathrm{e}^{\mathrm{j}\varphi}. \tag{15.529}$$

Similarly, we obtain for $n = 2$ and $m = 0$,

$$\underline{Y}_2^0(\theta, \varphi) = \sqrt{\frac{5}{4\pi}} P_2^0(\eta) = \sqrt{\frac{5}{4\pi}} P_2(\eta). \tag{15.530}$$

Substituting $P_2(\eta)$ in (15.522) gives

$$\underline{Y}_2^0(\theta, \varphi) = \frac{1}{2}\sqrt{\frac{5}{4\pi}} (3\cos^2\theta - 1). \tag{15.531}$$

For $n = 2$ and $m = 1$, it follows that

$$\underline{Y}_2^1(\theta, \varphi) = \sqrt{\frac{5}{24\pi}} P_2^1(\eta) \, \mathrm{e}^{\mathrm{j}\varphi}, \tag{15.532}$$

while with

$$P_2^1(\eta) = -(1 - \eta^2)^{\frac{1}{2}} \frac{\mathrm{d}}{\mathrm{d}\eta} P_2(\eta), = -\sqrt{1 - \eta^2} \frac{3}{2} \cdot 2\eta = -3\eta\sqrt{1 - \eta^2}, \tag{15.533}$$

(15.519) and (15.522) come into use.

Substituting (15.533) into (15.532), and employing $\eta = \cos\theta$, and $\cos^2\theta + \sin^2\theta = 1$, finally delivers as a result

$$\underline{Y}_2^1(\theta, \varphi) = -\sqrt{\frac{15}{8\pi}}\, \cos\theta \sin\theta\, \mathrm{e}^{\mathrm{j}\varphi}\,. \tag{15.534}$$

Problem 9.9.

Proposed Approach

According to (9.54) and Table 9.1, the directional characteristic is determined by the *Fourier* transform of its source-strength load with $\Omega = -\beta\cos\theta$, namely,

$$\underline{p}_\to(r, \theta) = \frac{\mathrm{j}\omega\varrho}{4\pi}\, \frac{\mathrm{e}^{-\mathrm{j}\beta r_0}}{r_0} \int_{-\infty}^{+\infty} \underline{q}'(x)\, \mathrm{e}^{+\mathrm{j}(\beta\cos\theta)x}\, \mathrm{d}x\,. \tag{15.535}$$

Since the source-strength load is a *Gaussian*-like function, the integration extends to $\pm\infty$.

Task **(a)** Substitution of the source-strength-load function, $\underline{q}'(x)$, in (9.75), which is a *Gaussian* function, into (15.535), results in

$$\underline{p}_\to(r, \theta) = \underbrace{\frac{\mathrm{j}\omega\varrho}{4\pi}\, \frac{\mathrm{e}^{-\mathrm{j}\beta r_0}}{r_0}\, \underline{q}'_0}_{\text{const.}} \int_{-\infty}^{+\infty} \mathrm{e}^{-x^2/2h^2}\, \mathrm{e}^{+\mathrm{j}(\beta\cos\theta)x}\, \mathrm{d}x\,, \tag{15.536}$$

where $\underline{q}'_0$ is a constant.

The integral represents the *Fourier* transform in the form of

$$\underline{G}(\Omega) = \int_{-\infty}^{+\infty} g(x)\, \mathrm{e}^{-\mathrm{j}\Omega x}\, \mathrm{d}x\,, \tag{15.537}$$

with $g(x) = \mathrm{e}^{-x^2/2h^2}$—a *Gaussian* function—and $\Omega = -\beta\cos\theta$, leading to

$$g(x) = \mathrm{e}^{-x^2/2h^2} \quad\circ\!\!-\!\!\bullet\quad \sqrt{2\pi}\, h\, \mathrm{e}^{-h^2\Omega^2/2} = \underline{G}(\Omega)\,. \tag{15.538}$$

Substitution of (15.538) into (15.536), with $\Omega = -\beta\cos\theta$, yields

$$\underline{p}_\to(r, \theta) = \frac{\mathrm{j}\omega\varrho}{4\pi}\, \frac{\mathrm{e}^{-\mathrm{j}\beta r_0}}{r_0}\, \underline{q}'_0\, \sqrt{2\pi}\, h\, \mathrm{e}^{-(h\beta\cos\theta)^2/2}\,. \tag{15.539}$$

This leads to the directional characteristic

$$\Gamma = \frac{|\underline{p}_\to(r, \theta)|}{|\underline{p}_\to(r, 0)|} = \mathrm{e}^{-(h\beta\cos\theta)^2/2}\,. \tag{15.540}$$

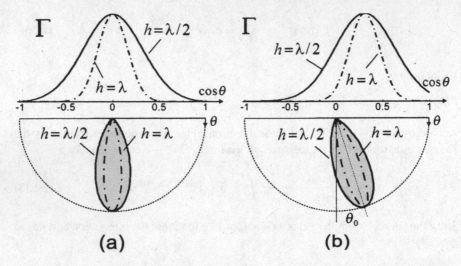

Fig. 15.49 Directional characteristic of a line source with a *Gaussian*-shaped source-strength load as in (9.74), with $h = \lambda/2$ and $h = \lambda$, respectively. (**a**) Solution without phase shift (**b**) Solution where the line source is modified by a phase offset as in (15.544)

Figure 15.49a illustrates the directional characteristics for $h = \lambda/2$ and $h = \lambda$ in both *Cartesian* ($-1 \leq \cos\theta \leq 1$) and polar ($0 \leq \theta \leq \pi$) representation. With $h = \lambda/2$, and $\beta = 2\pi/\lambda$, $h\beta = \pi$, and the directional-characteristic of the source of finite length, the expression reduces to a negligible value at $\cos\theta = \pm 1$.

Task (**b**) When the source-strength load in (9.75) is modified by a phase shift in (9.76) with θ_0 being a specific phase-shift angle, take (9.76) with a variable substitution of $\Omega_0 = \beta \cos\theta_0$. This produces

$$\phi = (\beta \cos\theta_0)\, x = \Omega_0\, x. \tag{15.541}$$

Due to the *Fourier*-transform's property for a phase shift, (15.537) becomes

$$\int_{-\infty}^{+\infty} [\, g(x)\, e^{j\,\Omega_0\, x}\,]\, e^{-j\Omega x}\, dx = \int_{-\infty}^{+\infty} g(x)\, e^{-j(\Omega - \Omega_0)\, x}\, dx$$
$$= \underline{G}(\Omega - \Omega_0), \tag{15.542}$$

where $\underline{G}(\Omega)$ is determined in (15.538) and simplifying renders

$$g(x)\, e^{j\,\Omega_0\, x} \,\circ\!\!\!-\!\!\!\bullet\, \underline{G}[\Omega - \Omega_0]. \tag{15.543}$$

Similar to the derivations conducted in the solution for *Task* (**a**), the directional characteristic changes due to a modified phase shift when substituting $\Omega_0 = \beta \cos\theta_0$. Thereupon, we finally arrive at

$$\Gamma = \frac{|\underline{p}_\rightarrow(r, \theta)|}{|\underline{p}_\rightarrow(r, 0)|} = e^{-[h(\beta \cos \theta - \beta \cos \theta_0)]^2 / 2} = e^{-(h\beta)^2 (\cos \theta - \cos \theta_0)^2 / 2}. \quad (15.544)$$

Figure 15.49b illustrates the modified directional characteristic with a small phase offset, namely, $\theta_0 = 2\pi/5$, for $h = \lambda/2$, and $h = \lambda$.

15.10 Chapter 10

Problem 10.1.

Proposed Approach

According to *Fraunhofer*'s approximation, we write the sound pressure as

$$\underline{p}_\rightarrow(r, \theta, \varphi) = \frac{j \omega \varrho}{2\pi r_0} e^{-j\beta r_0}$$
$$\cdot \int_{-\infty}^{\infty} \int_{-\infty}^{\infty} v(x, y) \, e^{j(\beta \cos \theta)x} \, e^{j(\beta \cos \varphi)y} \, dx \, dy. \quad (15.545)$$

Given that $v(x, y) = v_0 \cdot \psi_x(x) \cdot \psi_y(y)$ holds, the preceding expression becomes

$$\underline{p}_\rightarrow(r, \theta, \varphi) = \frac{j \omega \varrho v_0}{2\pi r_0} e^{-j\beta r_0} \int_{-\infty}^{\infty} \psi_x(x) \, e^{j(\beta \cos \theta)x} dx$$
$$\cdot \int_{-\infty}^{\infty} \psi_y(y) e^{j(\beta \cos \varphi)y} \, dy, \quad (15.546)$$

so that the directional characteristic is given by

$$\Gamma(\theta, \varphi) = \underbrace{\int_{-\infty}^{\infty} \psi_x(x) \, e^{j(\beta \cos \theta)x} \, dx}_{\Gamma(\theta)} \cdot \underbrace{\int_{-\infty}^{\infty} \psi_y(y) \, e^{j(\beta \cos \varphi)y} \, dy}_{\Gamma(\varphi)}, \quad (15.547)$$

which leads to

$$\Gamma(\theta, \varphi) = \Gamma(\theta) \cdot \Gamma(\varphi). \quad (15.548)$$

Problem 10.2.

Proposed Approach

Since the cardioid directional characteristic along the y–direction is given, and the line array consisting of these elementary sources is arranged along the x–direction, the source strength of the entire set-up is separable like in (10.30). The directional characteristic is then determinable for two orthogonal directions (angular variables) separately as two factors—compare (10.29).

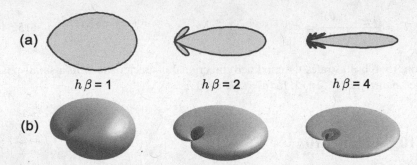

(a)

$h\beta = 1$ $h\beta = 2$ $h\beta = 4$

(b)

Fig. 15.50 Directional characteristics, Γ, of a line array with cardioid elements for $h\beta = 1, 2, 4$
(a) Directional patterns concerning the elevation, $|\Gamma(\theta)|$ **(b)** Three dimensional directional patterns,
$|\Gamma(\theta, \varphi)|$

For x–direction we obtain

$$\Gamma(\theta) = \text{si}(h\,\beta \cos\theta)\,, \tag{15.549}$$

and for the y–direction the appropriate result is

$$\Gamma(\varphi) = \frac{1 + \cos\varphi}{2}\,. \tag{15.550}$$

In summary, we consequently arrive at

$$\Gamma(\theta, \varphi) = \Gamma(\theta)\,\Gamma(\varphi) = \frac{1}{2}\,\text{si}(h\,\beta \cos\theta)\,(1 + \cos\varphi)\,. \tag{15.551}$$

Figure 15.50 illustrates the directional characteristics for both (15.549) and
(15.551). Figure 15.47 shows the directional profiles in the azimuthal plane.

Problem 10.3.

Proposed Approach

Given the radius, R, of a circular piston membrane, we consider a differential element, $\mathrm{d}A$, in location a, at a distance to the piston center—as depicted in Fig. 15.51a.
The plot shows that $\mathrm{d}A = a\,\mathrm{d}\varphi\,\mathrm{d}a$ holds. To cover the entire circular piston, the radial variable, a, and the angular variable, φ, have to cover $0 \le a \le R$, and $-\pi \le \mathrm{d}\varphi \le \pi$, respectively.

We now show that the solution resorts to *Rayleigh*'s integral as follows,

$$\underline{p} = \frac{j\,\omega\,\varrho\,v_0}{2\,\pi} \int_A \frac{e^{-j\beta r}}{r}\,\mathrm{d}A = \frac{j\,\omega\,\varrho\,v_0}{2\,\pi} \int_0^R \int_{-\pi}^{\pi} \frac{e^{-j\beta r}}{r}\,a\,\mathrm{d}\varphi\,\mathrm{d}a\,. \tag{15.552}$$

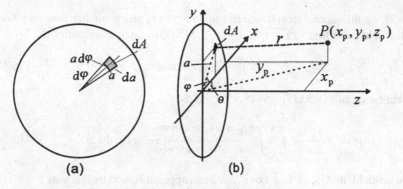

Fig. 15.51 Circular piston with a differential area element on it in both polar- and *Cartesian* coordinates (**a**) The differential area element, dA, on a circular piston, corresponds to a differential angular change, $d\varphi$, and a differential radial change, da, at radial distance, a, from the piston center, therefore $dA = a\, d\varphi\, da$ (**b**) Position of the observation point, $P(x_p, y_p, z_p)$, in relation to the position of the differential area element, dA

For convenience, the circular piston is placed in the x/y–plane as sketched in Fig. 15.51b.

In the following steps we first determine the distance, r, from an observation point, $P(x_p, y_p, z_p)$, on the membrane to the differential area element, dA, at position (x, y), on the surface of the piston—see in Fig. 15.51b. Thereby mind that

$$a^2 = x^2 + y^2, \qquad x_p^2 + y_p^2 + z_p^2 = r_0^2. \tag{15.553}$$

The distance, r, is given by the relation

$$
\begin{aligned}
r^2 &= (x_p - x)^2 + (y_p - y)^2 + z_p^2 \\
&= (x_p^2 + y_p^2 + z_p^2) + (x^2 + y^2) - 2x_p\, x - 2y_p\, y \\
&= r_0^2 + a^2 - 2\, \underbrace{x_p}_{r_0 \cos\theta}\, \underbrace{x}_{a\cos\varphi} - 2y_p\, y\,.
\end{aligned}
\tag{15.554}
$$

The current problem is radially symmetric. Therefore, the solution reduces to the x/z–plane with $y_p = 0$. Furthermore, $a \ll r_0$ means that we deal with the far-field. Thus, we write,

$$r^2 \approx r_0^2 - 2a\, r_0 \cos\theta \, \cos\varphi, \tag{15.555}$$

or, in a modified way,

$$r \approx \sqrt{r_0^2 - 2a\, r_0 \cos\theta \, \cos\varphi} = r_0 \sqrt{1 - 2\frac{a}{r_0}\cos\theta\,\cos\varphi}. \tag{15.556}$$

For $a \ll r_0$, the second term inside the square-root is much smaller than one. Given this far-field condition, $\sqrt{1 - \eta} \approx 1 - \eta/2$, (15.556) further simplifies to

$$r \approx r_0 - a \cos \theta \cos \varphi. \tag{15.557}$$

A substitution of (15.557) into (15.552) yields

$$\underline{p} = \frac{j \omega \varrho v_0}{2 \pi} \int_0^R \int_{-\pi}^{\pi} \frac{e^{-j \beta (r_0 - a \cos \theta \cos \varphi)}}{r_0 - a \cos \theta \cos \varphi} a \, d\varphi \, da \,. \tag{15.558}$$

In the far-field, that is, $r_0 \gg a \cos \theta \cos \varphi$, an approximation by ignoring $\cos \theta \cos \varphi$ in the distance tèrm in the denominator of the above equation while retaining it in the phase term, results in

$$\underline{p} \approx \frac{j \omega \varrho v_0}{2 \pi} \frac{e^{-j \beta r_0}}{r_0} \int_0^R \int_{-\pi}^{\pi} e^{j \beta a \cos \theta \cos \varphi} a \, d\varphi \, da. \tag{15.559}$$

To resolve this double integral, consider the classical integral

$$\int_{-\pi}^{\pi} e^{j \gamma \cos \varphi} \, d\varphi = 2 \pi \, \mathrm{J}_0(\gamma) \,, \tag{15.560}$$

where $\mathrm{J}_0(\gamma)$ is the 0th-order *Bessel* function of the 1st kind with $\gamma = \beta a \cos \theta$.

Substitution of γ and (15.560) into (15.559) yields

$$\underline{p} \approx j \omega \varrho v_0 \frac{e^{-j \beta r_0}}{r_0} \int_0^R a \, \mathrm{J}_0(\overbrace{\beta a \cos \theta}^{\gamma}) \, da \,. \tag{15.561}$$

To solve this radial integral further, another relevant integral is useful, namely,

$$\int_0^b x \, \mathrm{J}_0(x) \, dx = b \, \mathrm{J}_1(b) \,. \tag{15.562}$$

Now consider a variable substitution, $\epsilon = \beta \cos \theta$, and $\gamma = a \epsilon$, $d\gamma = \epsilon \, da$, for the case that $a = 0$, $\gamma = 0$, and $a = R$, $\gamma = \epsilon R$. The radial integral in (15.561) is then expressible as

$$\int_0^R a \, \mathrm{J}_0(a \epsilon) \, da = \frac{1}{\epsilon^2} \int_0^{\epsilon R} \gamma \, \mathrm{J}_0(\gamma) \, d\gamma = \frac{1}{\epsilon^2} \epsilon R \, \mathrm{J}_1(\epsilon R) \,, \tag{15.563}$$

where $\mathrm{J}_1(\gamma)$ is the 1st-order *Bessel*-function of the 1st kind.

A substitution of (15.563) into (15.561) yields

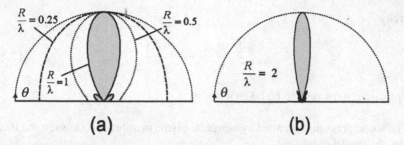

Fig. 15.52 Directional characteristics, Γ, of a circular piston (a) $R/\lambda = 0.25$, 0.5, 1 (b) $R/\lambda = 2$

$$\underline{p} \approx j\,\omega\,\varrho\,v_0\,\frac{e^{-j\beta r_0}}{r_0}\,\frac{R}{\beta \cos\theta}\,\mathbf{J}_1(\beta R \cos\theta). \tag{15.564}$$

In this expression, the directional characteristic comes out as

$$\Gamma(\theta) \propto \frac{\mathbf{J}_1(\beta R \cos\theta)}{\cos\theta}, \tag{15.565}$$

This expression arrives at its at its maximum when $\theta = 90°$ since $\cos\theta = 0$, due to which the argument of $\mathbf{J}_1(\gamma)$ is also small. Note that $\mathbf{J}_1(\gamma) \approx \gamma/2$ when $\gamma \ll 1$, which produces the maximum in (15.565) as follows,

$$\Gamma_{max} \approx \frac{\beta R \cos\theta}{2\cos\theta} = \frac{\beta R}{2}. \tag{15.566}$$

By normalizing the directional characteristic to its maximum, the normalized characteristic becomes

$$\Gamma(\theta) = 2\,\frac{\mathbf{J}_1(R\,\beta\,\cos\theta)}{R\,\beta\,\cos\theta} = 2\,\frac{\mathbf{J}_1(2\,\pi\,\frac{R}{\lambda}\,\cos\theta)}{2\,\pi\,\frac{R}{\lambda}\,\cos\theta}. \tag{15.567}$$

Compare (15.567) also with (10.10). Figure 15.52 illustrates different directional characteristics as in (15.567) for $R/\lambda = 0.25$, 0.5, 1, and 2, respectively. The figure indicates that, on the one hand, the directional characteristic of a circular piston of a given radius, R, becomes narrower with decreasing wavelength, λ, that is, increasing frequency. On the other hand, they also become narrower with an increasing radius at a given frequency/wavelength, λ.

Problem 10.4.

Proposed Approach

To find the radiation resistance of a circular piston membrane, that is,

$$r_{rad} = \mathrm{Re}\{\underline{Z}_{rad}\}, \tag{15.568}$$

whereby

$$\underline{Z}_{\text{rad}} = \frac{\underline{F}}{\underline{v}} = \frac{\int_{A_0} \underline{p} \, dA}{\underline{v}} . \tag{15.569}$$

the following steps need to be taken.

The sound pressure generated by a circular piston membrane follows from (10.26), that is, the *Rayleigh* integral,

$$\underline{p}_{\rightarrow}(l_1, \Theta_1) = \frac{j\omega\varrho}{2\pi} \int_{l_2=0}^{R} \int_{\Theta_2=0}^{2\pi} \underline{v} \, \frac{e^{-j\beta a}}{a} \, l_2 \, d\Theta_2 \, dl_2 , \tag{15.570}$$

where $a = \sqrt{l_1^2 + l_2^2 - 2l_1 l_2 \cos\Theta_2}$—modified from b in (10.26).

To find the force acting on the membrane, we perform the following integration,

$$\underline{F} = \int_{l_1=0}^{R} \int_{\Theta_1=0}^{2\pi} \underline{p}_{\rightarrow}(l_1, \Theta_1) \, l_1 \, d\Theta_1 \, dl_1 . \tag{15.571}$$

Substituting (15.570) into (15.571) yields

$$\underline{F} = \frac{j\omega\varrho\underline{v}}{2\pi} \int_{\Theta_2=0}^{2\pi} \int_{l_2=0}^{R} \int_{\Theta_1=0}^{2\pi} \int_{l_1=0}^{R} \frac{e^{-j\beta a}}{a} \, l_1 \, dl_1 \, d\Theta_1 \, l_2 \, dl_2 \, d\Theta_2 . \tag{15.572}$$

It is necessary to perform the integration by only including the contributions from the differential areas determined by l_1 and l_2—again according to Fig. 10.10. Therefore, we integrate over the circle with a maximum distance between the differential areas, a. For the current example, l_2 is chosen, and the distance between the differential areas is given by $2 l_2 \cos\theta$, which is the maximum distance in a circle determined by the radius l_2. We extend the bounds of (15.572) to cover the new circle.

The variable, a, still represents the distance between the two areas, and θ is now the angle created by l_2 and a. To find the contribution of pressure from one area to the other one, and then vice versa, we split the bounds of Θ_1. These modifications finally result in (15.573), viz,

$$\underline{F} = \frac{j\omega\varrho\underline{v}}{2\pi} \int_{\Theta_2=0}^{2\pi} \int_{l_2=0}^{R} 2 \int_{-\frac{\pi}{2}}^{\frac{\pi}{2}} \int_{0}^{2 l_2 \cos\theta} \frac{e^{-j\beta a}}{a} \, a \, da \, d\theta \, l_2 \, dl_2 \, d\Theta_2 . \tag{15.573}$$

Solving the double integral

$$\int_{-\frac{\pi}{2}}^{\frac{\pi}{2}} \int_{0}^{2 l_2 \cos\theta} l_2 \, e^{-j\beta a} da \, d\theta = \int_{-\frac{\pi}{2}}^{\frac{\pi}{2}} \frac{l_2}{j\beta} \left(1 - e^{-2j\beta l_2 \cos\theta}\right) d\theta , \tag{15.574}$$

and using *Euler*'s identity for yet another substitution, renders

$$\int_{-\frac{\pi}{2}}^{\frac{\pi}{2}} \frac{l_2}{j\beta} \left[1 - \cos(2\beta l_2 \cos\theta) + j\sin(2\beta l_2 \cos\theta)\right] d\theta. \tag{15.575}$$

The following identities are needed for the *Bessel* function of the 1st kind and of 0th-order, that is,

$$\pi J_0(x) = \int_{-\frac{\pi}{2}}^{\frac{\pi}{2}} \cos(x\cos\theta)\, d\theta, \tag{15.576}$$

and the *Struve* function of the 0th order, which is

$$\pi H_0(x) = \int_{-\frac{\pi}{2}}^{\frac{\pi}{2}} \sin(x\cos\theta)\, d\theta. \tag{15.577}$$

Substitution of (15.576) and (15.577) into the integral in (15.575) leads to

$$\int_{-\frac{\pi}{2}}^{\frac{\pi}{2}} \frac{l_2}{j\beta} \left[1 - \cos(2\beta l_2 \cos\theta) + j\sin(2\beta l_2 \cos\theta)\right] d\theta$$

$$= \frac{l_2\pi}{j\beta}\left[1 - J_0(2\beta l_2) + j H_0(2\beta l_2)\right]. \tag{15.578}$$

Further, substitution of (15.578) into (15.573) and using $\beta = \omega/c$ provides

$$\underline{F} = \varrho c \underline{v} \int_{\Theta_2=0}^{2\pi} \int_{l_2=0}^{R} l_2 \left[1 - J_0(2\beta l_2) + j H_0(2\beta l_2)\right] dl_2\, d\Theta_2. \tag{15.579}$$

and integration over Θ_2 yields

$$\underline{F} = 2\pi \varrho c \underline{v} \int_{l_2=0}^{R} l_2\left[1 - J_0(2\beta l_2) + j H_0(2\beta l_2)\right] dl_2. \tag{15.580}$$

Again two important identities are relevant here, that is, the *Bessel* function of the 1st kind and the 1st order,

$$\int_0^b x\, J_0(x)\, dx = b\, J_1(b), \tag{15.581}$$

and the *Struve* function of the 1st order, which reads as follows,

$$\int_0^b x\, H_0(x)\, dx = b\, H_1(b). \tag{15.582}$$

A change of variables for $x = 2\beta R$ and $\mathrm{d}l_2 = \mathrm{d}x/(2\beta)$ yields

$$\int_0^{2\beta R} \frac{x}{2\beta} \left[1 - \mathbf{J}_0(2\beta x) + \mathrm{j}\,\mathbf{H}_0(2\beta x)\right] \frac{\mathrm{d}x}{2\beta}, \tag{15.583}$$

and is followed by applying the identities

$$\frac{1}{4\beta^2} \left[\frac{(2\beta R)^2}{2} - 2\beta\,R\,\mathbf{J}_1(2\beta x) + 2\beta\,R\,\mathrm{j}\,\mathbf{H}_1(2\beta x)\right]. \tag{15.584}$$

Rearranging with the area of the circular piston, $A_0 = \pi R^2$, returns

$$\frac{A_0}{2\pi} \left[1 - \frac{\mathbf{J}_1(2\beta R)}{\beta R} + \mathrm{j}\frac{\mathbf{H}_1(2\beta R)}{\beta R}\right]. \tag{15.585}$$

Substituting this expression into (15.580) yields

$$\underline{F} = A_0\,\varrho\,\mathrm{c}\,\underline{v} \left[1 - \frac{\mathbf{J}_1(2\beta R)}{\beta R} + \mathrm{j}\frac{\mathbf{H}_1(2\beta R)}{\beta R}\right]. \tag{15.586}$$

From the definition of $\underline{Z}_{\mathrm{rad}}$ follows our final result, which is

$$r_{\mathrm{rad}} = \mathrm{Re}\{\underline{Z}_{\mathrm{rad}}\} = \mathrm{Re}\left\{\frac{\underline{F}}{\underline{v}}\right\} = A_0\,\varrho\,\mathrm{c}\left[1 - \frac{\mathbf{J}_1(2\beta R)}{\beta R}\right]. \tag{15.587}$$

Problem 10.5.

Proposed Approach

The sound pressure radiated by the nth zone is given by (10.16) as

$$\underline{p}_{\rightarrow,\,n\text{th zone}} = \pm 2\,\varrho\,\mathrm{c}\,\underline{v}\,\mathrm{e}^{-\mathrm{j}\beta r_0}. \tag{15.588}$$

For the 1st (odd-numbered) zone adjacent to the center of the opening, we thus get

$$\underline{p}_{\rightarrow,\,1\text{th zone}} = 2\,\varrho\,\mathrm{c}\,\underline{v}\,\mathrm{e}^{-\mathrm{j}\beta r_0}, \tag{15.589}$$

whereby the 2nd (even-numbered) zone from the center of the opening, contributes

$$\underline{p}_{\rightarrow,\,2\text{nd zone}} = -2\,\varrho\,\mathrm{c}\,\underline{v}\,\mathrm{e}^{-\mathrm{j}\beta r_0}. \tag{15.590}$$

The reference point is located at a distance of $r_0 = 3\,\lambda$, perpendicular to the center of the opening surface, and the edge length of the quadratic opening is $a = 6.7\lambda$. This prompts the determination of the number of circular zones with circular radii, r_n—see Fig. 10.7,

$$R_n^2 = r_n^2 + r_0^2 = \left(r_0 + n \cdot \frac{\lambda}{2}\right)^2. \tag{15.591}$$

Task **(a)** Equation (15.591) leads to

$$r_n^2 = 2\, r_0\, n \cdot \frac{\lambda}{2} + n^2 \frac{\lambda^2}{4}, \quad \text{or} \quad r_n = \sqrt{n\, r_0 \cdot \lambda + n^2 \frac{\lambda^2}{4}}. \tag{15.592}$$

and a substitution of $r_0 = 3\lambda$ into (15.592) results in

$$r_n = \lambda \sqrt{3n + \frac{n^2}{4}}, \quad \text{for} \quad n = 1, 2\dots. \tag{15.593}$$

Equation (15.593) allows to determine the zones for the orders $n = 1, 2, \ldots$, up the edge length $a/\sqrt{2} = 4.74\lambda$, namely,

$$\left\{ \begin{aligned} r_1 &= \lambda \sqrt{3 + \tfrac{1}{4}} = 1.803\lambda \\[4pt] r_2 &= \lambda \sqrt{6 + 1} = 2.646\lambda \\[4pt] r_3 &= \lambda \sqrt{9 + \tfrac{9}{4}} = 3.354\lambda \\[4pt] r_4 &= \lambda \sqrt{12 + \tfrac{16}{4}} = 4\lambda \\[4pt] r_5 &= \lambda \sqrt{15 + \tfrac{25}{4}} = 4.609\lambda \\[4pt] r_6 &= \lambda \sqrt{18 + \tfrac{36}{4}} = 5.196\lambda\ . \end{aligned} \right. \tag{15.594}$$

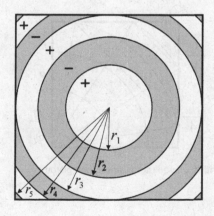

Fig. 15.53 *Huygens-Fresnel* zones on a square piston—schematic given the wavelength

Task **(b)** From the above calculations of radii $r_1, \ldots r_6$, we plot the *Fresnel* zones given by the quadratic edge length $a = 6.7\,\lambda$—shown in Fig. 15.53.

Task **(c)** The area of the A_{zone} with radius, r_1, is $A_{\text{zone}} = \pi r_1^2 = 10.21\lambda^2$. Figure 15.53 shows that the three most inner zones are undisturbed. However, the zones 4–6 are constricted by the edges, and thus contribute partially to the resulting total sound pressure, viz,

$$\underline{p}_{\rightarrow,\text{total}} = 2\,\varrho\,c\,\underline{v}\,e^{-j\frac{2\pi}{\lambda}r_0}\,\frac{A_1 - A_2 + A_3 - A_4 + A_5 - A_6}{A_{\text{zone}}}. \tag{15.595}$$

To determine their partial contributions, we calculate the areas A_4, A_5, A_6. These partial areas are approximately estimated by considering the areas in one of the four corners as shown in Fig. 15.54. We perform this estimation, as follows,

$$A_k \approx \frac{4\,B_k}{\pi\,(r_k^2 - r_{k-1}^2)}, \qquad k = 4, 5, 6 \tag{15.596}$$

by calculating the trapezoid areas, $\frac{1}{2}(b_{k-1} + b_k) \cdot d_k$, with

$$\frac{b_k}{2} = r_k \sin\frac{\phi_k}{2}, \quad \text{and} \quad d_k = r_k \cos\frac{\phi_k}{2} - r_{k-1}\cos\frac{\phi_{k-1}}{2}, \tag{15.597}$$

and, further,

$$B_k = \left(\overbrace{r_{k-1}\sin\frac{\phi_{k-1}}{2} + r_k\sin\frac{\phi_k}{2}}^{(b_{k-1}+b_k)/2}\right)\left(\overbrace{r_k\cos\frac{\phi_k}{2} - r_{k-1}\cos\frac{\phi_{k-1}}{2}}^{d_k}\right)$$

$$+ \frac{r_k^2}{2}(\phi_k - \sin\phi_k) - \frac{r_{k-1}^2}{2}(\phi_{k-1} - \sin\phi_{k-1}). \tag{15.598}$$

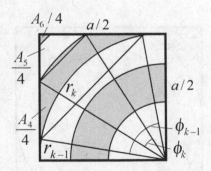

Fig. 15.54 Partial areas within one-quarter of the square piston

In the case of A_6, these calculations produce $r_5 = a/\sqrt{2}$ and $\phi_6 \approx 0$, so that we end with

$$A_6 \approx 4\, \frac{r_5 \sin \frac{\phi_5}{2}(\frac{a}{\sqrt{2}} - r_5 \cos \frac{\phi_5}{2}) - \frac{r_5^2}{2}(\phi_5 - \sin \phi_5)}{\pi\,(r_6^2 - r_5^2)}. \tag{15.599}$$

Equations (15.596)–(15.599) produce $A_4 = 0.488\,A_{zone}$, $A_5 = 0.139\,A_{zone}$, and $A_6 = 0.005\,A_{zone}$. According to (15.595), summarizing the contributions from all six zones, with $r_0 = 3\lambda$ and $\beta = 2\pi/\lambda$, provides our final result as

$$\begin{aligned} \underline{p}_{\rightarrow,total} &\approx 2\,\varrho c \underline{v}\, e^{-j\frac{2\pi}{\lambda}r_0}\,(1 - 1 + 1 - 0.488 + 0.139 - 0.005) \\ &= 1.292\,\varrho c \underline{v}\, e^{-j6\pi}. \end{aligned} \tag{15.600}$$

15.11   Chapter 11

Problem 11.1.

Proposed Approach

The extended *Euler* equation for porous media is written as

$$-\frac{\partial p}{\partial x} = \varrho\,\frac{\partial v_i}{\partial t} + \sigma\,\Xi\, v_i, \tag{15.601}$$

and the continuity equation as

$$-\frac{\partial v_i}{\partial x} = \kappa\,\frac{\partial p}{\partial t}. \tag{15.602}$$

Task **(a)** Taking the partial derivative with respect to x on both sides of (15.601) results in

$$-\frac{\partial^2 p}{\partial x^2} = \varrho\,\frac{\partial^2 v_i}{\partial x\,\partial t} + \sigma\,\Xi\,\frac{\partial v_i}{\partial x}, \tag{15.603}$$

and taking the partial derivative with respect to t on both sides of (15.602) results in

$$\frac{\partial^2 v_i}{\partial x\,\partial t} = -\kappa\,\frac{\partial^2 p}{\partial t^2}. \tag{15.604}$$

Substitution of (15.604) and (15.602) into (15.603) causes

$$\frac{\partial^2 p}{\partial x^2} = \varrho\,\kappa\,\frac{\partial^2 p}{\partial t^2} + \sigma\,\Xi\,\kappa\,\frac{\partial p}{\partial t}. \tag{15.605}$$

This is the 2nd-order wave equation.

In complex notation, the above equation reads

$$\frac{\partial^2 \underline{p}}{\partial x^2} = \varrho \kappa \, (j\omega)^2 \underline{p} + j\omega \sigma \, \Xi \, \kappa \, \underline{p}$$
$$= (j\omega \varrho + \sigma \, \Xi)(j\omega \kappa) \, \underline{p}. \tag{15.606}$$

The second line of this equation, viz,

$$\frac{\partial^2 \underline{p}}{\partial x^2} - (j\omega \varrho + \sigma \, \Xi)(j\omega \kappa) \, \underline{p} = 0. \tag{15.607}$$

is known as the *Helmholtz* equation for porous media.

A comparison of (15.607) with (11.4) leads to

$$\underline{\gamma} = \sqrt{(j\omega \varrho + \sigma \, \Xi)(j\omega \kappa)}. \tag{15.608}$$

Task **(b)** One of the (forward propagating) solutions of the above equation is

$$\underline{p}_{i\rightarrow}(x) = \hat{\underline{p}}_{\rightarrow} e^{-\underline{\gamma} x}. \tag{15.609}$$

Substitution of this solution into the extended *Euler* equation (15.601) yields

$$\underline{\gamma} \, \underline{p}_{i\rightarrow}(x) = (\varrho j\omega + \sigma \, \Xi) \underline{v}_{i\rightarrow}(x). \tag{15.610}$$

The characteristic impedance, a field impedance, then becomes

$$\underline{Z}_c = \frac{\underline{p}_{i\rightarrow}(x)}{\underline{v}_{i\rightarrow}(x)} = \frac{j\omega \varrho + \sigma \, \Xi}{\underline{\gamma}}. \tag{15.611}$$

Taking (11.17) into account, a substitution of (15.608) into (15.611) leads to

$$\underline{Z}_c = \frac{j\omega \varrho + \sigma \, \Xi}{\sqrt{(j\omega \varrho + \sigma \, \Xi)(j\omega \kappa)}} = \sqrt{\frac{j\omega \varrho + \sigma \, \Xi}{j\omega \kappa}} = \sqrt{\varrho_{eq}(\omega) \, K}, \tag{15.612}$$

where $K = 1/\kappa$.

Problem 11.2.

Proposed Approach

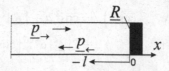

Fig. 15.55 Coordinates for *Kundt's* tube

To solve this problem, we generalize the fundamental method by considering the standing-wave-ratio method without losses. Our discussion is straightforward and also applicable to other methods, such as the *transfer-function* method.

In the coordinate system defined in Fig. 15.55, the forward propagating and reflected wave components are

$$\underline{p}_{\rightarrow}(l) = \underline{\hat{p}}\, e^{(\breve{\alpha}+j\beta)l}\,, \tag{15.613}$$

$$\underline{p}_{\leftarrow}(l) = \underline{R}\,\underline{\hat{p}}\, e^{-(\breve{\alpha}+j\beta)l}\,, \tag{15.614}$$

where $\underline{R} = |\underline{R}|\, e^{j\phi_R}$.

The total pressure field becomes

$$\underline{p}(l) = \underline{p}_{\rightarrow}(l) + \underline{p}_{\leftarrow}(l) = \underline{\hat{p}}\,[\, e^{(\breve{\alpha}+j\beta)l} + \underline{R}\, e^{-(\breve{\alpha}+j\beta)l}\,]\,, \tag{15.615}$$

so that we get

$$\underline{p}(l) = \underline{\hat{p}}\, e^{\breve{\alpha}l}[\, e^{j\beta l} + \underline{R}\, e^{-2\breve{\alpha}l} e^{-j\beta l}\,]\,. \tag{15.616}$$

The pressure at the tube position at a distance l, from the surface of the material under test becomes

$$\underline{p}(l) = \underline{\hat{p}}\, e^{\breve{\alpha}l}\,[\cos\beta l + j\,\sin\beta l$$
$$+ |\underline{R}|\, e^{-2\breve{\alpha}l}\cos(\phi_R - \beta l) + j\,|\underline{R}|\, e^{-2\breve{\alpha}l}\sin(\phi_R - \beta l)]\,. \tag{15.617}$$

With $a = \beta l$ and $b = \phi_R - \beta l$, the magnitude—the so-called *envelope*—of the sound-pressure signal, $p_{\text{env}}(l) = |\underline{p}(l)|$, results in

$$p_{\text{env}}(l) = \hat{p}\, e^{\breve{\alpha}l}\,\cdot$$
$$\sqrt{(\cos a + |\underline{R}|\, e^{-2\breve{\alpha}l}\cos b)^2 + (\sin a + |\underline{R}|\, e^{-2\breve{\alpha}l}\sin b)^2}\,. \tag{15.618}$$

Applying both

$$\left(\cos a + |\underline{R}|\,e^{-2\breve{\alpha}l}\cos b\right)^2 = \cos^2 a + |\underline{R}|^2\,e^{-4\breve{\alpha}l}\cos^2 b$$
$$+ 2|\underline{R}|\,e^{-2\breve{\alpha}l}\cos a\cos b, \qquad (15.619)$$

and

$$\left(\sin a + \underline{R}\,e^{-2\breve{\alpha}l}\sin b\right)^2 = \sin^2 a + |\underline{R}|^2\,e^{-4\breve{\alpha}l}\sin^2 b$$
$$+ 2|\underline{R}|\,e^{-2\breve{\alpha}l}\sin a\sin b, \qquad (15.620)$$

results in the total sound pressure as

$$p_{\mathrm{env}}(l) = \hat{p}\,e^{\breve{\alpha}l}$$
$$\sqrt{1 + |\underline{R}|^2 e^{-4\breve{\alpha}l} + 2|\underline{R}|\,e^{-2\breve{\alpha}l}(\cos a\cos b + \sin a\sin b)}. \quad (15.621)$$

Keeping in mind that $\cos(a-b) = \cos a\,\cos b + \sin a\,\sin b$, substitution of $a = \beta l$ and $b = \phi_R - \beta l$ into (15.621) yields

$$p_{\mathrm{env}}(l) = \hat{p}\,e^{\breve{\alpha}l}\sqrt{1 + |\underline{R}|^2\,e^{-4\breve{\alpha}l} + 2|\underline{R}|\,e^{-2\breve{\alpha}l}\cos(2\beta l - \phi_R)}.. \quad (15.622)$$

The argument inside the square root reaches maxima when $2\beta l_{\max} - \phi_R = 2n\pi$, that is, at

$$p(l_{\max}) = \hat{p}\,e^{\breve{\alpha}l}\sqrt{1 + |\underline{R}|^2\,e^{-4\breve{\alpha}l} + 2|\underline{R}|\,e^{-2\breve{\alpha}l}}$$
$$= \hat{p}\,e^{\breve{\alpha}l_{\max}}\left(1 + |\underline{R}|\,e^{-2\breve{\alpha}l_{\max}}\right), \qquad (15.623)$$

and reaches minima for $2\beta l_{\max} - \phi_R = [(2n+1)/2]\,\pi$, namely,

$$p(l_{\min}) = \hat{p}\,e^{\breve{\alpha}l}\sqrt{1 + |\underline{R}|^2\,e^{-4\breve{\alpha}l} - 2|\underline{R}|\,e^{-2\breve{\alpha}l}}$$
$$= \hat{p}\,e^{\breve{\alpha}l_{\min}}\left(1 - |\underline{R}|\,e^{-2\breve{\alpha}l_{\min}}\right). \qquad (15.624)$$

The standing-wave ratio thus results in

$$S = \frac{p(l_{\max})}{p(l_{\min})} = \frac{e^{\breve{\alpha}l_{\max}}(1 + |\underline{R}|\,e^{-2\breve{\alpha}l_{\max}})}{e^{\breve{\alpha}l_{\min}}(1 - |\underline{R}|\,e^{-2\breve{\alpha}l_{\min}})}, \qquad (15.625)$$

leading to the following solution for $|\underline{R}|$,

$$|\underline{R}| = \frac{S\,e^{\breve{\alpha}(l_{\min}-l_{\max})} - 1}{S\,e^{-\breve{\alpha}(l_{\max}+l_{\min})} + e^{-2\breve{\alpha}l_{\max}}}. \qquad (15.626)$$

When $\check{\alpha} \to 0$, the tube is considered lossless, and the above equation reduces to (15.347).

Problem 11.3.

Proposed Approach

Similar to (7.70), we express the reflectance, $\underline{R}$, at the front surface—shown in Fig. 11.14. Given the separation, s, and the distance to the surface, l, of the micro phone, M_2, we obtain

$$\underline{R} = \frac{e^{j\beta s} - \underline{H}_{12}}{\underline{H}_{12} - e^{-j\beta s}} \, e^{-j2\beta l}, \tag{15.627}$$

where

$$\underline{H}_{12} = \frac{\underline{P}_2}{\underline{P}_1} \tag{15.628}$$

is the transfer function between the pressure spectra, $\underline{P}_1$ and $\underline{P}_2$. It is given by the *Fourier* integral, (9.55) or (9.56), of the pressure signals, $p_1(t)$, and $p_2(t)$. These are as measured by the two microphones, M_1, and M_2. $\underline{P}_1$ and $\underline{P}_2$ are both functions of frequency.

The surface impedance of the porous specimen evolves from the reflectance, $\underline{R}$, as follows,

$$\underline{Z}_s = \varrho c \, \frac{1 + \underline{R}}{1 - \underline{R}}. \tag{15.629}$$

As shown in Fig. 11.14, the porous material of thickness, d, is fitted inside the tube against a rigid termination. Denote the total sound pressure at this position, $\underline{p}_3$, and the respective velocity, $\underline{v}_3$. The microphone M_3 measures the total sound pressure.

The total sound pressure, $\underline{p}_m$, and the particle velocity, $\underline{v}_m$, at the front surface of the porous sample is

$$\begin{bmatrix} \underline{p}_m \\ \underline{v}_m \end{bmatrix} = \begin{bmatrix} \cos(\underline{\gamma} d) & j\underline{Z}_c \sin(\underline{\gamma} d) \\ j \sin(\underline{\gamma} d)/\underline{Z}_c & \cos(\underline{\gamma} d) \end{bmatrix} \begin{bmatrix} \underline{p}_3 \\ \underline{v}_3 \end{bmatrix}. \tag{15.630}$$

Thus, at the rigid termination, $\underline{v}_3 = 0$, the surface impedance, $\underline{Z}_s$, follows from the above equation as

$$\underline{Z}_s = \frac{\underline{p}_m}{\underline{v}_m} = \frac{\underline{Z}_c}{j} \frac{\cos(\underline{\gamma} d)}{\sin(\underline{\gamma} d)} = j\underline{Z}_c \tan(\underline{\gamma} d), \tag{15.631}$$

where the surface impedance, $\underline{Z}_s$, is given in (15.629) via (15.627) through experimentally measured pressure spectra from the two microphones in front of the sample material, $\mathbf{M}_1$ and $\mathbf{M}_2$.

From (15.630), it also follows that

$$\underline{P}_m = \underline{P}_3 \cos(\underline{\gamma} d), \tag{15.632}$$

The complex propagation coefficient, consequently, results as follows

$$\underline{\gamma} = \cos^{-1}(\underline{H}_{3m})/d, \tag{15.633}$$

where

$$\underline{H}_{3m} = \frac{\underline{P}_m}{\underline{P}_3}. \tag{15.634}$$

$\underline{P}_2$, and $\underline{P}_3$ are also provided by the sound-pressure signals via the *Fourier* integral—compare (9.55) or (9.56). The respective sound-pressure signal, $\underline{p}_3$, is provided by microphone, $\mathbf{M}_3$. $\underline{P}_3$ is a function of frequency as well. Since microphone $\mathbf{M}_2$ is positioned at a certain distance, l, off the surface, we write

$$\underline{H}_{3m} = \frac{\underline{P}_m}{\underline{P}_2} \frac{\underline{P}_2}{\underline{P}_3} = \frac{\underline{P}_m}{\underline{P}_2} \underline{H}_{32}, \tag{15.635}$$

with

$$\underline{H}_{32} = \frac{\underline{P}_2(f)}{\underline{P}_3(f)}. \tag{15.636}$$

The sound pressure at the surface of the porous test-sample, $\underline{P}_m$ is the result of superposition of the incident and reflected components in the frequency domain. Thus we have

$$\underline{P}_m = \underline{P}_{m\rightarrow} + \underline{P}_{m\leftarrow} = \underline{P}_{m\rightarrow} + \underline{R}\,\underline{P}_{m\rightarrow}, \tag{15.637}$$

where the sound pressure, $\underline{P}_2$, at microphone $\mathbf{M}_2$, is the result of a superposition of two components as well, that is,

$$\underline{P}_2 = \underline{P}_{2\rightarrow} + \underline{P}_{2\leftarrow} = \underline{P}_{m\rightarrow}\left(e^{j\beta l} + \underline{R}\,e^{-j\beta l}\right), \tag{15.638}$$

The ratio of (15.637) and (15.638) leads to

$$\frac{\underline{P}_m}{\underline{P}_2} = \frac{\underline{P}_{m\rightarrow}(1+\underline{R})}{\underline{P}_{m\rightarrow}(e^{j\beta l} + \underline{R}\,e^{-j\beta l})} = \frac{1+\underline{R}}{e^{j\beta l} + \underline{R}\,e^{-j\beta l}}. \tag{15.639}$$

Substitution of (15.639) into (15.635) and into (15.633) renders the propagation coefficient, namely,

$$\underline{\gamma} = \frac{1}{d} \cos^{-1}\left(\frac{1+\underline{R}}{e^{j\beta l} + \underline{R}\,e^{-j\beta l}}\,\underline{H}_{32}\right), \qquad (15.640)$$

where the transfer function, $\underline{H}_{32}$, is measured via the microphones $\mathbf{M_2}$ and $\mathbf{M_3}$, through the pressure spectra—as in (15.636)—while $\underline{R}$ is determined using (15.627), via the microphones $\mathbf{M_1}$ and $\mathbf{M_2}$, via the pressure spectra—as in (15.628).

In summary, using three microphones, two in front of the sample-material's surface and one in its back, the propagation coefficient and the characteristic impedance are determinable by measurement.

Note that knowing the thickness of the tested material, d, is not necessary in (15.632) and (15.640) as long as only the characteristic impedance is of interest because of

$$\underline{\gamma} d = \cos^{-1}\left(\frac{1+\underline{R}}{e^{j\beta l} + \underline{R}\,e^{-j\beta l}}\,\underline{H}_{32}\right). \qquad (15.641)$$

Substitution of (15.641) into (15.631) yields the characteristic impedance, $\underline{Z}_c$, without necessarily knowing d, provided that the microphone separation, s, and the distance, l, from microphone $\mathbf{M_2}$ to the surface of the material sample under test are known.

Problem 11.4.

Proposed Approach

In Medium 1, as illustrated in Fig. 15.56, the incident (forward) and the reflected sound-pressure components are

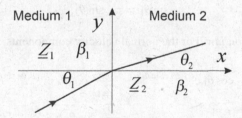

Fig. 15.56 Coordinates and sketch of the interface at two different media

$$\begin{cases} \underline{p}_{1\rightarrow} = \hat{p}_{1\rightarrow}\, e^{-j\beta_1(x\cos\theta_1 + y\sin\theta_1)} \\ \underline{p}_{1\leftarrow} = \hat{p}_{1\leftarrow}\, e^{j\beta_1(x\cos\theta_1 - y\sin\theta_1)} \quad. \end{cases} \tag{15.642}$$

The corresponding velocity components are

$$\begin{cases} \underline{v}_{x1\rightarrow} = \dfrac{\hat{p}_{1\rightarrow}}{\underline{Z}_1}\cos\theta_1\, e^{-j\beta_1(x\cos\theta_1 + y\sin\theta_1)} \\ \underline{v}_{x1\leftarrow} = -\dfrac{\hat{p}_{1\leftarrow}}{\underline{Z}_1}\cos\theta_1\, e^{j\beta_1(x\cos\theta_1 - y\sin\theta_1)} \quad. \end{cases} \tag{15.643}$$

In Medium 2, we express the refracted-wave components as

$$\underline{p}_{2\rightarrow} = \hat{p}_2\, e^{-j\beta_2(x\cos\theta_2 + y\sin\theta_2)}, \tag{15.644}$$

and

$$\underline{v}_{x2\rightarrow} = \dfrac{\hat{p}_2}{\underline{Z}_2}\cos\theta_2\, e^{-j\beta_2(x\cos\theta_2 + y\sin\theta_2)}. \tag{15.645}$$

We now state the relevant boundary conditions according to the continuity of the sound pressure and the velocity's normal components at $x = 0$. From (15.642) and (15.644), it follows that

$$\hat{p}_{1\rightarrow}\, e^{-j\beta_1(y\sin\theta_1)} + \hat{p}_{1\leftarrow}\, e^{-j\beta_1(y\sin\theta_1)} = \hat{p}_2\, e^{-j\beta_2(y\sin\theta_2)}. \tag{15.646}$$

This condition is valid everywhere at the boundary $x = 0$, and for arbitrary y. Thus it follows for $y = 0$,

$$\hat{p}_{1\rightarrow} + \hat{p}_{1\leftarrow} = \hat{p}_2, \tag{15.647}$$

and

$$\beta_1\sin\theta_1 = \beta_2\sin\theta_2, \tag{15.648}$$

which is the *refraction law*. For the normal velocity components, we have

$$\dfrac{\hat{p}_{1\rightarrow}}{\underline{Z}_1}\cos\theta_1\, e^{-j\beta_1(y\sin\theta_1)} - \dfrac{\hat{p}_{1\leftarrow}}{\underline{Z}_1}\cos\theta_1\, e^{-j\beta_1(y\sin\theta_1)}$$
$$= \dfrac{\hat{p}_2}{\underline{Z}_2}\cos\theta_2\, e^{-j\beta_2(y\sin\theta_2)}, \tag{15.649}$$

and for $y = 0$,

$$(\hat{p}_{1\rightarrow} - \hat{p}_{1\leftarrow})\frac{\cos\theta_1}{\underline{Z}_1} = \hat{p}_2 \frac{\cos\theta_2}{\underline{Z}_2}. \tag{15.650}$$

Substitution of (15.647) into (15.650) yields

$$(\hat{p}_{1\rightarrow} - \hat{p}_{1\leftarrow})\frac{\cos\theta_1}{\underline{Z}_1} = (\hat{p}_{1\rightarrow} + \hat{p}_{1\leftarrow})\frac{\cos\theta_2}{\underline{Z}_2}. \tag{15.651}$$

According to the refraction law in (15.648), we obtain with $\sin^2 a + \cos^2 a = 1$,

$$\cos\theta_2 = \sqrt{1 - \left(\frac{\beta_1}{\beta_2}\sin\theta_1\right)^2}. \tag{15.652}$$

Substitution of (15.652) into (15.651) further yields

$$(\hat{p}_{1\rightarrow} - \hat{p}_{1\leftarrow})\frac{\cos\theta_1}{\underline{Z}_1} = (\hat{p}_{1\rightarrow} + \hat{p}_{1\leftarrow})\frac{1}{\underline{Z}_2}\sqrt{1 - \left(\frac{\beta_1}{\beta_2}\sin\theta_1\right)^2}. \tag{15.653}$$

Division of both sides by $\hat{p}_{1\rightarrow}$ results in

$$\left(1 - \frac{\hat{p}_{1\leftarrow}}{\hat{p}_{1\rightarrow}}\right)\frac{\cos\theta_1}{\underline{Z}_1} = (1 + \frac{\hat{p}_{1\leftarrow}}{\hat{p}_{1\rightarrow}})\frac{1}{\underline{Z}_2}\sqrt{1 - \left(\frac{\beta_1}{\beta_2}\sin\theta_1\right)^2}, \tag{15.654}$$

and substitution of $\underline{R} = \hat{p}_{1\leftarrow}/\hat{p}_{1\rightarrow}$ into (15.654) leads to

$$(1 - \underline{R})\frac{\cos\theta_1}{\underline{Z}_1} = (1 + \underline{R})\frac{1}{\underline{Z}_2}\sqrt{1 - \left(\frac{\beta_1}{\beta_2}\sin\theta_1\right)^2}. \tag{15.655}$$

A few re-arrangements result in

$$\underline{R} = \frac{\underline{Z}_2\cos\theta_1 - \underline{Z}_1\sqrt{1 - \left(\frac{\beta_1}{\beta_2}\sin\theta_1\right)^2}}{\underline{Z}_2\cos\theta_1 + \underline{Z}_1\sqrt{1 - \left(\frac{\beta_1}{\beta_2}\sin\theta_1\right)^2}}. \tag{15.656}$$

In view of the conditions for total reflection, $|\underline{R}| = 1$, (15.656) takes the form

$$\underline{R} = \frac{a - b}{a + b}, \tag{15.657}$$

and, if the expression inside the square root becomes imaginary,

$$\left(\frac{\beta_1}{\beta_2} \sin \theta_1\right)^2 > 1, \tag{15.658}$$

which is the condition for total reflection.

With $\beta = \omega/c$ it follows that

$$\sin \theta_1 > \frac{c_1}{c_2}. \tag{15.659}$$

We recognize that total reflection is only possible when

$$c_2 > c_1. \tag{15.660}$$

Once the reflectance is known and total reflection does not occur, there exists a *reflected wave*. The reflected component is given by

$$\underline{P}_{1\leftarrow} = \hat{p}_{1\rightarrow} \frac{\underline{Z}_2 \cos \theta_1 - \underline{Z}_1 \sqrt{1 - \left(\frac{\beta_1}{\beta_2} \sin \theta_1\right)^2}}{\underline{Z}_2 \cos \theta_1 + \underline{Z}_1 \sqrt{1 - \left(\frac{\beta_1}{\beta_2} \sin \theta_1\right)^2}} \, e^{j\beta_1 (x \cos \theta_1 - y \sin \theta_1)}. \tag{15.661}$$

Further, there is a *transmitted wave*. It is marked by the following expression (15.631),

$$\underline{P}_{2\rightarrow} = \hat{p}_2 \, e^{-j\beta_2 (x \cos \theta_2 + y \sin \theta_2)}, \tag{15.662}$$

where

$$\hat{p}_2 = \hat{p}_{1\rightarrow}(1 + \underline{R}). \tag{15.663}$$

Substitution of (15.652) and (15.663) into (15.661) ends with

$$\underline{P}_{1,\leftarrow} = \frac{2 \, \hat{p}_{1\rightarrow} \, \underline{Z}_2 \cos \theta_1 \, e^{-j \left(x \sqrt{\beta_2^2 - \beta_1^2 \sin^2 \theta_1} + y \beta_1 \sin \theta_1\right)}}{\underline{Z}_2 \cos \theta_1 + \underline{Z}_1 \sqrt{1 - \left(\frac{\beta_1}{\beta_2} \sin \theta_1\right)^2}}. \tag{15.664}$$

To find the *degree of absorption*, we apply the term $\alpha = 1 - |\underline{R}|^2$, as follows,

$$\alpha = \frac{4 \underline{Z}_1 \underline{Z}_2 \cos \theta_1 \sqrt{1 - \left(\frac{\beta_1}{\beta_2} \sin \theta_1\right)^2}}{\left[\underline{Z}_2 \cos \theta_1 + \underline{Z}_1 \sqrt{1 - \left(\frac{\beta_1}{\beta_2} \sin \theta_1\right)^2}\right]^2}. \tag{15.665}$$

The *maximum absorption coefficient* is reached for $\alpha = 1$. According to (15.665), the condition for this to happen is given by

$$\frac{Z_2}{Z_1} = \frac{\sqrt{1 - \left(\frac{\beta_1}{\beta_2}\sin\theta_1\right)^2}}{\cos\theta_1}.$$ (15.666)

Problem 11.5.

Proposed Approach

The reflection coefficient at the boundary between air and a wall is a function of the angle of sound incidence. The following expression denotes the situation, namely,

$$R(\theta_1) = \frac{Z_{\text{wall}}\cos\theta_1 - Z_{\text{w,air}}}{Z_{\text{wall}}\cos\theta_1 + Z_{\text{w,air}}}.$$ (15.667)

With $\alpha(\theta_1) = 1 - |R(\theta_1)|^2$, the absorption coefficient is given.

The degree of absorption of a wall depends on whether the ratio $|Z_{\text{wall}}|/Z_{\text{w,air}}$ is greater or less than one. If $|Z_{\text{wall}}| < Z_{\text{w,air}}$, normal (perpendicular) incidence produces the highest degree of absorption. With an increasing incident angle, the degree of absorption decreases monotonically—see Fig. 11.6. For $|Z_{\text{wall}}| > Z_{\text{w,air}}$, there exist an optimal angle between $0°$ and $90°$ where the relatively highest degree of absorption is achieved. With the incident angle increasing from $0°$ to $90°$, the degree of absorption rises first until it reaches the peak value, afterward it falls. The optimal angle is given by

$$\theta_{1_{\max}} = \cos^{-1}\left(\frac{Z_{\text{w,air}}}{|Z_{\text{wall}}|}\right).$$ (15.668)

15.12   Chapter 12

Problem 12.1.

Proposed Approach

Figure 15.57 illustrates major steps for designing the ceiling boundary, primarily making use of 1st-order mirror images of the speaker, **S**. In practice, the ceiling is optimized iteratively, often by use of computer-aided calculations. In the figure, some ceiling segments are labeled by a, b, c, d... . The mirror-image lines, M_b, M_c, M_d determine the 1st-order mirror sources. The mirror-image lines point perpendicularly to the extended lines of the ceiling segments, L_a, L_b, L_c, L_d... .

1. A conceptual reflective plane, a, is positioned midway between the speaker, **S**, and the mirror source $S_a^{(1)}$ via the mirror-image line, M_a. The position of the speaker's mouth defines its height.

2. At the intended height of the wall, a, find an angle-bisecting line such that a reflecting screen, b, directs the speaker's sound toward the audience. Use its extended line, L_b, to find the mirror source, $S_b^{(1)}$, via the mirror-image

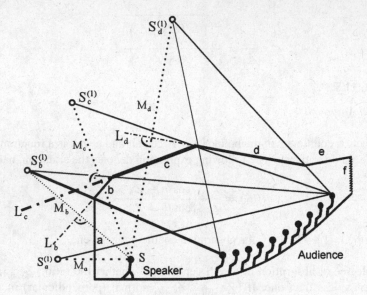

Fig. 15.57 A reasonable ceiling boundary addressing the audience coverage of the auditorium sketched in Fig. 12.7

line, M_b. Check its coverage of the audience area to find the length of the screen, b.

3. At the end of reflecting screen, b, connect another reflecting screen, c, by using its extended line, L_c, to find the mirror, $S_c^{(1)}$ via the mirror-image line M_c. Check its coverage of the audience area to find the length of the screen, c.

4. At the end of the reflecting screen, c, connect to the next reflecting screen, d, by using its extended line, L_d, to find the mirror source, $S_d^{(1)}$, via the mirror image line M_d. Check its coverage of the audience area to find the length of the screen, d.

5. Part of the ceiling, segment e, is slightly tilted such that the effect of the source, **S**, is limited to the last row of the audience. The sound reflected from e moves on to the rear wall.

6. Install an absorbent or sound-diffusing rear wall, f—or an adequate material to cover this wall—to eliminate potential reflections.

Problem 12.2.

Proposed Approach

The room dimensions, $l_x = 25\,\text{m}$, $l_y = 16\,\text{m}$, $l_z = 8\,\text{m}$, determine the volume, V. Each of the sub-areas of the interior surfaces, A_i, possesses a specific degree of absorption—here $\bar{\alpha}_1 = 0.2$, $\bar{\alpha}_2 = 0.7$, $\bar{\alpha}_3 = 0.4$—resulting in the total equivalent absorption area.

The average degree of absorption results from

$$\overline{\alpha} = \frac{\sum_i A_i \overline{\alpha}_i}{\sum_i A_i}$$

$$= (\underbrace{l_x l_y \overline{\alpha}_1}_{\text{floor}} + \underbrace{l_x l_y \overline{\alpha}_2}_{\text{ceiling}} + \underbrace{2 l_x l_z \overline{\alpha}_3}_{\text{y--wall}} + \underbrace{2 l_y l_z \overline{\alpha}_3}_{\text{x--wall}})/1456\,\text{m}^2$$

$$= (80 + 280 + 160 + 102.4)\,\text{m}^2/1456\,\text{m}^2 = 0.427, \qquad (15.669)$$

with a total surface area of $A = 1456\,\text{m}^2$.

Task (**a**) According to (12.24) the equivalent-absorption area is determined by applying $\overline{\alpha}$ in (15.669)

$$A_\alpha = \overline{\alpha} \cdot A + \underbrace{8\,\overline{\alpha} \cdot (V/m)}_{\text{air absorption}} = 0.427(1456 + 25.69)\,\text{m}^2 = 648\,\text{m}^2 . \qquad (15.670)$$

The reverberation time is calculated using *Sabine*'s formula in (12.27) as follows, with the room volume of $V = 3200\,\text{m}^3$ given.[3]

$$T_{60,\text{S}} \approx 0.163 \cdot (\text{s/m}) \cdot \frac{V}{A_\alpha} = 0.163 \cdot (\text{s/m}) \cdot \frac{(25 \cdot 16 \cdot 8)\,\text{m}^3}{648\,\text{m}^2} = 0.805\,\text{s}. \qquad (15.671)$$

Since the overall absorption resulting from (15.669) is relatively high, the *Eyring* formula in (12.26) is more adequate, namely,

$$T_{60,\text{E}} = 0.163 \cdot (\text{s/m}) \cdot \frac{V}{-\ln(1 - 0.427)\,A}, \qquad (15.672)$$

and, with the current quantities inserted,

$$T_{60,\text{E}} = 0.163 \cdot (\text{s/m}) \cdot \frac{3200\,\text{m}^3}{(811 + 25.6)\,\text{m}^2} = 0.623\,\text{s}. \qquad (15.673)$$

Task (**b**) When the room is occupied by an audience of 150 people, assuming an equivalent absorption area of $0.5\,\text{m}^2$/person, the reverberation time using *Eyring*'s

---

[3] Note that in this book we exclusively use quantity equations where the italic letter symbols stand for physical quantities, that is, numerical values times their units, for example, $T_{60} = 1\,\text{s}$. In practice, sometimes numerical-value equations are in use. There the symbols denote numerals that are exclusively associated with specific units, for example, $\text{T}_{60} = 0.163\frac{\text{V}}{\text{A}}\,\text{s}$. This equation requires to insert V in cubic meters, and A in square meters only. In science, the usage of numerical-value equations is strongly discouraged.

formula, (15.672), becomes

$$T_{60,\mathrm{E}} = 0.163 \cdot (\mathrm{s/m}) \cdot \frac{V}{(836.6 + 150 \cdot 0.5)\,\mathrm{m}^2}$$

$$= 0.163 \cdot (\mathrm{s/m}) \cdot \frac{3200\,\mathrm{m}^3}{911.6\,\mathrm{m}^2} = 0.572\,\mathrm{s}. \tag{15.674}$$

while the approximate reverberation time obtained by the *Sabine* formula results as

$$T_{60,\mathrm{S}} \approx 0.163 \cdot (\mathrm{s/m}) \cdot \frac{V}{A_\alpha}$$

$$= 0.163 \cdot (\mathrm{s/m}) \cdot \frac{3200\,\mathrm{m}^3}{(648 + 150 \cdot 0.5)\,\mathrm{m}^2} = 0.721\,\mathrm{s}. \tag{15.675}$$

Task (c) The critical radius is the distance at which the direct-energy density equals the diffuse-field energy density. According to (12.37), the following holds for a 0th-order spherical source,

$$r_{\mathrm{c,s}} = \sqrt{\frac{A_\alpha}{16\,\pi}}. \tag{15.676}$$

Specific directional characteristics of sound sources make the critical distances depending on these. For example, a dipole sound source enlarges the critical distance by a direction-related energy gain, G, compared to that of a monopole source.

$$G \int_0^{2\pi} \int_0^{\pi} \left(\overline{P}_{\max} \cos\theta\right)^2 (r\,\sin\theta\,\mathrm{d}\theta)(r\,\mathrm{d}\varphi) = \int \overline{P}_{\max}^2\,\mathrm{d}A_{\mathrm{s}}, \tag{15.677}$$

whereby we have

$$G\,2\,\pi\,r^2\,\overline{P}_{\max}^2 \int_0^{\pi} \cos^2\theta\,\sin\theta\,\mathrm{d}\theta = 4\,\pi\,r^2\,\overline{P}_{\max}^2, \tag{15.678}$$

$$G \left[-\frac{1}{3}\cos^3\theta \right]_0^{\pi} = 2, \quad \rightarrow \quad G = 3. \tag{15.679}$$

The critical distance of a dipole sound source, $r_{\mathrm{c,d}}$, is then determined using the $r_{\mathrm{c,s}}$ of a spherical sound source in (15.676) and the energy-related gain calculated in (15.679), as

$$r_{\mathrm{c,d}} = \sqrt{G}\,r_{\mathrm{c,s}} = \sqrt{3}\sqrt{\frac{A_\alpha}{16\,\pi}} = (1.73 \cdot 3.59)\,\mathrm{m} = 6.21\,\mathrm{m}. \tag{15.680}$$

Problem 12.3.

Proposed Approach

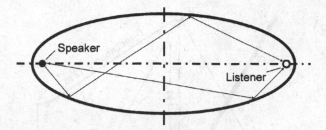

Fig. 15.58 Reasonable boundary for a "whisper gallery" in form of an ellipse with the sound source at one of the focus points, and the receiver at the other one

An ellipse-shaped boundary as shown in Fig. 15.58 is capable of generating optimal sound propagation from one point in space to another one. The sound source, for example, a speaker, is positioned at one of the two focus points, while the receiver, for example, a listener, is located at the other one.

Problem 12.4.

Proposed Approach

An angle of 120° between the horizontal wall, W_1, and the oblique wall, W_2, is given. The height of the source is 5 m, the height of the listener's mouth is 1.4 m. The horizontal distance between the source, S, and the listener, R, is 8 m—see Fig. 12.15.

Task (**a**) Figure 15.59 illustrates the following steps in schematic way.

1. Find the mirror sound source, $S_a^{(1)}$, along the perpendicular line, L_1, with respect to the horizontal wall, W_1—Fig. 15.59a. Its height is given in Fig. 12.15 as 5 m.

2. Connect a straight line between this 1st-order mirror source, $S_a^{(1)}$, to the receiver, R, for estimating the mirror source distance.

3. Find another mirror sound source, $S_b^{(1)}$, along the line, L_2, with respect to the oblique wall, W_2, perpendicular to W_2—see Fig. 15.59a.

4. Connect a straight line between this 1st-order mirror source, $S_b^{(2)}$, to the receiver, R, for estimating the mirror source distance.

5. Find one 2nd-order mirror sound source, $S_a^{(2)}$, of the 1st-order sound source, $S_a^{(1)}$, along the perpendicular line, L_3, with respect to the extended line of the oblique wall, W_2—see Fig. 15.59b.

6. Connect a straight line between this 2nd-order mirror source, $W_a^{(2)}$, to the receiver, for estimating the 2nd-order mirror source distance.

7. Find another 2nd-order mirror sound source, $S_b^{(2)}$, of the 1st-order sound source, $S_b^{(1)}$, along the perpendicular line, L_4, with respect to the extended line of the horizontal wall, W_1—see Fig. 15.59b.

8. Connect a straight line between this 2nd-order mirror source, $S_b^{(2)}$, to the receiver, R, for estimating the 2nd-order mirror source distance.

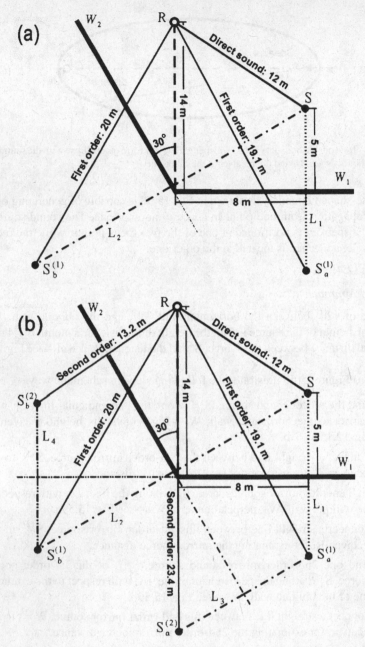

Fig. 15.59 1st- and 2nd-order mirror sources for a sound source and a receiver in front of an oblique wall and a horizontal wall (**a**) 1st-order mirrored sources, $S_a^{(1)}$ and $S_b^{(1)}$ (**b**) 2nd-order mirrored sources, $S_a^{(2)}$ and $S_b^{(2)}$, derived from the 1st-order ones in (**a**)

Task **(b)** Assuming a sound speed of c = 343 m/s, the reflection coefficient, $R = (1 - \overline{\alpha})^{1/2}$, is determined by the degree of absorption, $\overline{\alpha} = 0.5$. We determine the distance, d, between a source and the receiver with

$$d = \sqrt{l_1^2 + l_2^2 - 2\,l_1\,l_2 \cos\theta}. \tag{15.681}$$

Table 15.5 lists the distances from each source—the primary, 1st-order, and the 2nd-order ones—to the receiver, and their arrival-time delays. For the 1st-order and 2nd-order mirror sources, wall absorption is taken into account once or twice, according to their propagation paths. The distance ratio, d_0/d, is used to estimate the relative pressure reduction due to its $1/r$ relationship. The amplitudes of each individual reflection, Column 6, result from the product of the reflection coefficients, R, Column 4, and the distance ratios d_0/d, Column 5. Figure 15.60 plots the room impulse response calculated with these data.

Table 15.5 Time, distance, and amplitude relations between the primary source/image sources and the receiver

Sources	Distance d [m]	Delay time [ms]	R	d_0/d	Amplitude
S	12.0	35	1.0	1.0	1.0
$S_a^{(1)}$	19.1	55.6	0.707	0.628	0.44
$S_b^{(1)}$	20.5	60	0.707	0.585	0.41
$S_a^{(2)}$	23.4	68.2	0.5	0.513	0.26
$S_b^{(2)}$	13.2	38.5	0.5	0.909	0.45

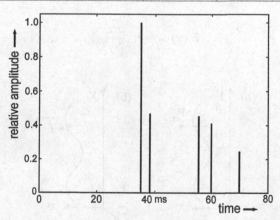

Fig. 15.60 Impulse response consisting of the direct sound pulse, two 1st-order, and two 2nd-order reflections. Their temporal positions and relative amplitudes are listed in Table 15.5

15.13   Chapter 13

Problem 13.1.

Proposed Approach
An infinitely expanded thin plate serves well as a model for a single-leaf wall where both sides of the thin plate move together so that any bending only travels along the $y-$ and $z-$directions. The plate is described by a mass load [mass per unit area], m'', and a bending-stiffness layer B' [bending-stiffness per unit width]. The following discussion considers an elementary section of unit width of this plate—as illustrated in Fig. 13.4.

   Task (**a**) The normal wave propagation of the bending wave in the plate is along the $x-$direction. Therefore, the normal velocities, v_x, are equal on both sides of the plate—see Fig. 15.61a. In other words, both sides of the thin plate move in synchrony.

   By considering a mass load, m'', of the elementary plate while ignoring the bending-stiffness, B', at this time, the pressures at the two sides due to the mass load build up a pressure difference which amounts to

$$\underline{p}_m = \underline{p}_1 - \underline{p}_2 = j\omega m'' \underline{v}_x. \tag{15.682}$$

By considering the bending-stiffness, but now ignoring the mass load, a bending moment or *torque* [per unit width], T', has to be applied to bend the thin plate as shown in Fig. 15.61b. The resulting flexural curvature (...*flexion*) is proportional to the bending moment, namely,

$$T'(y) = -B' \frac{\partial^2 \xi_x}{\partial y^2}, \tag{15.683}$$

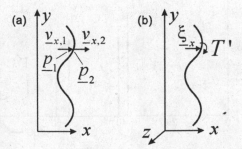

Fig. 15.61 Bending wave with a plate (**a**) Velocity normals and pressures on both sides of the plate with a bending mass load "m", at a point of discussion (**b**) Displacement and bending moment, T', at a point of discussion

where the negative sign results from taking T' positive for power flow in positive x–direction. Yet, here this direction of T' is opposite to the one that generates a positive flexion. This bending moment per unit width, dz, corresponds to a shear force over an elementary distance along the y–direction (see Fig. 15.62), which is,

$$\partial T' = F'_{s,x}\,\partial y, \quad \text{or} \quad F'_{s,x} = \frac{\partial T'}{\partial y}. \tag{15.684}$$

This shear force per unit width, $F'_{s,x}$, corresponds to the difference of the pressures at both sides of the plane, which causes the bending of the plate, that is,

$$\partial F'_{s,x} = p_b\,\partial y, \quad \text{namely,} \quad p_b = \frac{\partial F'_{s,x}}{\partial y}. \tag{15.685}$$

This bending pressure difference, when substituting (15.684), and then (15.683) in turn in (15.685), becomes

$$p_b = \frac{\partial F'_{s,x}}{\partial y} = \frac{\partial^2 T'}{\partial y^2} = -B'\frac{\partial^4 \xi_x}{\partial y^4}, \tag{15.686}$$

and, rewritten for harmonic excitation,

$$\underline{p}_b = -B'\frac{\partial^4 \underline{\xi}_x}{\partial y^4} = -\frac{B'}{j\omega}\frac{\partial^4 \underline{v}_x}{\partial y^4}. \tag{15.687}$$

Consequently, as the plate is featured with a mass load and a bending-stiffness, the mass-relevant pressure, $\underline{p}_m$, and the bending pressure, $\underline{p}_b$, have to counterbalance each other with $\underline{p}_m = \underline{p}_b$ according to (15.682) and (15.687), which leads to

$$j\omega\, m''\,\underline{v}_x + \frac{B'}{j\omega}\frac{\partial^4 \underline{v}_x}{\partial y^4} = 0. \tag{15.688}$$

This is the bending-wave equation of free-oscillation plates, featured by the mass load, m'' and the bending-stiffness,[4] B'.

Task (b) In fulfilling the boundary condition for the normal components of the particle velocities, $\underline{v}_{1,x}$ and $\underline{v}_{2,x}$, on both side of a (wall) plate, we have

$$\underline{v}_{1,x} = \underline{v}_{2,x} = \underline{v}_{b,x}, \tag{15.689}$$

where $\underline{v}_{b,x}$ is the normal component of the bending-wave velocity.

---

[4] Real plates are also furnished with some damping loss, often of a small amount. Yet, for simplicity, we ignore resistive damping here.

Fig. 15.62 Bending wave along a plate. A shear force, $F'_{s,x}$ acts upon a differential element, ∂y, of the plate. This corresponds to a bending moment, $\partial T'$

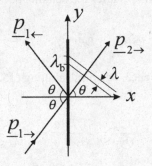

Fig. 15.63 Obliquely incident plane wave onto a plate along the reflected and transmitted components

For the wavelength in air, λ, with an incidence angle, θ, and the excited bending-wavelength, λ_b as shown in Fig. 15.63, it follows

$$\lambda = \lambda_b \sin \theta, \tag{15.690}$$

The phase coefficient thus is

$$\beta = \frac{\beta_b}{\sin \theta}, \tag{15.691}$$

where the angle θ represents the incident angle, the reflected angle, and that of the radiated wave on the opposite side.

We write the corresponding sound pressures as

$$\begin{cases} \underline{p}_{1\rightarrow} \ e^{-j\beta(x\cos\theta + y\sin\theta)} & \text{(incident)} \\ \underline{p}_{1\leftarrow} \ e^{-j\beta(-x\cos\theta + y\sin\theta)} & \text{(reflected)} \\ \underline{p}_{2\rightarrow} \ e^{-j\beta(x\cos\theta + y\sin\theta)} & \text{(transmitted)} \end{cases}$$

At the plate's surface, $x = 0$, considering (15.691), the pressure difference between the two opposite sides becomes

$$\underline{p}_1 - \underline{p}_2 = (\underline{p}_{1\rightarrow} + \underline{p}_{1\leftarrow} - \underline{p}_{2\rightarrow})\, e^{-j\beta y \sin\theta}. \tag{15.692}$$

Taking this pressure difference as an excitation of the bending wave of the plate,

$$(\underline{p}_{1\rightarrow} + \underline{p}_{1\leftarrow} - \underline{p}_{2\rightarrow})\,e^{-j\beta_b y} =$$
$$\left(j\omega m'' + \frac{B'}{j\omega}\beta^4 \sin^4\theta\right)\underline{v}_{x\rightarrow}\,e^{-j\beta_b y},\qquad (15.693)$$

with

$$\underline{v}_{x\rightarrow} = \underline{v}_{2\rightarrow}\cos\theta,\qquad (15.694)$$

this leads to

$$\underline{p}_{1\rightarrow} + \underline{p}_{1\leftarrow} - \underline{p}_{2\rightarrow} = \left(j\omega m'' + \frac{B'}{j\omega}\beta^4 \sin^4\theta\right)\underline{v}_{2\rightarrow}\cos\theta.\qquad (15.695)$$

and, with

$$\underline{v}_{2\rightarrow} = \frac{\underline{p}_{2\rightarrow}}{\varrho c},\qquad (15.696)$$

we also get

$$\underline{v}_{2\rightarrow} = \underline{v}_{1\rightarrow} + \underline{v}_{1\leftarrow} = \frac{1}{\varrho c}(\underline{p}_{1\rightarrow} - \underline{p}_{1\leftarrow}).\qquad (15.697)$$

A substitution of (15.696) into (15.697) renders

$$\underline{p}_{1\leftarrow} = \underline{p}_{1\rightarrow} - \underline{p}_{2\rightarrow}.\qquad (15.698)$$

and a further substitution of (15.696) and (15.698) into (15.695) yields

$$2\underline{p}_{1\rightarrow} - 2\underline{p}_{2\rightarrow} = \left(j\omega m'' + \frac{B'}{j\omega}\beta^4 \sin^4\theta\right)\underline{p}_{2\rightarrow}\frac{\cos\theta}{\varrho c},\qquad (15.699)$$

resulting in a pressure ratio of

$$\frac{\underline{p}_{1\rightarrow}}{\underline{p}_{2\rightarrow}} = 1 + j\left(\omega m'' - \frac{B'}{\omega}\beta^4 \sin^4\theta\right)\frac{\cos\theta}{2\varrho c}.\qquad (15.700)$$

By expressing this ratio in logarithmic terms, we finally obtain the so-called *sound-transmission loss*, R, as

$$R = 20\lg\left|\frac{\underline{p}_{1\rightarrow}}{\underline{p}_{2\rightarrow}}\right| = 20\lg\left|1 + j\left(\omega m'' - \frac{B'}{\omega}\beta^4 \sin^4\theta\right)\frac{\cos\theta}{2\varrho c}\right|\ \text{dB}.\quad (15.701)$$

Task (c) From (15.701) with $\beta = \omega/c$, the sound-transmission loss becomes

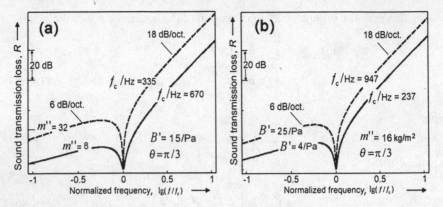

Fig. 15.64 Sound-transmission loss, R, as a function of frequency for a directed oblique incidence ($\theta = \pi/3$) to a single-leaf wall (**a**) Different R's for two different mass loads, $8/\mathrm{kg\,m^{-2}}$ and $32/\mathrm{kg\,m^{-2}}$, while keeping the bending-stiffness constant at $B' = 15/\mathrm{Pa}$ (**b**) R for two different values of the bending-stiffness, $25/\mathrm{Pa}$ and $4/\mathrm{Pa}$, while the mass load is constant at $m'' = 16/\mathrm{kg\,m^{-2}}$

$$R = 20\,\lg \left| 1 + \mathrm{j}\left(\omega\, m'' - \omega^3 \frac{B'}{c^4}\,\sin^4\theta \right) \frac{\cos\theta}{2\varrho c} \right| \ \mathrm{dB}\,. \tag{15.702}$$

Figure 15.64 illustrates the sound-transmission losses of a single-leaf wallboard with respect to the frequency—see also Fig. 13.7. Below the coincidence frequency, the transmission-loss curves show a 6 dB increase per frequency doubling—due to mass dominance—while above the coincidence frequency, the curves follow an 18 dB increase per frequency doubling—due to stiffness dominance—indicated by the term that includes ω^3 in (15.702).

In the figures, the curves are presented over the normalized frequency. For different mass loads or bending-stiffnesses,[5] the coincidence frequencies, f_c, vary according to (15.705).

Problem 13.2.

Proposed Approach

For the solution, start with (13.17) or (15.701).

Task (**a**) When the term in parentheses in (15.701) equals zero, the plate (wall) concerned puts the sound completely through, that is, $\underline{p}_{1\to} = \underline{p}_{2\to}$. This occurs when the component of the phase coefficient in the air parallel to the plate, $\beta\sin\theta$, is equal to the bending-phase coefficient, $\beta_b = \beta\sin\theta$. This phenomenon is called *trace matching*. With the term in parenthesis equaling zero, we get

---

[5] The bending-stiffness is quantified in physical units as $[\mathrm{N/m^2}]$, Pa (Pascal), or $[\mathrm{m^{-1}\,kg\,s^{-2}}]$, which is a stiffness $[\mathrm{N/m}]$ per unit length. In practice, many panels made of plastics possess bending stiffnesses in the MPa or $\mathrm{N/mm^2}$ range.

$$\omega\, m'' = \frac{B'}{\omega}\, \beta^4 \sin^4 \theta\,, \tag{15.703}$$

With $\beta = \omega/c$ and $\beta_b = \beta \sin \theta$, the above equation rewrites as

$$\frac{\omega}{c} \sin \theta = \beta \sin \theta = \beta_b = \sqrt{\omega}\, \sqrt[4]{\frac{m''}{B'}}\,, \tag{15.704}$$

This determines the coincidence frequency, f_c, to be

$$2\pi\, f_c = \sqrt{\frac{m''\, c^4}{B' \sin^4 \theta}}\,, \quad \text{or} \quad f_c = \frac{1}{2\pi} \sqrt{\frac{m''\, c^4}{B' \sin^4 \theta}}\,. \tag{15.705}$$

For $\sin \theta \to 1$, the coincidence frequency approaches the lower limiting frequency, namely,

$$f_{c,\text{lim}} = \frac{1}{2\pi} \sqrt{\frac{m''\, c^4}{B'}}\,. \tag{15.706}$$

Below this limiting coincidence frequency, $f_{c,\text{lim}}$, no trace matching is possible, because then the wavelength in air is larger than the bending wavelength. At the coincidence frequency, the sound wave in the air travels in parallel to the wallboard, that is $\sin \theta = 1$.

Task **(b)** For normal incidence, $\sin \theta \approx 0$, or for a wall which is extremely compliant with bending waves, $B' \approx 0$, (15.701) reduces to

$$R \approx 20 \lg \left(\frac{\omega\, m'' \cos \theta}{2\varrho c} \right)\,. \tag{15.707}$$

In other words, the sound-transmission loss depends only on the mass, m'', with 6 dB increase per a mass doubling. This is the so-called *mass law*.

Problem 13.3.

Proposed Approach

Figure 15.65 illustrates a double-leaf, yet infinitely extended wall consisting of two wallboards with mass loads, m_1'' and m_2''—as shown in Fig. 15.65a. Between the two wallboards, there is a lossless cavity with a compliance load, n'' [compliance per unit area]. The sound pressures on both sides of the double-leaf wall are $\underline{p}_1$ and $\underline{p}_2$.

For a plane wave hitting the wall perpendicularly in the sinistral half-space, Fig. 15.65b shows its mechanic equivalent circuit with circuit elements m_1'', m_2'', n'', and r'', [expressed per unit area]. The dexter wallboard radiates sound waves into the dexter half-space against the wave impedance, $Z_w = \varrho c$, with ϱ being the density of air, and c the speed of sound in air.

Fig. 15.65 Double-leaf wall
and its equivalent circuits (a)
Double-leaf wall
construction (b) Equivalent
mechanic circuit (c)
Equivalent electric circuit

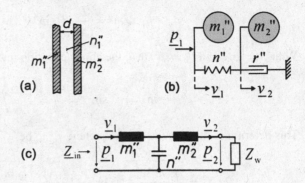

Using the electromechanic analogy #1, the so-called *impedance analogy*—compare Sect. 3.1, the field quantities become

$$\underline{F} \,\hat{=}\, \underline{u} \quad \text{and} \quad \underline{v} \,\hat{=}\, \underline{i} \tag{15.708}$$

For the current problem, we prefer to use the sound pressure instead of the force, that is,

$$\underline{p} \,\hat{=}\, \underline{u} \quad \text{and} \quad \underline{v} \,\hat{=}\, \underline{i} \tag{15.709}$$

This analogy leads to the equivalent electrical circuit as shown in Fig. 15.65c. In the sinistral half-space, the sound pressure on the surface of the wallboard consists of an incident and a reflected component, namely,

$$\underline{p}_1 = \underline{p}_{1\rightarrow} + \underline{p}_{1\leftarrow}, \tag{15.710}$$

The same is true for the particle velocity,

$$\underline{v}_1 = \underline{v}_{1\rightarrow} + \underline{v}_{1\leftarrow}. \tag{15.711}$$

Hence, for the wave impedance follows

$$\frac{\underline{p}_{1\rightarrow}}{\underline{v}_{1\rightarrow}} = Z_{\mathrm{w}} = -\frac{\underline{p}_{1\leftarrow}}{\underline{v}_{1\leftarrow}}, \tag{15.712}$$

which leads to

$$Z_{\mathrm{w}} \underline{v}_1 = \underline{p}_{1\rightarrow} - \underline{p}_{1\leftarrow}. \tag{15.713}$$

In addition to (15.710) and (15.713), we obtain

$$\underline{p}_{1\rightarrow} = \frac{1}{2} \left(\underline{p}_1 + Z_{\mathrm{w}} \underline{v}_1 \right). \tag{15.714}$$

With $\underline{p}_1 = \underline{Z}_1 \underline{v}_1$, the incident component of sound pressure now writes as

$$\underline{p}_{1\rightarrow} = \frac{1}{2}\left(\underline{Z}_1 + Z_{\mathrm{w}}\right)\underline{v}_1. \tag{15.715}$$

The wall radiates the sound wave into the dexter half-space, which is determined by the sound pressure, $\underline{p}_2 = \underline{p}_{2\rightarrow}$. The ratio of the incident sound pressure, $\underline{p}_{1\rightarrow}$, at the wall surface in the dexter half-space and the one at the wall surface in the sinistral half-space, $\underline{p}_{2\rightarrow}$, determines the sound-transmission loss, R, as follows,

$$R = 20\,\lg\left|\frac{\underline{p}_{1\rightarrow}}{\underline{p}_2}\right| = 20\,\lg\left|\frac{\underline{p}_{1\rightarrow}}{\underline{p}_{2\rightarrow}}\right|\quad\mathrm{dB}. \tag{15.716}$$

Because of $\underline{p}_2 = Z_{\mathrm{w}}\,\underline{v}_2$, we obtain

$$\frac{\underline{p}_{1\rightarrow}}{\underline{p}_2} = \frac{1}{2}\frac{\underline{Z}_1 + Z_{\mathrm{w}}}{Z_{\mathrm{w}}}\frac{\underline{v}_1}{\underline{v}_2}. \tag{15.717}$$

According to Fig. 15.65c, the input impedance, $\underline{Z}_1$, follows as

$$\begin{aligned}
\underline{Z}_1 &= j\,\omega\,m_1'' + \frac{\frac{1}{j\omega n''}(Z_{\mathrm{w}} + j\,\omega\,m_2'')}{\frac{1}{j\omega n''} + Z_{\mathrm{w}} + j\,\omega\,m_2''}\\
&= j\,\omega\,m_1'' + \frac{Z_{\mathrm{w}} + j\,\omega\,m_2''}{1 - \omega^2\,n''\,m_2'' + j\,\omega\,n''\,Z_{\mathrm{w}}}\\
&= \frac{Z_{\mathrm{w}}(1 - \omega^2\,m_1''\,n'') + j\,\omega\,(m_1'' + m_2'' - \omega^2\,n''\,m_1''m_2'')}{1 - \omega^2\,n''m_2'' + j\,\omega\,n''\,Z_{\mathrm{w}}}.
\end{aligned} \tag{15.718}$$

Note that

$$\omega_1 = \frac{1}{\sqrt{m_1''\,n''}}, \quad \omega_2 = \frac{1}{\sqrt{m_2''\,n''}}, \quad \text{and} \quad \omega_{\mathrm{d}} = \frac{1}{\sqrt{m_{\mathrm{d}}''\,n''}}, \tag{15.719}$$

with ω_{d} being the (angular) *drum frequency*, and

$$m_{\mathrm{d}}'' = \frac{m_1''\,m_2''}{m_1'' + m_2''}. \tag{15.720}$$

With $\omega_1 < \omega_{\mathrm{d}}$, and $\omega_2 < \omega_{\mathrm{d}}$, the input impedance, $\underline{Z}_1$, simplifies to

$$\underline{Z}_1 = \frac{Z_{\mathrm{w}}\left[1 - \left(\frac{\omega}{\omega_1}\right)^2\right] + j\,\omega\,m_{\mathrm{d}}''\left[1 - \left(\frac{\omega}{\omega_{\mathrm{d}}}\right)^2\right]}{\left[1 - \left(\frac{\omega}{\omega_2}\right)^2\right] + j\,\omega\,n''\,Z_{\mathrm{w}}}. \tag{15.721}$$

In a similar fashion as in Fig. 15.65c, it follows that

$$\frac{v_2}{v_1} = \frac{\frac{1}{j\omega n''}}{\frac{1}{j\omega n''} + j\omega m_2'' + Z_w} = \frac{1}{1 - \omega^2 m_2'' n'' + j\omega n'' Z_w}$$

$$= \frac{1}{1 - \left(\frac{\omega}{\omega_2}\right)^2 + j\omega n'' Z_w} , \tag{15.722}$$

A substitution of (15.721) and (15.722) into (15.717) yields

$$\frac{p_{1\rightarrow}}{p_2} = \frac{1}{2}\left\{\left[1 - \left(\frac{\omega}{\omega_1}\right)^2\right] + \left[1 - \left(\frac{\omega}{\omega_2}\right)^2\right]\right.$$

$$\left. + j\omega\left[\frac{m_d''}{Z_w}\left[1 - \left(\frac{\omega}{\omega_d}\right)^2\right] + n'' Z_w\right]\right\} , \tag{15.723}$$

and, we now express R as

$$R = 20\lg\left|\frac{p_{1\rightarrow}}{p_2}\right| \text{ dB} . \tag{15.724}$$

This expression reveals that at low frequencies, that is, $\omega < (\omega_1, \omega_2, \omega_d)$, the sound-transmission loss, R, increases with frequency at $6\,\text{dB/oct}$. For high frequencies, $\omega > \omega_d$, (15.723) reduces to

$$\frac{p_{1\rightarrow}}{p_2} \approx -j\omega^3 \frac{m_d''}{2\omega_d^2 Z_w} , \tag{15.725}$$

that is, R increases with frequency by $18\,\text{dB/oct}$. A resonance occurs when the imaginary part of (15.723) equals zero, viz,

$$\left(\frac{\omega_0}{\omega_d}\right)^2 = 1 + \frac{n''}{m_d''} Z_w^2 = 1 + \frac{Z_w^2}{\omega_d^2 (m_d'')^2} . \tag{15.726}$$

Considering ω_d in (15.719), the resonance frequency, f_0, becomes

$$f_0 = \frac{1}{2\pi}\sqrt{\omega_d^2 + \left(\frac{Z_w}{m_d''}\right)^2} . \tag{15.727}$$

Note that in practice when constructing walls, the mass load, m_d'', is made large enough to tune the drum-resonance frequency to be smaller than the resonance frequency ω_0. Assuming $m_1'' = m_2''$, it follows

$$\omega_d = \sqrt{2}\,\omega_1 = \sqrt{2}\,\omega_2 . \tag{15.728}$$

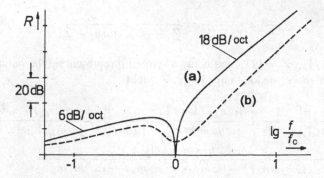

Fig. 15.66 Transmission loss as a function of frequency for a double-leaf wall (a) Directed oblique incidence. (b) Random incidence

At $\omega = \omega_d$, the pressure ratio becomes

$$\frac{\underline{p}_{1\to}}{\underline{p}_2} \approx 1, \tag{15.729}$$

which means that the wall transmits the incident sound completely through to the opposite side without any losses, that is, $R \approx 0\,\mathrm{dB}$. Figure 15.66 illustrates the sound-transmission loss, R, as function of frequency—compare Fig. 13.7.

Problem 13.4.

Proposed Approach

Using analogy #2 with $\underline{F} \hat{=} \underline{i}$ and $\underline{v} \hat{=} \underline{u}$, Fig. 15.67 illustrates an equivalent circuit of the dynamic absorber shown in Fig. 13.17. We consider a total admittance, $\underline{Y}$, of the branches as indicated in Fig. 15.67, and the other admittance of the single compliance, $\underline{Y}_n$, which shares the partial force, $\underline{F}'$ flowing through it, Consequently, the force ratio results as

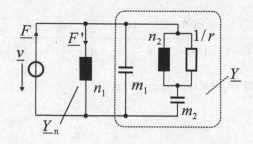

Fig. 15.67 Equivalent electric circuit of the dynamic absorber depicted in Fig. 13.17

$$\frac{F'}{F} = \frac{\underline{Y}_n}{\underline{Y}_n + \underline{Y}} = \frac{\underline{Z}}{\underline{Z}_n + \underline{Z}} = \frac{\underline{Z}}{j\omega n_1 + \underline{Z}}, \tag{15.730}$$

with $\underline{Z}_n = 1/\underline{Y}_n$, $\underline{Z} = 1/\underline{Y}$, and using a symbol $\parallel$ representing the parallel of two branch impedances, the total impedance, $\underline{Z}$, yields

$$\underline{Z} = \frac{1}{j\omega m_1} \left\| \left[\frac{1}{j\omega m_2} + \frac{j\omega n_2 \cdot (1/r)}{j\omega n_2 + 1/r} \right] = \frac{1}{j\omega m_1} \right\| \left[\frac{1 + j\omega n_2 r - \omega^2 m_2 n_2}{j\omega m_2 (1 + j\omega n_2 r)} \right]$$

$$= \frac{1 + j\omega n_2 r - \omega^2 m_2 n_2}{j\omega m_2 (1 + j\omega n_2 r) + j\omega m_1 - \omega^2 n_2 m_1 r - j\omega^3 m_1 m_2 n_2}$$

$$= \frac{1 + j\omega n_2 r - \omega^2 m_2 n_2}{-\omega^2 n_2 r (m_1 + m_2) + j\omega (m_1 + m_2 - \omega^2 m_1 m_2 n_2)}. \tag{15.731}$$

Substituting (15.731) into (15.730) results in

$$\frac{F'}{F} = \frac{\underline{Z}}{j\omega n_1 + \underline{Z}}$$

$$= \frac{1 - \omega^2 m_2 n_2 + j\omega n_2 r}{-j\omega^3 n_1 n_2 r m_s - \omega^2 n_1 (m_s - \omega^2 m_1 m_2 n_2) + 1 - \omega^2 m_2 n_2 + j\omega n_2 r}$$

$$= \frac{1 - \omega^2 m_2 n_2 + j\omega n_2 r}{\omega^4 m_1 n_1 m_2 n_2 - \omega^2 (m_2 n_2 + m_s n_1) + 1 + j\omega n_2 r (1 - \omega^2 m_s n_1)}$$

$$= \frac{1 - \frac{\omega^2}{\omega_2^2} + j\omega n_2 r}{\frac{\omega^4}{\omega_1^2 \cdot \omega_2^2} - \omega^2 \left(\frac{1}{\omega_2^2} + \frac{1}{\omega_3^2} \right) + 1 + j\omega n_2 r \left(1 - \frac{\omega^2}{\omega_3^2} \right)}, \tag{15.732}$$

where $m_s = m_1 + m_2$ and

$$\omega_1 = \frac{1}{\sqrt{m_1 n_1}}, \quad \omega_2 = \frac{1}{\sqrt{m_2 n_2}}, \quad \text{and} \quad \omega_3 = \frac{1}{\sqrt{m_s n_1}}. \tag{15.733}$$

Using $\Omega = \omega/\omega_3$, further simplification of (15.732) is possible,

$$\frac{F'}{F} = \frac{\frac{\omega_3^2}{\omega_2^2} - \Omega^2 + j\omega n_2 r \frac{\omega_3^2}{\omega_2^2}}{\Omega^4 \frac{\omega_3^2}{\omega_1^2} - \Omega^2 \left(\frac{\omega_3^2}{\omega_2^2} + 1 \right) + \frac{\omega_3^2}{\omega_2^2} + j\omega n_2 r \frac{\omega_3^2}{\omega_2^2} (1 - \Omega^2)}. \tag{15.734}$$

Using short forms of angular frequency ratios

$$x = \frac{\omega_2}{\omega_3}, \quad y = \frac{\omega_3^2}{\omega_1^2} = \frac{m_1}{m_s}, \tag{15.735}$$

the square of (15.734) is

$$\left|\frac{F'}{F}\right|^2 = \frac{(x^2 - \Omega^2)^2 + (\omega \, n_2 \, r \, x^2)^2}{\left[\, y \, \Omega^4 - (1 + x^2) \, \Omega^2 + x^2 \,\right]^2 + \left[\, \omega \, n_2 \, r \, x^2 \, (1 - \Omega^2) \,\right]^2}. \tag{15.736}$$

By defining a loss factor

$$d = n_2 \, r \, \omega_3 \, x^2 = n_2 \, r \, \frac{\omega_2^2}{\omega_3} = \frac{r}{m_2 \, \omega_3} = \frac{r}{Z_\mathrm{m}}, \tag{15.737}$$

with $Z_\mathrm{m} = m_2 \, \omega_3$, which is essentially the impedance ratio of damping loss versus mass, (15.736) delivers

$$\left|\frac{F'}{F}\right|^2 = \frac{(\Omega^2 - x^2) + d^2 \Omega^2}{\left[\, y \, \Omega^4 - (1 + x^2) \, \Omega^2 + x^2 \,\right]^2 + d^2 \Omega^2 (\Omega^2 - 1)^2}. \tag{15.738}$$

This expression facilitates the following discussion by featuring two special cases, viz,

– $d \to \infty$ (rigid coupling)

$$\left|\frac{F'}{F}\right|^2 \to \frac{1}{(\Omega^2 - 1)^2}, \tag{15.739}$$

implying a strong resonance as if without the dynamic absorber.

– $d \to 0$

$$\left|\frac{F'}{F}\right|^2 \to \frac{(\Omega^2 - x^2)}{\left[\, y \, \Omega^4 - (1 + x^2) \, \Omega^2 + x^2 \,\right]^2}, \tag{15.740}$$

indicating a transfer function with two poles and one zero.

When optimally tuned, the dynamic absorber features two resonances as illustrated in 15.68.

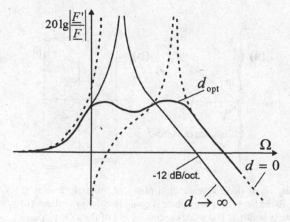

Fig. 15.68 Transfer function of the dynamic absorber of Fig. 13.17, for the cases of $d \to \infty$ and $0 < d_\mathrm{opt} \ll \infty$

Problem 13.5.

Proposed Approach

The problem is solvable using *Raleigh*'s integral in (10.2). The distant reference point in the far-field, r_0, with $r_0 \gg l$ makes *Fraunhofer*'s approximation applicable to the problem—see Fig. 15.69a.

The *Raleigh* integral applies to the determination of the sound pressure at the reference point.

$$\underline{p}(r) = \frac{j\omega\varrho}{2\pi} \int_{A_0} \underline{v}(x,y) \frac{e^{-j\beta r}}{r} \, dA. \tag{15.741}$$

Task (**a**) We approximate the distance from the reference point to a radiation point on the plate, $r = r_0 - x\cos\theta - y\cos\varphi$, as in (10.4), such that

$$\underline{p}(r) \approx \frac{j\omega\varrho}{2\pi} \frac{e^{-j\beta r_0}}{r_0}$$
$$\cdot \int_{x=-l/2}^{l/2} \int_{y\to-\infty}^{\infty} \underline{v}(x,y) \, e^{j\beta x\cos\theta} \, e^{j\beta y\cos\varphi} dx\, dy. \tag{15.742}$$

Task (**b**) To address its radiation characteristic, a certain symmetry is exploited. Indeed, by considering the y/z–plane also for $r_0 \gg l$, $\theta \to \pi/2$, and thus let $\cos\theta \to 0$, the above equation reduces to

$$\underline{p}(r) \approx \frac{j\omega\varrho}{2\pi} \frac{e^{-j\beta r_0}}{r_0} \int_{x=-l/2}^{l/2} dx \int_{y\to-\infty}^{\infty} \underline{v}(x,y) \, e^{j\beta y\cos\varphi} \, dy. \tag{15.743}$$

In case of a uniform velocity distribution on the plate, and considering the angular variable φ' in Fig. 15.69b and Θ, which is the angle between r_0 and the vector to

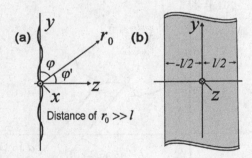

Fig. 15.69 Bending wave along a beam (thin plate) of infinitely length. (**a**) y/z–plane with a reference point, r_0, in the far-field. The x–plane is going into the y/z–plane. (**b**) x/y–plane showing the beam plate with its width, l. The z–axis is coming out of the x/y–plane

the elementary element in y/x–plane, we get $\cos\varphi \approx \sin\varphi' - \sin\Theta$—similar to Fig. 10.3. This leads to

$$\underline{p}(r) \approx \frac{j\omega\varrho l\,e^{-j\beta r_0}}{2\pi r_0}\,\underline{v}\int_{y\to-\infty}^{\infty}e^{j\beta\sin\varphi'\,y}\,e^{-j\beta\sin\Theta\,y}\,dy\,.\tag{15.744}$$

Replacement of the variables $\Omega_0 = \beta\sin\varphi'$, and $\Omega = \beta\sin\Theta$, results in

$$\underline{p}(r) \approx \frac{j\omega\varrho l\,e^{-j\beta r_0}}{2\pi r_0}\,\underline{v}\int_{y\to-\infty}^{\infty}e^{j\Omega_0 y}\,e^{-j\Omega y}\,dy\,.\tag{15.745}$$

Making use of the *Fourier* transform of $e^{j\omega_0 t}$ renders

$$\int_{-\infty}^{\infty}e^{j\omega_0 t}\,e^{-j\omega t}\,dt = \mathrm{FT}(e^{j\omega_0 t})\,,\tag{15.746}$$

or, symbolically,

$$e^{j\omega_0 t}\ \circ\!\!-\!\!\bullet\ 2\pi\delta(\omega-\omega_0)\,,\tag{15.747}$$

so that

$$\underline{p}(r) \approx \frac{j\omega\varrho l\,e^{-j\beta r_0}}{r_0}\,\underline{v}\,\delta\,(\beta\sin\Theta - \beta\sin\varphi')\,.\tag{15.748}$$

Problem 13.6.

Proposed Approach

The ASTM-E90 standard involves the sound-transmission losses from 100 Hz to 4 kHz in third-octave band steps. When locating the sound-transmission-class, STC,

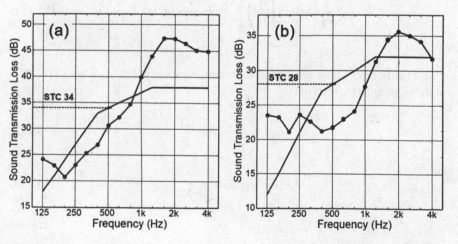

Fig. 15.70 Sound-transmission losses experimentally measured in standard test chambers as of Table 13.1. **(a)** Wallboard 1, the STC amounts to 34. **(b)** Wallboard 2, the STC amounts to 28

contour onto the sound-transmission-loss curve, the sum of deficiencies must not exceed 32 dB under the constraint that every single deficit must not exceed 8 dB. The plots in Fig. 15.70 illustrate the results for the measured sound-transmission losses.

15.14   Chapter 14

Problem 14.1.

Proposed Approach

We express the relation of the sound intensity and the sound power as

$$\overline{P}' = \int_A I(r)\mathrm{d}A, \tag{15.749}$$

where $A = 2\pi r^2$ represents the area at a distance r, assuming that the sound-power load, $\overline{P}'$, [sound power per unit length], is constant. This leads to the sound intensity

$$I(r) = \frac{\overline{P}'}{2\pi r^2}. \tag{15.750}$$

For a line source of length l, radiating a constant power load of incoherent sound, the sound intensity at the reference point, r_0, perpendicular to the middle of the line source—as sketched in Fig. 15.71—is given by the following expression,

$$I(r_0) = \int_{-l/2}^{l/2} I(r)\mathrm{d}x = \int_{-l/2}^{l/2} \frac{\overline{P}'}{2\pi r^2}\mathrm{d}x = \frac{\overline{P}'}{2\pi} \int_{-l/2}^{l/2} \frac{1}{r_0^2 + x^2}\mathrm{d}x$$

$$= \frac{\overline{P}'}{2\pi} \frac{1}{r_0} \left[\arctan\left(\frac{x}{r_0}\right) \right]_{-l/2}^{l/2} = \frac{\overline{P}'}{\pi r_0} \arctan\left(\frac{l}{2r_0}\right). \tag{15.751}$$

Task **(a)** For very long incoherent line sources, assume $l \gg r_0$. Using (15.751), the sound intensity comes out as

$$I(r_0) = \frac{\overline{P}'}{2r_0} \propto \frac{1}{r_0}, \tag{15.752}$$

since

$$\lim_{\gamma \to \infty} \arctan\gamma = \frac{\pi}{2}, \tag{15.753}$$

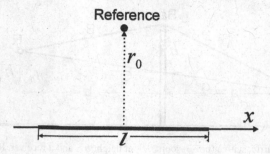

Fig. 15.71 Line source of length l. The reference point is by a distance of r_0 away from the middle point of the line source

Accordingly, the sound pressure level becomes

$$L = 10 \lg \frac{\overline{P}'}{2} - 10 \lg r_0, \qquad (15.754)$$

which indicates a 3-dB decrease per distance-doubling.

Task (**b**) For a very short incoherent line source, assuming $l \ll r_0$, we have

$$\arctan \gamma \approx \gamma, \quad \gamma \ll 1. \qquad (15.755)$$

Substituting (15.755) into (15.751) leads to

$$I(r_0) = \frac{\overline{P}' l}{2 \pi r_0^2} \propto \frac{1}{r_0^2}, \qquad (15.756)$$

which indicates a 6-dB drop per distance doubling.

Taking $\overline{P} = \overline{P}' l$ as the sound power, (15.756) provides us with the sound intensity for a point source radiating into the half-space.

Problem 14.2.

Proposed Approach

The insertion loss, D_i, of a noise-barrier is given in (14.9) as

$$D_i \approx 10 \lg (20.4 \, k_f + 3), \qquad (15.757)$$

where k_f is the *Fresnel number*, defined as

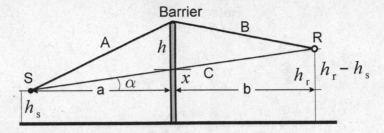

Fig. 15.72 Noise barrier with a (noise) source, S, at height h_s, and a receiver, R, at height h_r

$$k_f = \frac{(A + B) - C}{\lambda/2}, \tag{15.758}$$

and λ is the wavelength.

Figure 15.72 illustrates distances, A, B, C, a, b, and different heights.

Task **(a)** With $h_b = h_{barrier}$, $h_s = h_{sender}$, and $h_r = h_{receiver}$, the condition

$$a, b \gg h_b - \min(h_s, h_r) \tag{15.759}$$

is fulfilled. The difference in heights, h, is

$$h \approx h_b - h_r + x, \tag{15.760}$$

and the condition (15.759) leads to the following approximations,

$$A \approx \sqrt{a^2 + h^2}, \quad B \approx \sqrt{b^2 + h^2}, \quad C \approx a + b, \tag{15.761}$$

with

$$\tan \alpha = \frac{x}{a} = \frac{h_r - h_s}{a + b}, \tag{15.762}$$

so that

$$x = a \frac{h_r - h_s}{a + b}. \tag{15.763}$$

By substitution of (15.763) into (15.760), we obtain

$$h = h_b - h_r + a \frac{h_r - h_s}{a + b}, \tag{15.764}$$

from which, with (15.761), follows

$$A + B - C \approx a \sqrt{1 + \frac{h^2}{a^2}} + b \sqrt{1 + \frac{h^2}{b^2}} - a - b. \tag{15.765}$$

We now apply the series expansion

$$\sqrt{1 + x^2} \approx 1 + \frac{1}{2} x^2 - \frac{1}{2 \cdot 4} x^4 \cdots . \tag{15.766}$$

When ignoring the x^4-term and higher-order terms, due to condition (15.759), a further approximation of (15.765) leads to

$$A + B - C \approx a \left(1 + \frac{h^2}{2 a^2}\right) + b \left(1 + \frac{h^2}{2 b^2}\right) - a - b$$

$$\approx \frac{h^2}{2} \left(\frac{1}{a} + \frac{1}{b}\right) . \tag{15.767}$$

Substituting (15.767) into (15.757)–(15.758) finally results in the insertion loss as follows,

$$D_i = 10 \lg \left[3 + 0.06 \, f \, h^2 \left(\frac{1}{a} + \frac{1}{b}\right)\right], \tag{15.768}$$

$$h = h_b - h_r + a \frac{h_r - h_s}{a + b}, \tag{15.769}$$

where $\lambda = c/f$, $c = 343\,\text{m/s}$, and h are taken from (15.764).

Task (b) Determine the insertion loss for a 4-lane road with the following given parameters,

$$a = 14\,\text{m}, \, b = 20\,\text{m}, \, h_s = 0.6\,\text{m}, \, h_r = 2\,\text{m}, \, h_b = 4\,\text{m}, \, f = 500\,\text{Hz}.$$

By employing (15.764), we get

$$h = \frac{4}{m} - \frac{2}{m} + \frac{14}{m} \frac{1.4/\text{m}}{(14 + 20)/\text{m}} = 2.6\,\text{m}, \tag{15.770}$$

and substituting this result into (15.768) renders

$$D_i = 10 \lg \left[3 + \frac{0.06}{\text{s/m}} \frac{500}{\text{Hz}} \left(\frac{2.6}{m}\right)^2 \left(\frac{1}{14/\text{m}} + \frac{1}{20/\text{m}}\right)\right]$$

$$= 14.4\,\text{dB} . \tag{15.771}$$

Task (c) When the frequency is increased by a factor of two or more, that is, with $\gamma \geq 2$, the insertion loss, (15.769), becomes

$$D_i \approx 10 \lg(\gamma\, f) = 10 \lg \gamma + 10 \lg f, \tag{15.772}$$

indicating a 3-dB increase per frequency-doubling.

Problem 14.3.

Proposed Approach

A line-array of sound sources consist of an infinite number of point sources. With the assumption that the array radiates a sound power of $\overline{P}'$, such an array qualifies as a model of heavy-traffic road noise—see Fig. 14.15. The insertion loss results from comparing the sound intensities before and after the barrier was installed.

Task (a) Figure 15.73 illustrates a top view of the line array of noise sources. At a location, x, along the line source, the local insertion loss results as

$$dD_i = 10 \lg[3 + 20.4\, k_f(\lambda)] = 10 \lg \frac{d\, I_0(x)}{d\, I_b(x)}, \tag{15.773}$$

where $I_0(x)$ represents the sound intensity along with the line array without the noise barrier, while $I_b(x)$ is the one with the noise barrier. This leads to

$$dI_b(x) = \frac{dI_0(x)}{3 + 20.4\, k_f(\lambda)}. \tag{15.774}$$

With

$$I_{0,S} = \int_{-\infty}^{\infty} d\, I_0(x) dx, \quad \text{and} \quad I_{b,S} = \int_{-\infty}^{\infty} d\, I_0(x) dx, \tag{15.775}$$

the total noise contribution along the line source thus results in

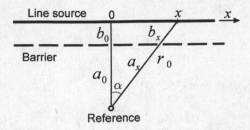

Fig. 15.73 Traffic-noise model. An infinitely long line-array with a noise barrier extending parallel to the array

$$D_{i,S} = 10 \lg \frac{I_{0,S}(x)}{I_{b,S}(x)}$$

$$= 10 \lg \left[\int_{-\infty}^{\infty} d\, I_0(x) dx \right] - 10 \lg \left[\int_{-\infty}^{\infty} d\, I_b(x) dx \right]. \quad (15.776)$$

Substitution of (15.774) into (15.776) gives

$$D_{i,S} = 10 \lg \left[\int_{-\infty}^{\infty} d\, I_0(x)\, dx \right] - 10 \lg \left[\int_{-\infty}^{\infty} \frac{d\, I_0(x)}{3 + 20.4\, k_f(\lambda)}\, dx \right]. \quad (15.777)$$

Consideration of Fig. 15.73 and the ratio of the sound powers, and substitution of (15.750) into (15.777) leads to

$$D_{i,S} = 10 \lg \left(\int_{-\infty}^{\infty} \frac{\overline{P}'}{2\pi} \frac{dx}{r_0^2 + x^2} \right)$$

$$- 10 \lg \left[\int_{-\infty}^{\infty} \frac{\overline{P}'}{2\pi} \frac{dx}{(r_0^2 + x^2)\,(3 + 20.4\, k_f)} \right]. \quad (15.778)$$

In case of a constant power load, $\overline{P}'$, the insertion loss reduces to

$$D_{i,S} = 10 \lg \left(\int_{-\infty}^{\infty} \frac{dx}{r_0^2 + x^2} \right) - 10 \lg \left[\int_{-\infty}^{\infty} \frac{dx}{(r_0^2 + x^2)\,(3 + 20.4\, k_f)} \right]$$

$$= 10 \lg \left(\int_{-\infty}^{\infty} \frac{dx}{r_0^2 + x^2} \right)$$

$$- 10 \lg \left\{ \int_{-\infty}^{\infty} \frac{dx}{\left[3(r_0^2 + x^2) + 20.4 \frac{h^2 r_0}{\lambda} \sqrt{r_0^2 + x^2} \left(\frac{1}{a_0} + \frac{1}{b_0} \right) \right]} \right\}, \quad (15.779)$$

where (15.768) comes into use. The integrals in the first and the second term become

$$I_{0,S} = \int_{-\infty}^{\infty} \frac{dx}{r_0^2 + x^2} = 2 \int_{0}^{\infty} \frac{dx}{r_0^2 + x^2}$$

$$= \frac{2}{r_0} \arctan \frac{x}{r_0} \Big|_0^{\infty} = \frac{\pi}{r_0} = \frac{\pi}{a_0 + b_0}. \quad (15.780)$$

and

$$I_{b,S} = \int_{-\infty}^{\infty} \frac{dx}{\left[3(r_0^2 + x^2) + 20.4 \frac{h^2 r_0}{\lambda} \sqrt{r_0^2 + x^2} \left(\frac{1}{a_0} + \frac{1}{b_0}\right)\right]}$$

$$= \frac{2}{3 r_0} \int_0^{\infty} \frac{d(x/r_0)}{\left[1 + \left(\frac{x}{r_0}\right)^2\right] + Q_b \sqrt{1 + \left(\frac{x}{r_0}\right)^2}}$$

$$= \frac{2}{3 r_0} \int_0^{\infty} \frac{dy}{1 + y^2 + Q_b \sqrt{1 + y^2}} \tag{15.781}$$

with

$$Q_b = 20.4 \frac{h^2}{3 \lambda} \left(\frac{1}{a_0} + \frac{1}{b_0}\right), \tag{15.782}$$

and

$$y = \frac{x}{r_0} = \sinh \alpha. \tag{15.783}$$

Note that the following holds, $\cosh^2 \alpha - \sinh^2 \alpha = 1$, and $dy = \cosh \alpha \, d\alpha$. A substitution of the variables into in (15.781) leads to

$$I_{b,S} = \frac{2}{3 r_0} \int_0^{\infty} \frac{dy}{1 + y^2 + Q_b \sqrt{1 + y^2}}$$

$$= \frac{2}{3 r_0} \int_0^{\infty} \frac{\cosh \alpha \, d\alpha}{\cosh^2 \alpha + Q_b \cosh \alpha} = \frac{2}{3 r_0} \int_0^{\infty} \frac{d\alpha}{\cosh \alpha + Q_b}. \tag{15.784}$$

The solution of this integral take on two different forms, depending on the Q_b^2-values as follows,

$$I_{b,S} = \frac{2}{3 r_0} \left(\frac{2}{\sqrt{1 - Q_b^2}} \arctan \frac{\sqrt{1 - Q_b^2}}{1 + Q_b^2}\right) \qquad \text{for } 1 > Q_b^2, \tag{15.785}$$

and

$$I_{b,S} = \frac{2}{3 r_0} \left(\frac{1}{\sqrt{Q_b^2 - 1}} \ln \frac{1 + Q_b + \sqrt{Q_b^2 - 1}}{1 + Q_b - \sqrt{Q_b^2 - 1}}\right) \qquad \text{for } 1 < Q_b^2, \tag{15.786}$$

where Q_b is already defined in (15.782).

Task (b) Employing the arrangement of Problem 14.2, with $h = 2.6 \,\mathrm{m}$, $f = 500 \,\mathrm{Hz}$, $a_0 = 14 \,\mathrm{m}$, and $b_0 = 20 \,\mathrm{m}$, the evaluation of (15.782) results in $Q_b^2 = 65.7 \gg 1$. We now simplify the Eq. (15.786) to

$$I_{b,S} \approx \frac{2}{3 r_0 Q_b} \ln(1 + 2 Q_b). \tag{15.787}$$

A further substitution, namely, of (15.787) into (15.779) considering (15.780), finally produces

$$D_{\mathrm{i,S}} = 10 \lg \frac{\pi}{a_0 + b_0} - 10 \lg \left[\frac{2}{3\, r_0\, Q_b} \ln(1 + 2\, Q_b) \right]$$

$$= 10 \lg \frac{3\,\pi\, Q_b}{2\, \ln(1 + 2\, Q_b)} = 18.1\,\mathrm{dB}\,. \tag{15.788}$$

Problem 14.4.

Proposed Approach

Figure 15.74 illustrates a segment of the tube. The sound power is determined by

$$P = \frac{A\, p^2}{\varrho\, c}\,, \tag{15.789}$$

with A being the cross-sectional area.

Task (**a**) In an elementary segment of the tube, $\mathrm{d}x$, the elementary power is

$$\mathrm{d}P = -S\, \mathrm{d}x\, p^2 \operatorname{Re}[\underline{Y}_{\mathrm{wall}}]\,, \tag{15.790}$$

where S is the circumference of the square tube. Substituting (15.789) into (15.790) leads to

$$\frac{\mathrm{d}P}{\mathrm{d}x} = -\frac{S\,\varrho\, c \operatorname{Re}[\underline{Y}_{\mathrm{wall}}]}{A}\, P\,. \tag{15.791}$$

The solution of this differential equation is

$$P = P_0\, \mathrm{e}^{-\frac{S\,\varrho\, c\, \operatorname{Re}[\underline{Y}_{\mathrm{wall}}]}{A} x} = P_0\, \mathrm{e}^{-\check{\alpha}_{\mathrm{p}} x}\,, \tag{15.792}$$

with

$$\check{\alpha}_{\mathrm{p}} = \frac{S\,\varrho\, c \operatorname{Re}[\underline{Y}_{\mathrm{wall}}]}{A}\,. \tag{15.793}$$

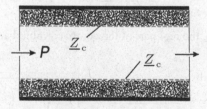

Fig. 15.74 Segment of a square tube lined with absorbent material on the interior tube wall. The characteristic impedance of the material is $\underline{Z}_{\mathrm{c}}$

Inside porous materials as sketched in Fig. 15.74, the characteristic impedance is as given in (11.34), that is,

$$\underline{Z}_c = \frac{1}{\sigma}\sqrt{\frac{j\omega\varrho + \sigma\,\Xi}{j\omega\,\kappa}} = \frac{\varrho c}{\sigma}\sqrt{1 - j\,\frac{\sigma\,\Xi}{\omega\,\varrho}}\,, \tag{15.794}$$

therefore, the admittance is given by

$$\underline{Y}_{\text{wall}} = \frac{\sigma}{\varrho c}\frac{1}{\sqrt{1 - j\,\frac{\sigma\,\Xi}{\omega\,\varrho}}}\,. \tag{15.795}$$

To evaluate the real part of the complex-valued admittance, evaluate first a complex function in the form of

$$\underline{f}(\underline{z}) = \frac{1}{\sqrt{a + jb}}, \quad \text{with} \quad \underline{z} = a + jb\,. \tag{15.796}$$

The real part of this function reads as follows,

$$\begin{aligned}
\text{Re}[\,\underline{f}(\underline{z})\,] &= \frac{1}{2}\Big[\,\underline{f}(\underline{z}) + \underline{f}^*(\underline{z})\,\Big] = \frac{1}{2}\Big[\,\underline{f}(\underline{z}) + \underline{f}(\underline{z}^*)\,\Big] \\
&= \frac{1}{2}\left[\frac{1}{\sqrt{a + jb}} + \frac{1}{\sqrt{a - jb}}\right].
\end{aligned} \tag{15.797}$$

Then, squaring (15.797) leads to

$$\begin{aligned}
\text{Re}^2[\,\underline{f}(\underline{z})\,] &= \frac{1}{4}\left[\frac{1}{a + jb} + \frac{1}{a - jb} + \frac{2}{\sqrt{a^2 + b^2}}\right] \\
&= \frac{1}{2}\left[\frac{a + \sqrt{a^2 + b^2}}{a^2 + b^2}\right],
\end{aligned} \tag{15.798}$$

$$\begin{aligned}
\text{Re}[\,\underline{f}(\underline{z})\,] &= \frac{1}{\sqrt{2}}\sqrt{\frac{a + \sqrt{a^2 + b^2}}{a^2 + b^2}} \\
&= \frac{1}{\sqrt{2}\sqrt{a^2 + b^2}}\sqrt{a^2 + \sqrt{a^2 + b^2}}\,.
\end{aligned} \tag{15.799}$$

By employing (15.799), the real part of the complex-valued admittance in (15.795) comes out as

$$\text{Re}[\,\underline{Y}_{\text{wall}}] = \frac{\sigma}{\sqrt{2}\,\varrho c\,\sqrt{1 + [\sigma\,\Xi/(\omega\,\varrho)]^2}}\sqrt{1 + \sqrt{1 + \left(\frac{\sigma\,\Xi}{\omega\,\varrho}\right)^2}}\,, \tag{15.800}$$

and

$$\check{\alpha}_p = \frac{S\,\sigma}{\sqrt{2}\,A\,\sqrt{1 + [\sigma\,\varXi/(\omega\,\varrho)]^2}}\,\sqrt{1 + \sqrt{1 + \left(\frac{\sigma\,\varXi}{\omega\,\varrho}\right)^2}}\,. \qquad (15.801)$$

The sound power level is determined as follows,

$$\begin{aligned} L_\Delta &= 10\lg\frac{P(x+x_\Delta)}{P(x)} = 10\lg\frac{e^{-\alpha_p\,(x+x_\Delta)}}{e^{-\alpha_p\,x}} \\ &= -10\,\alpha_p\,x_\Delta\,\lg e = -4.34\,\alpha_p\,x_\Delta\,. \end{aligned} \qquad (15.802)$$

Further evaluation requires calculation of $b = \sigma\,\varXi/(\omega\,\varrho)$ in (15.801) with $\varrho = 1.293\cdot10^{-3}$ (g/cm^3)

For cotton we get

$$b_{\text{cotton}} = \frac{\sigma\,\varXi}{\omega\,\varrho} = \frac{0.95\cdot1\,(\text{g cm}^{-3}\,\text{s}^{-1})}{2\,\pi\,1000\,(\text{s}^{-1})\cdot1.293\cdot10^{-3}\,(\text{g/cm}^3)} = 0.117\,, \qquad (15.803)$$

and for mineral wool

$$b_{\text{wool}} = \frac{\sigma\,\varXi}{\omega\,\varrho} = \frac{0.75\cdot100\,(\text{g cm}^{-3}\,\text{s}^{-1})}{2\,\pi\,1000\,(\text{s}^{-1})\cdot1.293\cdot10^{-3}\,(\text{g/cm}^3)} = 9.24\,. \qquad (15.804)$$

Consequently, the relevant damping coefficients are

$$\check{\alpha}_{\text{cotton}} = \frac{16/\text{cm}\cdot0.95\,\sqrt{1+\sqrt{1+(0.117)^2}}}{\sqrt{2}\cdot16^2/\text{cm}^2\cdot\sqrt{1+(0.117)^2}} = 0.945\,\text{cm}^{-1}, \qquad (15.805)$$

$$\check{\alpha}_{\text{wool}} = \frac{1/\text{cm}\cdot0.75\,\sqrt{1+\sqrt{1+(9.24)^2}}}{\sqrt{2}\cdot1/\text{cm}^2\cdot\sqrt{1+(9.24)^2}} = 0.183\,\text{cm}^{-1}\,. \qquad (15.806)$$

Thus, we finally arrive at

$$\frac{L_{\Delta,\,\text{cotton}}}{x_\Delta} = 4.101\ \text{dB/cm},$$

$$\frac{L_{\Delta,\,\text{wool}}}{x_\Delta} = 0.795\ \text{dB/cm}\,. \qquad (15.807)$$

Task **(b)** Inside the porous material, the complex propagation coefficient in (11.16), $\underline{\gamma}$, plays a dominate role concerning the sound attenuation, according to the expression

$$\underline{p}_{\text{i}}(x) = p_0\,e^{-\underline{\gamma}x}\,, \qquad (15.808)$$

where $\underline{p}_i$ is the sound pressure inside the porous material, and

$$\underline{\gamma} = \sqrt{(j\omega\varrho + \sigma\,\Xi)(j\omega\,\kappa)} = \frac{\omega}{c}\sqrt{-1+j\frac{\sigma\,\Xi}{\omega\,\varrho}}\,. \qquad (15.809)$$

Pursuing the real part of $\underline{\gamma}$ in similar fashion as in (15.796) and (15.799), yields

$$\underline{f}(\underline{z}) = \sqrt{a+jb}, \quad \text{with} \quad \underline{z} = a+jb\,, \qquad (15.810)$$

$$\mathrm{Re}[\,\underline{f}(\underline{z})] = \frac{1}{\sqrt{2}}\sqrt{a+\sqrt{a^2+b^2}}\,. \qquad (15.811)$$

Applying (15.810) and (15.811) to (15.809) leads to

$$\breve{\alpha} = \mathrm{Re}[\,\underline{\gamma}] = \frac{\omega}{c\sqrt{2}}\sqrt{-1+\sqrt{1+\left(\frac{\sigma\,\Xi}{\omega\,\varrho}\right)^2}}\,. \qquad (15.812)$$

The reduction level of the sound pressure thus involves $\breve{\alpha}$, namely,

$$L_\Delta = 20\lg\frac{\underline{p}_i(x+x_\Delta)}{\underline{p}_i(x)} = 20\lg\frac{\underline{p}_0 e^{-\breve{\alpha}(x+x_\Delta)}}{\underline{p}_0 e^{-\breve{\alpha}x}} = -8.68\,\breve{\alpha}\,x_\Delta\,. \qquad (15.813)$$

and we find

$$\breve{\alpha}_{\text{cotton}} = \frac{1}{\sqrt{2}}\frac{2\pi\,1000/\mathrm{s}^{-1}}{343\cdot100/(\mathrm{cm/s})}\sqrt{-1+\sqrt{1+(0.117)^2}}$$
$$= 0.0107\,\mathrm{cm}^{-1}, \qquad (15.814)$$

$$\breve{\alpha}_{\text{wool}} = \frac{1}{\sqrt{2}}\frac{2\pi\,1000/\mathrm{s}^{-1}}{343\cdot100/(\mathrm{cm/s})}\sqrt{-1+\sqrt{1+(9.23)^2}}$$
$$= 0.383\,\mathrm{cm}^{-1}, \qquad (15.815)$$

and, consequently,

$$\frac{L_{\Delta,\,\text{cotton}}}{x_\Delta} = -8.68\,\breve{\alpha} = -0.093\ \mathrm{dB/cm}\,,$$

$$\frac{L_{\Delta,\,\text{wool}}}{x_\Delta} = -8.68\,\breve{\alpha} = -3.32\ \mathrm{dB/cm}\,. \qquad (15.816)$$

Thus, the material thickness, d, takes on the following values,

$$d = \begin{cases} 105 \text{ cm} & \text{for cotton,} \\ 3.1 \text{ cm} & \text{for mineral wool .} \end{cases} \tag{15.817}$$

Problem 14.5.

Proposed Approach

The sound speed, in this case, is a function of the height, that is, $c = c(z)$. From (14.19) it follows

$$S(\theta, z) = \frac{c}{\sin \theta} - c_{\text{trace}} = 0 . \tag{15.818}$$

Next, determine $\mathrm{d}\,\theta / \mathrm{d}\, z$, and then its relation to $x(z)$. It is helpful to consider the derivative of a function $F(x, y) = 0$ and, with the general rule

$$\frac{\mathrm{d}\, x}{\mathrm{d}\, y} = \frac{-\frac{\partial F}{\partial y}}{\frac{\partial F}{\partial x}} . \tag{15.819}$$

we obtain with the chain rule,

$$\frac{\mathrm{d}\, F}{\mathrm{d}\, y} = \frac{\partial F}{\partial x} \frac{\mathrm{d}\, x}{\mathrm{d}\, y} + \frac{\partial F}{\partial y} = 0 \tag{15.820}$$

since $F(x, y) = 0$.

Task **(a)** Apply (15.819) to the current problem, where $S(\theta, z) = 0$ as in (15.818),

$$\frac{\mathrm{d}\, \theta}{\mathrm{d}\, z} = \frac{-\frac{\partial S}{\partial z}}{\frac{\partial S}{\partial \theta}} . \tag{15.821}$$

From (15.818), it follows that

$$\frac{\partial S}{\partial z} = \frac{1}{\sin \theta} \frac{\partial c}{\partial z} = \frac{1}{\sin \theta} c'(z), \quad \text{with} \quad c'(z) = \frac{\partial c}{\partial z} . \tag{15.822}$$

and

$$\frac{\partial S}{\partial \theta} = \frac{-c(z)}{\sin^2 \theta} \cos \theta = \frac{-c(z)}{\sin \theta \tan \theta} . \tag{15.823}$$

Substituting (15.822) and (15.823) into (15.821) results in

$$\frac{\mathrm{d}\, \theta}{\mathrm{d}\, z} = \frac{c'(z)}{c(z)} \tan \theta . \tag{15.824}$$

and, from (14.20)—as shown in Fig. 14.17—we get

$$\theta = \arctan \frac{dx}{dz}. \tag{15.825}$$

Differentiation (15.825) leads to

$$\frac{d\theta}{dz} = \frac{d}{dz}\left(\arctan \frac{dx}{dz}\right) = \frac{1}{1 + \left(\frac{dx}{dz}\right)^2} \frac{d^2x}{dz^2}. \tag{15.826}$$

Equaling (15.826) to (15.824) with substitution of (15.825) results in

$$\frac{1}{1 + \left(\frac{dx}{dz}\right)^2} \frac{d^2x}{dz^2} = \frac{c'(z)}{c(z)} \tan\left(\arctan \frac{dx}{dz}\right) = \frac{c'(z)}{c(z)} \frac{dx}{dz}. \tag{15.827}$$

This indicates that the sound-ray function, $x(z)$, is described by a nonlinear differential equation of the 2nd order, that is,

$$\frac{1}{1 + \left(\frac{dx}{dz}\right)^2} \frac{d^2x}{dz^2} = \frac{c'(z)}{c(z)} \frac{dx}{dz}. \tag{15.828}$$

Assignment of

$$g(z) = \frac{dx}{dz} \tag{15.829}$$

simplifies the solution of (15.828) to

$$\frac{1}{1 + g^2(z)} \frac{dg(z)}{dz} = \frac{c'(z)}{c(z)} g(z). \tag{15.830}$$

Rearranging (15.830) leads to

$$\frac{d\,g(z)}{[1 + g^2(z)]g(z)} = \frac{c'(z)}{c(z)} dz. \tag{15.831}$$

Integrating both sides of (15.831) renders

$$\int \frac{d\,g(z)}{[1 + g^2(z)]g(z)} = \int \frac{c'(z)}{c(z)} dz + K_0. \tag{15.832}$$

Note that

$$\int \frac{d\,g(z)}{[1 + g^2(z)]g(z)} = \frac{1}{2} \ln\left[\frac{g^2(z)}{1 + g^2(z)}\right] = \ln\left[\frac{g(z)}{\sqrt{1 + g^2(z)}}\right], \tag{15.833}$$

and

$$\int \frac{c'(z)}{c(z)}\, dz = \ln c(z). \tag{15.834}$$

Substituting (15.831) and (15.834) into (15.832) results in

$$\ln\left[\frac{g(z)}{\sqrt{1+g^2(z)}}\right] = \ln c(z) + \ln K_1, \tag{15.835}$$

leading to

$$\frac{g(z)}{\sqrt{1+g^2(z)}} = K_1\, c(z). \tag{15.836}$$

Solving this expression for $g(z)$ brings us

$$g^2(z) = K_1^2 c^2(z)[1 + g^2(z)],$$
$$g^2(z)\,[1 - K_1^2 c^2(z)] = K_1^2\, c^2(z), \tag{15.837}$$

and, consequently

$$g(z) = \pm \frac{K_1\, c(z)}{\sqrt{1 - K_1^2\, c^2(z)}} = \pm \frac{c(z)}{\sqrt{1/K_1^2 - c^2(z)}}. \tag{15.838}$$

By using (15.829), rearranging (15.838) and replacing an internal variable, $z \to \mu$ lead to the integral

$$x(z) = \pm \int_{z_1}^{z} \frac{c(\mu)}{\sqrt{1/K_1^2 - c^2(\mu)}}\, d\mu + K_2, \tag{15.839}$$

where K_1, K_2 are determined by the boundary conditions at the source location, x_0, z_0. The sound ray, $x(z)$, starts at the source position,

$$x(z_0) = x_0, \tag{15.840}$$

and it proceeds with a given initial slope of

$$\frac{d\,x}{d\,z}(z_0) = m. \tag{15.841}$$

$$x(z) = x_0 \pm \int_{z_0}^{z} \frac{c(\mu)}{\sqrt{1/K_1^2 - c^2(\mu)}}\, d\,\mu, \tag{15.842}$$

so that

$$m = \pm \frac{c(z_0)}{\sqrt{1/K_1^2 - c^2(z_0)}}. \tag{15.843}$$

Squaring and rearranging yields

$$\frac{1}{K_1^2} = c^2(z_0)\left(1 + \frac{1}{m^2}\right). \tag{15.844}$$

Substituting (15.844) into (15.842) results in

$$x(z) = x_0 \pm \int_{z_0}^{z} \frac{c(\mu)}{\sqrt{c^2(z_0)(1 + 1/m^2) - c^2(\mu)}}\, d\mu, \tag{15.845}$$

and

$$\frac{dx(z)}{dz} = \pm \frac{c(z)}{\sqrt{c^2(z_0)(1 + 1/m^2) - c^2(z)}}. \tag{15.846}$$

Note that $c(z) > 0$ it is always true. Therefore it follows that

$$\left|\frac{dx(z)}{dz}\right| > 0, \tag{15.847}$$

which means that the tangent of the function $x(z)$ never becomes perpendicular in Fig. 14.17, unless for $m = 0$.

Task **(b)** Consider the practically relevant case of

$$c(z) = c_0 - \xi z \tag{15.848}$$

due to temperature gradient near the ground with $\xi \geq 0$ or $\xi \leq 0$, and $x_0 = 0$.

Substituting (15.848) into the major part of (15.845) yields

$$I = \int_{z_0}^{z} \frac{c_0 - \xi \mu}{\sqrt{(c_0 - \xi z_0)^2(1 + 1/m^2) - (c_0 - \xi \mu)^2}}\, d\mu. \tag{15.849}$$

Further, substituting

$$\eta = c_0 - \xi \mu, \quad d\mu = -\frac{1}{\xi}d\eta, \tag{15.850}$$

and

$$\gamma^2 = (c_0 - \xi z_0)^2 \left(1 + \frac{1}{m^2}\right), \tag{15.851}$$

into (15.849) leads to

$$-\frac{1}{\xi}\int\frac{\eta}{\sqrt{\gamma^2-\eta^2}}d\eta=\frac{1}{\xi}\sqrt{\gamma^2-\eta^2}.\qquad(15.852)$$

Next, pursue (15.849) in the form of (15.852), viz

$$I=\frac{1}{\xi}\sqrt{(c_0-\xi z_0)^2\left(1+\frac{1}{m^2}\right)-(c_0-\xi\mu)^2}\Bigg|_{z_0}^{z}$$

$$=\frac{1}{\xi}\left[\sqrt{(c_0-\xi z_0)^2\left(1+\frac{1}{m^2}\right)-(c_0-\xi z)^2}\right.$$

$$\left.-\sqrt{(c_0-\xi z_0)^2\left(1+\frac{1}{m^2}\right)-(c_0-\xi z_0)^2}\right]$$

$$=\frac{1}{\xi}\left[\sqrt{(c_0-\xi z_0)^2\left(1+\frac{1}{m^2}\right)-(c_0-\xi z)^2}-\left|\frac{(c_0-\xi z_0)}{m}\right|\right].\quad(15.853)$$

Recalling $x_0=0$, and substituting (15.853) into (15.849) and then into (15.845), results in

$$x(z)=\pm\frac{1}{\xi}\left[\sqrt{(c_0-\xi z_0)^2\left(1+\frac{1}{m^2}\right)-(c_0-\xi z)^2}\right.$$

$$\left.-\left|\frac{(c_0-\xi z_0)}{m}\right|\right],\qquad(15.854)$$

or, in an alternative form, as

$$\left(x(z)\pm\frac{1}{\xi}\left|\frac{(c_0-\xi z_0)}{m}\right|\right)^2+\left(z-\frac{c_0}{\xi}\right)^2=\left(\frac{c_0}{\xi}-z_0\right)^2\left(1+\frac{1}{m^2}\right).\quad(15.855)$$

This form in (15.855) represents exactly a circle equation, $(x-X_0)^2+(z-Z_0)^2=r^2$, centered around

$$(X_0,Z_0)=\left(\mp\frac{1}{\xi}\left|\frac{(c_0-\xi z_0)}{m}\right|,\ \frac{c_0}{\xi}\right),\qquad(15.856)$$

with a radius of

$$r=\left|\frac{c_0}{\xi}-z_0\right|\sqrt{1+\frac{1}{m^2}}.\qquad(15.857)$$

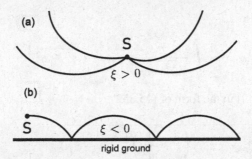

Fig. 15.75 Sound propagation under different environmental conditions. Different temperature profiles make the sound speed change with the height from the ground. The sound source is indicated by "S" **(a)** $\xi > 0$ ... The sound rays generally show convex curvatures. **(b)** $\xi < 0$ ... The sound rays show concave curvatures generally

The value of c_0/ξ, is always very large compared to the maximum wavelength. The attenuation due to wave propagation causes this high value. This is an indication for the circles's radius to be positive for $\xi > 0$, which means that the sound rays take on convex curvatures. However, the radius assumes a negative value when $\xi < 0$. The sound rays take on concave curvatures. Figure 15.75 illustrates these two cases. For $\xi < 0$, the sound rays' concave form is affected by a reflective ground.

Task **(c)** Determination of the acoustic horizon for $\xi > 0$. Under this environmental condition, the temperature decreases with increasing height—about $1°/100\,\text{m}$. The sound rays become convex-curved. Figure 15.76 illustrates the situation indicating that the source near the ground are no longer audible beyond a certain distance X_0—compare (15.856).

This effect requires that the circle center's height equals the circle's radius. Equaling (15.857) to Z_0 in (15.856) leads to

$$\frac{c_0}{\xi} = \left| \frac{c_0}{\xi} - z_0 \right| \sqrt{1 + \frac{1}{m^2}}, \tag{15.858}$$

$$1 = \left(1 - \frac{\xi z_0}{c_0}\right)^2 \left(1 + \frac{1}{m^2}\right), \tag{15.859}$$

$$\frac{1}{m^2} = \frac{1}{\left(1 - \frac{\xi z_0}{c_0}\right)^2} - 1. \tag{15.860}$$

Consequently, we now have

$$\left| \frac{1}{m} \right| = \sqrt{\frac{1}{\left(1 - \frac{\xi z_0}{c_0}\right)^2} - 1}. \tag{15.861}$$

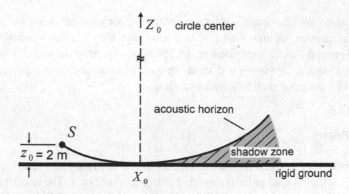

Fig. 15.76 Acoustic horizon for the temperature profile given by $\xi > 0$

Substituting (15.861) into the x–coordinate of the circle center in (15.856) leads to

$$X_0 = \frac{1}{\xi} \left| c_0 - \xi z_0 \right| \sqrt{\frac{1}{\left(1 - \frac{\xi z_0}{c_0}\right)^2} - 1} \,. \tag{15.862}$$

By employing the given data, $\xi = 1 \cdot 10^{-2}/\text{s}$, $c_0 = 343.6\,\text{m/s}$, and $z_0 = 2\,\text{m}$, we obtain

$$X_0 = \frac{1}{10^{-2}/\text{s}^{-1}} \left| \frac{343.6}{\text{m/s}} - \frac{2}{\text{m}} \cdot \frac{10^{-2}}{\text{s}^{-1}} \right| \sqrt{\frac{1}{\left[1 - \frac{(2/\text{m}) \cdot (10^{-2}/\text{s}^{-1})}{343.6/(\text{m/s})}\right]^2} - 1}$$

$$= 370.72\,\text{m}, \tag{15.863}$$

and

$$Z_0 = \frac{c_0}{\xi} = 34360\,\text{m} \tag{15.864}$$

as the starting point of the shadow zone.

Application of (15.860) leads to

$$\frac{1}{m^2} = \frac{1}{\left[1 - \frac{(2/\text{m}) \cdot (10^{-2}/\text{s}^{-1})}{343.6/(\text{m/s})}\right]^2} - 1 = 1.164 \cdot 10^{-4}. \tag{15.865}$$

Subsequently, usage of (15.857) finally renders the radius as

$$r = \left| \frac{343.6/(\text{m/s})}{10^{-2}/\text{s}^{-1}} - \frac{2}{\text{m}} \right| \sqrt{1 + 1.164 \cdot 10^{-4}} = 34360\,\text{m}. \tag{15.866}$$

Note that under weather conditions with *inversion*, for example, on a hot summer evening after sunset, we may find $\xi < 0$. Sound rays then become concave-curved toward the ground—as illustrated in Fig. 15.75b. Consequently, the sound is not heard over a long-range, except in a few discrete spots at discrete distances—provided that the ground is reflective such as a water surface.

Problem 14.6.

Proposed Approach

The microphone of the feed-forward device as shown in Fig. 14.13a receives a noise signal, $n(t)$, which is then polarity-inverted with a gain of $G = 1$. The resulting noise-control signal is delayed by $\tau_\Delta = 57.5 \cdot 10^{-6}$ s and subsequently fed into the noise field again. The sum of the original noise signal plus the control signal is

$$n_\Sigma(t) = n(t) - G\,n(t - \tau_\Delta).\tag{15.867}$$

We now apply the *Fourier* transform to this sum as follows,

$$N_\Sigma(f) = N(f)\left(1 - G\,e^{-j2\pi f \tau_\Delta}\right).\tag{15.868}$$

Thus, the resulting, modified noise field becomes

$$\begin{aligned} N_\Sigma(f) &= N(f)\left(1 - e^{-j2\pi f \tau_\Delta}\right)\\ &= N(f)\left[1 - \cos(2\pi f \tau_\Delta) + j\,\sin(2\pi f \tau_\Delta)\right]. \end{aligned}\tag{15.869}$$

To find the -3 dB limiting frequency, we consider the magnitude of (15.869),

$$20\lg\sqrt{[1 - \cos(2\pi f_1 \tau_\Delta)]^2 + \sin^2(2\pi f_1 \tau_\Delta)} = -3\,\text{dB},\tag{15.870}$$

namely,

$$2 - 2\cos(2\pi f_1 \tau_\Delta) = \frac{1}{2}, \quad \rightarrow \quad \cos(2\pi f_1 \tau_\Delta) = \frac{3}{4} = 0.75,\tag{15.871}$$

and we finally arrive at the following result,

$$f_1 = \frac{\cos^{-1}(0.75)}{2\pi\tau_\Delta} = \frac{41.4°}{360° \cdot 57.5 \cdot 10^{-6}\,\text{s}} = 2\,\text{kHz}.\tag{15.872}$$

Chapter 16
Appendices

16.1 Complex Notation of Sinusoidal Signals

An arbitrary sinusoidal signal can be written as

$$z(t) = \hat{z} \cos(\omega t + \phi), \tag{16.1}$$

with the three free parameters amplitude, $\hat{z}$, angular frequency, ω, and (zero)phase angle, ϕ. If the frequency is known and fixed, that is, for a monofrequent signal, only two free parameters are left—amplitude and phase angle.

Arithmetic operations with sinusoidal signals, such as addition, multiplication, differentiation and integration, are complicated in the representation as written above. There is thus a demand for more simple arithmetics to deal with these signals, particularly their amplitudes and phase angles. This can be accomplished in different ways. In this book we use the common symbolic representation of sinusoidal signals by means of so-called *complex amplitudes*.

To this end, the representation above is expanded by a complex imaginary part, but this operation is immediately inverted by forming the real part of the expression again. We thus write

$$z(t) = \mathrm{Re}\,\{\hat{z}\,\cos(\omega_0 t + \phi) + \mathrm{j}\,\hat{z}\,\sin(\omega_0 t + \phi)\}. \tag{16.2}$$

By applying *Euler*'s formula, $\mathrm{e}^{\mathrm{j}\phi} = \cos\phi + \mathrm{j}\sin\phi$, this expression can be rewritten as

$$z(t) = \mathrm{Re}\,\{\hat{z}\,\mathrm{e}^{\mathrm{j}(\omega_0 t + \phi)}\} = \mathrm{Re}\,\{\hat{z}\,\mathrm{e}^{\mathrm{j}\phi}\mathrm{e}^{\mathrm{j}\omega_0 t}\} = \mathrm{Re}\,\{\underline{z}\,\mathrm{e}^{\mathrm{j}\omega_0 t}\}. \tag{16.3}$$

with the term $\underline{z} = \hat{z}\,\mathrm{e}^{\mathrm{j}\phi}$ being the complex amplitude as mentioned above. The original real representation of the sinusoidal signal can be retrieved from the complex amplitude by multiplication with $\mathrm{e}^{\mathrm{j}\omega_0 t}$ and subsequent forming of the real part, that is, by applying the operator $\mathrm{Re}\,\{\dots\}$—see (16.3).

© Springer-Verlag GmbH Germany, part of Springer Nature 2021
N. Xiang and J. Blauert, *Acoustics for Engineers*,
https://doi.org/10.1007/978-3-662-63342-7_16

The most relevant rules for calculations with complex amplitudes are given below. For details please refer to the pertinent literature.

Addition and *Substraction*,

$$\underline{z}_1 + \underline{z}_2 = (\hat{z}_1 \cos\phi_1 + \hat{z}_2 \cos\phi_2) + j(\hat{z}_1 \sin\phi_1 + \hat{z}_2 \sin\phi_2)$$

$$\underline{z}_1 - \underline{z}_2 = (\hat{z}_1 \cos\phi_1 - \hat{z}_2 \cos\phi_2) + j(\hat{z}_1 \sin\phi_1 - \hat{z}_2 \sin\phi_2) \qquad (16.4)$$

Multiplication and *Division*,

$$\underline{z}_1\,\underline{z}_2 = (\hat{z}_1\,\hat{z}_2)\,e^{j(\phi_1+\phi_2)}, \qquad \underline{z}_1/\underline{z}_2 = (\hat{z}_1/\hat{z}_2)\,e^{j(\phi_1-\phi_2)} \qquad (16.5)$$

Integration and *Differentiation*,

$$\int \underline{z}\,dt = \frac{1}{j\omega}\,\underline{z}, \qquad \frac{d\underline{z}}{dt} = j\omega\,\underline{z} \qquad (16.6)$$

The rules for integration and differentiation become clear after multiplication of the complex amplitude, $\underline{z}$, with the factor $e^{j\omega t}$, which means reinserting the time dependency.

16.2 Complex Notation of Power and Intensity

Consider a force, $F_z(t)$, exciting a one-dimensional motion of $v_z(t)$ along a path z, for example at the mechanic port of an electro-mechanic transducer. The input energy can be written as follows, whereby from now on the index z is omitted for simplicity,

$$W = \int F(t)\,dz = \int F(t)\,\frac{dz}{dt}\,dt = \int F(t)\,v(t)\,dt. \qquad (16.7)$$

The transferred instantaneous power, $P(t)$, is than given by

$$P(t) = \frac{d}{dt}\int F(t)\,v(t)\,dt = F(t)\,v(t). \qquad (16.8)$$

Let both the force and the particle velocity be sinusoidal time functions, namely,

$$F(t) = \hat{F}\cos(\omega t + \phi_F) \quad \text{and} \quad v(t) = \hat{v}\cos(\omega t + \phi_v). \qquad (16.9)$$

This leads to the expression

$$P(t) = \hat{F}\cos(\omega t + \phi_F)\,\hat{v}\cos(\omega t + \phi_v). \qquad (16.10)$$

Execution of the multiplication with $\cos\alpha\,\cos\beta = \frac{1}{2}[\cos(\alpha+\beta) + \cos(\alpha-\beta)]$ renders

$$P(t) = \underbrace{\frac{\hat{F}\,\hat{v}}{2}\,[\cos(2\,\omega t + \phi_F + \phi_v)]}_{\text{alternating}} + \underbrace{\frac{\hat{F}\,\hat{v}}{2}\,[\cos(\phi_F - \phi_v)]}_{\text{constant}}. \qquad (16.11)$$

When determining the time average of the transmitted power, $\overline{P}$, the first part, which is alternating with double frequency, does not contribute. The average power is solely given by the second part, namely,

$$\overline{P} = \frac{\hat{F}\,\hat{v}}{2}\,[\cos(\phi_F - \phi_v)]. \qquad (16.12)$$

This average power is also called *active power* or *resistive power*.

For a complex notation with the complex amplitudes of the force and the particle velocity, that is

$$\underline{F} = \hat{F}\,e^{j(\omega t + \phi_F)} \quad \text{and} \quad \underline{v} = \hat{v}\,e^{j(\omega t + \phi_v)}, \qquad (16.13)$$

these notations lead to a *complex power*, $\underline{P}$, as follows,

$$\underline{P} = \frac{1}{2}\,[\underline{F}\,\underline{v}*] \quad \text{with} \quad \underline{v}* = \hat{v}\,e^{-j(\omega t + \phi_v)}, \qquad (16.14)$$

where the term $\underline{v}*$ is called the complex conjugate of $\underline{v}$. Some elaboration finally results in

$$\underline{P} = \overline{P} + j\,Q = \underbrace{\frac{\hat{F}\,\hat{v}}{2}\cos(\phi_F - \phi_v)}_{\text{active power}} + j\,\underbrace{\frac{\hat{F}\,\hat{v}}{2}\sin(\phi_F - \phi_v)}_{\text{reactive power}}. \qquad (16.15)$$

The real part of $\underline{P}$ is the active power, $\overline{P}$, as noted above. The imaginary part, Q, is called *reactive power* and has no direct physical relevance. Please note that by taking the complex conjugate of the particle velocity, $v*$, in (16.14) and not that of the force, we have chosen that the reactive power of mass is counted positive.

What holds for the power, also holds for the intensity, which is power per area. *Complex intensity* thus results as

$$\underline{I} = \frac{1}{2}\,[\underline{p}\,\underline{v}*]. \qquad (16.16)$$

Consequently, we denote the *active intensity*, $\mathrm{Re}\{\underline{I}\}$ also as $\overline{I}$ in this book.

16.3   Supplementary Textbooks for Self Study

The current textbook suffices as the sole teaching material for an introductory course given by experienced academic teachers who are able to provide specific explanations and stress relevant topics according to the prior knowledge and special interests of their students.

This book is also suitable for self study. In this case, however, we suggest that the reader may use further textbooks in parallel. Some suggestions are given in the following. The list only contains books in the English language and does not claim to be complete.

Blackstock, D.T. (2000), *Fundamentals of Physical Acoustics*, John Wiley & Sons, Hoboken, New York

Cremer, L. and Heckl, M. (1973), *Structure-Borne Sound – Structural Vibrations and Sound Radiation at Audio Frequencies*, translated by Ungar, E. E., Springer-Verlag, New York, Heidelberg, Berlin

Crocker, M. and Arenas, J.P. (2020) *Engineering Acoustics – Noise and Vibration Control* John Wiley & Sons, Hoboken, New York

Finch, R.D. (2005), *Introduction to Acoustics*, Pearson Prentice Hall, Upper Saddle River, New Jersey

Ford, R.D. (1970), *Introduction to Acoustics*, Elsevier, Amsterdam–London–New York, etc.

Ginsberg, J. (2018) *Acoustics– A Textbook for Engineers and Physicists, Vol. 1: Fundamentals, Vol. 2: Applications* Springer, New York etc.

Kinsley, L.E., Frey, A.R., Coppens, A.B. & Sanders, J.V. (2000), *Fundamentals of Acoustics*, 4th ed., John Wiley & Sons, Hoboken, New York

Kuttruff, H. (2007), *Acoustics – An Introduction*, Taylor & Francis, London–New York

Möser, M. (2004), *Engineering Acoustics – An Introduction to Noise Control*, Springer, Berlin–Heidelberg–New York

Pierce, A.D. (2019), *Acoustics – An Introduction to Its Physical Principles and Applications*, 3rd Ed. Springer, Cham, Switzerland

Raichel D.R. (2006), *The Science and Applications of Acoustics*, 2nd ed. Springer, Berlin–Heidelberg–New York

Randall, R.H. (1951), *An Introduction to Acoustics*, Reprint 2005, Dover Publications, Mineola, New York

Rossing, Th., Moore, F.R., & Wheeler, P.A. (2002), *The Science of Sound*, 3rd ed. Addison–Wesley, New York, etc.

Vorländer, M. (2008), *Auralization – Fundamentals of Acoustics, Modelling, Simulation, Algorithms and Acoustic Virtual Reality*, Springer, Berlin, Heidelberg

16.4 Letter Symbols, Notations, and Units

Roman-Letter Symbols

a	Acceleration
A	Area
A	Equivalent-absorption area
$A_\perp$	Effective area (area perpendicular to particle velocity)
b	Breadth, width, also: substitution for $\beta d \cos \delta$
B	Magnetic-flux density
B'	Bending stiffness
c	Sound-propagation speed in the free field (speed of sound)
c_b	Propagation speed of a bending wave
c	Sound-propagation speed in an exponential horn
c_p	Specific heat capacity at constant pressure
c_v	Specific heat capacity at constant volume
C	Capacitance
C'	Capacitance load (capacitance per length)
D	Dielectric-displacement density
D_i	Insertion loss
e	Piezoelectric coefficient
E	Electric field strength
f	Frequency
f_c	Coincidence frequency
F	Force
$\underline{g}$	Product of sound pressure and radius, $\underline{g} = \underline{p}\, r$
G	Electric conductance
G'	Conductance load (conductance per length)
h	Length, height
$h(t)$	Impulse response
$\underline{H}(\omega)$	Transfer function
i	Electric current
I	Sound intensity (sound power per area)
j	Unit of the imaginary numbers ($j^2 = -1$)
J_N	*Bessel* function of the first kind, order N
k	stiffness
k_f	*Fresnel* number
K	Compression module, $K = 1/\kappa$
l	Length
L	Inductance
L'	Inductance load (inductance per length)
L_I	Sound-intensity level
L_p	Sound-pressure level
L_P	Sound-power level
m	Mass
m'_a	Acoustic-mass load (acoustic mass per length)
M	Magnetic-field-transducer coefficient
n	Compliance
n'_a	Acoustic-compliance load (acoustic compliance per length)
N	Electric-field-transducer coefficient

Roman-Letter Symbols ... Continued

p	Sound pressure
P	Power
P_n	Legendre polynomial
P_n^m	Legendre function
$\underline{P}$	Active (resistive) power
q	Volume velocity
q_0	Source strength
q_m	Mirror-source strength
q'	Source-strength load (volume-velocity per length)
Q	Sharpness-of-resonance factor (quality factor), also: reactive power
Q_{el}	Electric charge
r	Damping (fluid damping), also: distance, also: radius
r_c	Critical distance
r_{rad}	Radiation resistance
R	Sound-reduction index, also: electric resistance, also: radius
$\underline{R}$	Reflectance
R'	Electric resistance load (resistance per length)
R_I	Isolation index
s	Strain
$\underline{s}$	Complex frequency, $\underline{s} = \breve{\alpha} + j\omega$
S	Standing-wave ratio, also: area of a wall, window, etc.
T	Reverberation time, also: temperature, also: torch, also: period duration
T'	Torch per width
T_{ip}	Transfer coefficient of a transducer, driven as sender
T_{pu}	Transfer coefficient of a transducer, driven as receiver (sensitivity)
T_{pp}	Sound-pressure-transfer factor
T_{up}	Transfer coefficient of a transducer, driven as a sender
u	Electric voltage
U	Perimeter
v	Particle velocity
V	Volume
W	Energy, or work
W'	Energy load (energy per length)
W''	Energy density (energy per volume)
x_{ff}	Far-field distance
Y	Admittance, $Y = 1/Z$
Y_a	Acoustic admittance
Y_f	Field admittance
Y_{mech}	Mechanic admittance (mobility)
Z	Impedance
Z_0	Terminating acoustic impedance of a tube
Z_a	Acoustic impedance
Z_f	Field impedance
Z_L	Line impedance (acoustic impedance of a tube)
Z_{mech}	Mechanic impedance
Z_{rad}	Radiation impedance
Z_w	Characteristic field impedance of a medium (wave impedance)
Z_{wall}	Wall impedance (surface impedance)

Greek-Letter Symbols

α	Degree of absorption
$\bar{\alpha}$	Degree of absorption for diffuse sound incidence
$\breve{\alpha}$	Attenuation coefficient
β	Phase coefficient (in physics often called angular wave number, k)
$\underline{\gamma}$	Complex propagation coefficient, $\underline{\gamma} = \alpha + \mathrm{j}\,\beta$
Γ	Directional characteristics of a sound source or receiver
δ	Damping coefficient, also *Kronecker* symbol in $\delta(t)$ ... *Dirac* impulse
ε	Dielectric permittivity
ε_0	Permittivity of the vacuum
ϵ	Flare coefficient
η	Ratio of the specific heat capacities, $c_\mathrm{p}/c_\mathrm{v}$, usually denoted γ
Θ	Angle of oblique sound incidence
θ	Elevation angle in a spherical coordinate system, also normal (stretching) stress
κ	Volume compressibility, $\kappa = \kappa_- + \kappa_\sim \approx \kappa_-$
λ	Wave length
μ_0	Permeability of the vacuum
μ_d	Dipole moment
$\varXi$	Flow resistivity
ξ	Particle displacement
ρ	Distance, also: position
ϱ	(Mass-)Density, $\varrho = \varrho_- + \varrho_\sim \approx \varrho_-$
σ	Electric polarization, also: porosity
ς	Mouth correction of a *Helmholtz* resonator
τ	Time interval
τ_ph	Phase delay
ν	Number of turns of a coil
ν_z	Number of *Huygens-Fresnel* zones
ϕ	Phase angle of a sinusoidal signal
φ	Azimuth, i.e. horizontal angle in a spherical coordinate system
Φ	Magnetic flux, also: vector potential
Ψ	Logarithmic-frequency interval
χ	Structure factor for porous media
ω	Angular frequency
Ω	Spherical angle

Specific Mathematical Notations and Terms

$\hat{z}$	Amplitude, peak value
$\underline{z}$	Complex amplitude—note that the $^\wedge$ on top of z is omitted
$\overline{\underline{z}}$	Time average, e.g., used for active power and active intensity
$\vec{z}$	Vector
$\mid \vec{z} \mid,\ z$	Magnitude of a vector
$\mid \underline{z} \mid,\ z$	Magnitude of a complex amplitude
$\mid \vec{\underline{z}} \mid,\ z$	Magnitude of a complex vector
$\mathrm{Re}\,\{\underline{z}\}$	Real-part operator, $\mathrm{Re}\,\{\underline{z}\} = \mathrm{Re}\,\{a + \mathrm{j}b\} = a$
z_-	Steady component of a function $z(t) = z_- + z_\sim$
$z_\sim$	Alternating component of a function $z(t) = z_- + z_\sim$
$\underline{z}_\rightarrow$	Forward-propagating wave component
$\underline{z}_\leftarrow$	Backward-propagating wave component
z_rms	Root of the time average of $z(t)^2$ (**r**oot **m**ean **s**quare)

For periodic functions of period duration T we have

$$z_\mathrm{rms} = \sqrt{\tfrac{1}{T}\int_0^\mathrm{T} z(t)^2 \mathrm{d}t}\ , \text{ of sinusoidal functions } z_\mathrm{rms} = \hat{z}/\sqrt{2}$$

$\circ\!\!\!-\!\!\!\bullet$	... *Fourier* transform
$\propto$	... Proportional
Coefficient	... Multiplier of dimension $\neq$ one
Degree	... Multiplier of dimension one and values of 0–1 (0–100%)
Factor	... Multiplier of dimension one

Units*⁾

Basic SI-units

m	*Meter* ... unit of length
$\mathrm{kg} = \mathrm{Ns}^2/\mathrm{m}$	*Kilogram* ... unit of mass
s	*Second* ... unit of time
A	*Ampere* ... unit of electric current

Some often used SI-derived units

$\mathrm{Hz} = 1/\mathrm{s}$	*Hertz* ... unit of frequency
$\mathrm{N} = (\mathrm{kg\ m})/\mathrm{s}^2$	*Newton* ... unit of force
$\mathrm{Pa} = \mathrm{N}/\mathrm{m}^2$	*Pascal* ... unit of pressure
$\mathrm{W} = \mathrm{V\ A} = (\mathrm{N\ m})/\mathrm{s}$	*Watt* ... unit of power
$\mathrm{V} = \mathrm{W}/\mathrm{A}$	*Volt* ... unit of electric potential difference, voltage
$\mathrm{C} = \mathrm{A\ s}$	*Coulomb* ... unit of electric charge
$\Omega = \mathrm{V}/\mathrm{A} = \mathrm{V}^2/\mathrm{W}$	*Ohm* ... unit of electric resistance to direct current
$\mathrm{F} = \mathrm{C}/\mathrm{V}$	*Farad* ... unit of capacitance
$\mathrm{H} = (\mathrm{V\ s})/\mathrm{A}$	*Henry* ... unit of self-inductance
$\mathrm{T} = \mathrm{N}/(\mathrm{A\ m})$	*Tesla* ... unit of magnetic flux

*⁾ All units used in this book are consistent with the SI-system (*système international d'unités*)

Index

A

Absorbent area
  equivalent, 198, 200
Absorber
  dynamic, 222
  Helmholtz, 180
  membrane, 180
  micro-perforated panel, 180, 181
  porous, 177, 178
  resonance, 180, 220
Absorption, 168, 191, 198, 205, 231, 234
  area, equivalent, 215
  coefficient, 111
  degree of, 111, 176, 189, 195, 199
  in air, 199
  resonance, 179
Acceleration, 5, 69
  meter, 85
Acoustics
  applied, 2
  architectural, 215, 220
  communication, 4
  definition of, 2
  electro-, 3, 4
  engineering, 2, 4
  geometrical, 185
  horizon, 227
  impedance, 27
  list of specialized fields, 3
  physical, 3, 4
  physiological, 3
  psycho-, 3
  room, 186
  statistical, 193
  technical, 2
Active, 43
  retro-, 43, 46
Adiabatic compression, 99, 169
Admittance
  mechanic, 21
Air absorption, 199
Air-gap, 73
  fictive, 76
Ampere's law, 73
Amplitude
  complex, 399
Analogy, 32
  dynamic, 32
  electroacoustic, 31, 33
  electromechanic, 31, 49
  impedance, 32
  mobile, 32
Anemometer
  hot-wire, 93
Appollonian circle, 176
Architectural acoustics, 215, 220
Area
  effective, 43, 52
  effective radiation, 6
  equivalent absorptive, 198
  function, 118
  of constant phase, 117, 118
Array
  line (linear), 144
Arrival time, 186
Attenuation, 8
Audience, 178
Auditory

© Springer-Verlag GmbH Germany, part of Springer Nature 2021
N. Xiang and J. Blauert, *Acoustics for Engineers*,
https://doi.org/10.1007/978-3-662-63342-7

Printed in the United States
by Baker & Taylor Publisher Services